Data-Intensive Computing in Smart Microgrids

Data-Intensive Computing in Smart Microgrids

Editor

Herodotos Herodotou

MDPI • Basel • Beijing • Wuhan • Barcelona • Belgrade • Manchester • Tokyo • Cluj • Tianjin

Editor
Herodotos Herodotou
Cyprus University of Technology
Cyprus

Editorial Office
MDPI
St. Alban-Anlage 66
4052 Basel, Switzerland

This is a reprint of articles from the Special Issue published online in the open access journal *Energies* (ISSN 1996-1073) (available at: https://www.mdpi.com/journal/energies/special_issues/data-intensive_computing_smart_microgrids).

For citation purposes, cite each article independently as indicated on the article page online and as indicated below:

LastName, A.A.; LastName, B.B.; LastName, C.C. Article Title. *Journal Name* **Year**, *Volume Number*, Page Range.

ISBN 978-3-0365-1627-1 (Hbk)
ISBN 978-3-0365-1628-8 (PDF)

Contents

About the Editor

Herodotos Herodotou is an Assistant Professor in the Department of Electrical Engineering and Computer Engineering and Informatics at the Cyprus University of Technology. He received his PhD in Computer Science from Duke University. His PhD dissertation work received the ACM SIGMOD Jim Gray Doctoral Dissertation Award Honorable Mention, as well as the Outstanding PhD Dissertation Award in Computer Science at Duke. Before joining CUT, he held research positions at Microsoft Research, Yahoo! Labs, and Aster Data, as well as software engineering positions at Microsoft and RWD Technologies. His research interests are in large-scale data processing and database systems. In particular, his work focuses on ease-of-use, manageability, and automated tuning of both centralized and distributed data-intensive computing systems. In addition, he is interested in applying database techniques in other areas, like smart power grid/microgrids, maritime informatics, and scientific and social computing.

Editorial

Introduction to the Special Issue on Data-Intensive Computing in Smart Microgrids

Herodotos Herodotou

Department of Electrical Engineering, Computer Engineering and Informatics, Cyprus University of Technology, Limassol 3036, Cyprus; herodotos.herodotou@cut.ac.cy

Received: 28 April 2021; Accepted: 6 May 2021; Published: 9 May 2021

Microgrids have recently emerged as the building block of a smart grid combining distributed renewable energy sources, energy storage devices, and load management in order to improve power system reliability, enhance sustainable development, and reduce carbon emissions. At the same time, the rapid advancements in sensor and metering technologies, wireless and network communication, as well as cloud and fog computing are leading to the collection and accumulation of large amounts of data (e.g., device status data, energy generation data, consumption data).

The application of big data analysis techniques (e.g., forecasting, classification, clustering) on such data can optimize the power generation and operation in real time by accurately predicting electricity demands, discovering electricity consumption patterns, and developing dynamic pricing mechanisms. The efficient and intelligent analysis of the data will enable smart microgrids to detect and restore from failures quickly, respond to electricity demand swiftly, supply more reliable and economical energy, and enable customers to have more control over their energy use. Overall, data-intensive analytics can provide effective and efficient decision support for all of the producers, operators, customers, and regulators in smart microgrids in order to achieve holistic smart energy management, including energy generation, transmission, distribution, and demand-side management.

We are pleased to present the Special Issue of *Energies* (ISSN 1996-1073) entitled "Data-Intensive Computing in Smart Microgrids", which contains an assortment of relevant research contributions. This Special Issue received a total of 14 manuscript submissions, out of which nine were selected for publication after a rigorous peer-review process. In the following, we provide a brief overview of the eight research articles and one review manuscript featured in this Special Issue.

In the article by Ahmad et al. [1], a new machine learning-based methodology is proposed for short-term electricity load forecasting. The methodology consists of three main steps: (i) feature selection using a combination of an XGBoost and a decision tree feature selector, (ii) redundancy removal using recursive feature elimination, and (iii) forecasting using support vector machine (SVM) and extreme learning machine (ELM). The hyper-parameters of SVM are tuned with the grid search algorithm while the hyper-parameters of ELM are tuned with a genetic algorithm.

Masood et al. [2] investigate three narrowband power line communication (NB-PLC) channel modeling techniques over medium voltage networks for advanced metering infrastructure. A transmission line theory model expresses the network as a two-port network to examine the characteristics of sending and receiving NB-PLC signals. A multipath signal propagation model incorporates the effect of multipath signals to determine the NB-PLC transfer function. Finally, a Simulink model considers the values of the network to examine the characteristics of NB-PLC signals.

In the article by Bukhsh et al. [3], the energy demand of a fog (micro) data center is evaluated by calculating the amount of energy required to run computing devices based on the number of requests as well as the amount of energy needed for cooling down the devices. The fog-based model can be used to ensure real-time energy management service provision for a green community. Various scenarios are simulated for fulfilling the energy demand from the utility and/or the use of renewable energy resources.

Ayub et al. [4] propose a framework for short- and medium-term electricity forecasting using a combination of deep learning and supervised machine learning techniques. First, Random Forest and XGBoost are used for identifying the most important features. Next, recursive feature elimination is performed to remove any redundant features. Finally, load forecasting is performed with support vector machines (SVMs) and a hybrid of convolutional neural networks (CNNs) and gated recurrent units (GRUs). Two meta-heuristic algorithms, namely, grey wolf optimization and earth worm optimization, are utilized to tune the hyper-parameters of SVMs and CNNs-GRUs, respectively.

Aslam et al. [5] address the problem of electricity theft detection in smart grids by proposing a modeling technique that combines deep learning (namely, long short-term memory (LSTM) neural networks and UNet) with ensemble learning (namely, adaptive boosting). In addition, the authors propose a new class balancing mechanism based on the interquartile minority oversampling technique to overcome the high-dimensional imbalanced data that are often used in these studies.

Qureshi et al. [6] propose an energy-efficient resource allocation strategy for real-time data-intensive smart systems that periodically offload computational and data-intensive loads to the cloud. The proposed strategy utilizes a mathematical model and a group of algorithms for minimizing the data files transfer overhead to computing resources by selecting appropriate pairs of computing and storage resources.

In the article by Ullah et al. [7], an energy optimization strategy is proposed to minimize operation cost and carbon emissions in a smart grid integrated with renewable energy sources. First, probability density functions are suggested to predict the behavior of wind and solar energy sources. To overcome uncertainty in power produced by wind and solar sources, demand response programs (DRPs) are proposed with the involvement of residential, commercial, and industrial consumers. Finally, incentive-based payments are introduced as price offered packages for executing the DRPs.

Ali et al. [8] propose a new optimization-based energy management framework that adapts consumer power usage patterns using real-time pricing signals and generation from utility and photovoltaic-battery systems in order to minimize electricity cost, reduce carbon emission, and mitigate peak power consumption. A hybrid genetic ant colony optimization algorithm is proposed to solve the complete scheduling model with or without photovoltaic and photovoltaic-battery systems.

Ullah et al. [9] provide a comprehensive review of various automatic generation control (AGC) models in diverse configurations of electric power systems. An in-depth analysis of various control methods used to mitigate the AGC issues is provided. The application of fast-acting energy storage devices, high voltage direct current interconnections, and flexible AC transmission systems devices in the AGC systems are also investigated. Furthermore, AGC systems employed in different renewable energy generation systems as well as in different configurations of microgrids/smart grids are discussed and compared.

Finally, we would like to thank all the authors who have contributed their papers for this Special Issue. We are sincerely grateful to the reviewers who contributed to assembling such a high-quality Special Issue. We are also indebted to the *Energies* editors, editorial office, and the publishing and production teams for their assistance in the preparation and publication of this Special Issue.

Conflicts of Interest: The author declares no conflict of interests.

References

1. Ahmad, W.; Ayub, N.; Ali, T.; Irfan, M.; Awais, M.; Shiraz, M.; Glowacz, A. Towards Short Term Electricity Load Forecasting Using Improved Support Vector Machine and Extreme Learning Machine. *Energies* **2020**, *13*, 2907. [CrossRef]
2. Masood, B.; Khan, M.A.; Baig, S.; Song, G.; Rehman, A.U.; Rehman, S.U.; Asif, R.M.; Rasheed, M.B. Investigation of Deterministic, Statistical and Parametric NB-PLC Channel Modeling Techniques for Advanced Metering Infrastructure. *Energies* **2020**, *13*, 3098. [CrossRef]
3. Bukhsh, R.; Javed, M.U.; Fatima, A.; Javaid, N.; Shafiq, M.; Choi, J.-G. Cost Efficient Real Time Electricity Management Services for Green Community Using Fog. *Energies* **2020**, *13*, 3164. [CrossRef]

4. Ayub, N.; Irfan, M.; Awais, M.; Ali, U.; Ali, T.; Hamdi, M.; Alghamdi, A.; Muhammad, F. Big Data Analytics for Short and Medium-Term Electricity Load Forecasting Using an AI Techniques Ensembler. *Energies* **2020**, *13*, 5193. [CrossRef]
5. Aslam, Z.; Javaid, N.; Ahmad, A.; Ahmed, A.; Gulfam, S.M. A Combined Deep Learning and Ensemble Learning Methodology to Avoid Electricity Theft in Smart Grids. *Energies* **2020**, *13*, 5599. [CrossRef]
6. Qureshi, M.S.; Qureshi, M.B.; Fayaz, M.; Zakarya, M.; Aslam, S.; Shah, A. Time and Cost Efficient Cloud Resource Allocation for Real-Time Data-Intensive Smart Systems. *Energies* **2020**, *13*, 5706. [CrossRef]
7. Ullah, K.; Ali, S.; Khan, T.A.; Khan, I.; Jan, S.; Shah, I.A.; Hafeez, G. An Optimal Energy Optimization Strategy for Smart Grid Integrated with Renewable Energy Sources and Demand Response Programs. *Energies* **2020**, *13*, 5718. [CrossRef]
8. Ali, S.; Khan, I.; Jan, S.; Hafeez, G. An Optimization Based Power Usage Scheduling Strategy Using Photovoltaic-Battery System for Demand-Side Management in Smart Grid. *Energies* **2021**, *14*, 2201. [CrossRef]
9. Ullah, K.; Basit, A.; Ullah, Z.; Aslam, S.; Herodotou, H. Automatic Generation Control Strategies in Conventional and Modern Power Systems: A Comprehensive Overview. *Energies* **2021**, *14*, 2376. [CrossRef]

Review

Automatic Generation Control Strategies in Conventional and Modern Power Systems: A Comprehensive Overview

Kaleem Ullah [1,*], Abdul Basit [1], Zahid Ullah [2], Sheraz Aslam [3] and Herodotos Herodotou [3,*]

1 US-Pakistan Center for Advanced Study in Energy, University of Engineering and Technology Peshawar, Peshawar 25000, Pakistan; abdul.basit@uetpeshawar.edu.pk

2 University of Management and Technology Lahore, Sialkot Campus, Sialkot 51310, Pakistan; zahid.ullah@skt.umt.edu.pk

3 Department of Electrical Engineering, Computer Engineering and Informatics, Cyprus University of Technology, 3036 Limassol, Cyprus; sheraz.aslam@cut.ac.cy

* Correspondence: kaleemullah@uetpeshawar.edu.pk (K.U.); herodotos.herodotou@cut.ac.cy (H.H.)

Academic Editor: André Madureira

Received: 27 March 2021; Accepted: 19 April 2021; Published: 22 April 2021

Abstract: Automatic generation control (AGC) is primarily responsible for ensuring the smooth and efficient operation of an electric power system. The main goal of AGC is to keep the operating frequency under prescribed limits and maintain the interchange power at the intended level. Therefore, an AGC system must be supplemented with modern and intelligent control techniques to provide adequate power supply. This paper provides a comprehensive overview of various AGC models in diverse configurations of the power system. Initially, the history of power system AGC models is explored and the basic operation of AGC in a multi-area interconnected power system is presented. An in-depth analysis of various control methods used to mitigate the AGC issues is provided. Application of fast-acting energy storage devices, high voltage direct current (HVDC) interconnections, and flexible AC transmission systems (FACTS) devices in the AGC systems are investigated. Furthermore, AGC systems employed in different renewable energy generation systems are overviewed and are summarized in tabulated form. AGC techniques in different configurations of microgrid and smart grid are also presented in detail. A thorough overview of various AGC issues in a deregulated power system is provided by considering the different contract scenarios. Moreover, AGC systems with an additional objective of economic dispatch is investigated and an overview of worldwide AGC practices is provided. Finally, the paper concludes with an emphasis on the prospective study in the field of AGC.

Keywords: automatic generation control; single/multi-area power system; intelligent control methods; microgrid; smart grid; renewable energy sources; virtual inertial control; demand response; soft computing control methods

1. Introduction

The power system primarily ensures its stable and safe operation by maintaining a constant equilibrium between power production and load demands. Such an equilibrium is maintained to hold the frequency of the network within the appropriate limits as defined by the grid codes. In real-time operation, there is often a persistent variation in the load demand that creates a relative balance between the load and generation. This relative balance always influences the system frequency and subsequently affects the security of the power system operation. The frequency of the power system is mainly controlled using two control loops, namely primary and secondary. The primary control loop prevents instant variations in the frequency before triggering the frequency protection switches. It is provided through the governor droops that typically give rise to the steady-state error. Secondary control, also termed automatic generation control (AGC) or load frequency control (LFC), is implemented to regulate the system frequency to its nominal value in the power system network. The AGC operates with the objectives of: (a) regulating the steady-state system frequency to its nominal value, (b) maintaining the tie-line power flow close to its scheduled value, and (c) keep overshoot and settling time within the acceptable ranges. To accomplish these objectives, a linear equation called area control error (ACE) is used, which is associated with two main focal variables, namely frequency variance, and tie-line power exchange. AGC uses ACE as a monitoring signal and sets it to zero in the event of any disturbance or variation in system loads.

The modern power infrastructure is currently undergoing significant transformations due to: (a) the integration of large-scale renewable energy generation sources, (b) introduction of new concepts in the power system like microgrids and smart grids and deregulation phenomena, and (c) digitalization's of the control structures in the power system network. The integration of increasingly intermittent renewable energy sources (RESs) such as wind and photovoltaic power plants results in uncertainty of active power productions, thereby determines the frequency variations. Furthermore, multiple physical, environmental, and operational constraints are emerging with numerous micro-energy sources that has shaped the operation of power systems in a different way and therefore poses various challenges to the secure power system operation. Besides this, the modern power grid is currently transitioning from a vertically integrated scenario to a deregulated power system, which requires the unbundling of power system elements into horizontal and vertical parts. This has increased the rivalry among utilities, resulting in free transmission access and thus raising the probability of transmission congestions. Moreover, the advent of various complicated power system elements with different capacities and consumption rates has further changed the dynamics of the power system network. In such circumstances, analysis of frequency control strategies is essential to cope with the above-mentioned challenges in the power system network. In other words, an AGC service is required that not only guarantees reliability, security, and economic operation of the system, but also offers an innovative and enhanced control version that meets all the requirements of the future power system. In this respect, different control approaches for AGC are proposed in the literature, which are divided into four categories: (a) classical controls, which are mainly based on conventional proportional-integral-derivative (PID) controllers; (b) modern controls including model predictive controls, adaptive and sliding mode controls, optimal controls, and digital controls; (c) intelligent control such as fuzzy logic and artificial neural networks; and (d) soft computing control approaches. In this respect, many literature surveys are published on different control philosophies of AGC operation regarding parametric variations/uncertainties, non-linearities, and different system load characteristics.

For instance, the authors in [1,2] briefly overviewed different frequency control schemes for AGC problems based on classical controls, optimal controls, adaptive and self-tunning, and artificial intelligence control methods. A comparative study is provided that summarizes the advantages and disadvantages of each proposed method based on different system load characteristics and generation profiles of various energy sources. However, the use of AGC in non-conventional power systems including the smart grid, and the complications related to the operation of AGC in non-linear power system models are not explored. In [3] the authors attempted to offer a complete survey on AGC regarding its application in conventional power systems, distribution generation systems, and a different configuration of the microgrid system. Further, the analysis also revealed the application of different energy storage devices, flexible AC transmission systems (FACTS) devices, and high voltage direct current (HVDC) links in the AGC system. However, the surveys still lack some contemporary regulation schemes and optimization approaches for AGC, such as deep learning and reinforcement learning approaches. The same approach regarding the review on AGC is observed in [4], which further described various challenges associated with the integration of fast energy storage systems,

FACTS devices, wind-diesel power plant, and photovoltaic (PV) systems into the AGC schemes. The authors in [5] provided an in-depth literature review on different perspectives and challenges of AGC in both traditional and future power systems. Here, the use of AGC in the diverse configuration of the power system is firstly reviewed and applications of different intelligent and modern control methods are enlightened. In [6], the authors presented a detailed literature review on different methodologies for integrating battery energy storage systems (BESSs) into AGC schemes due to their rapid response and high energy density in terms of BES scale, ownership responsibility, and collaboration with the system operator. Various challenges in the AGC of future smart power systems are reviewed in [7], where different AGC methods incorporating the BESS-based electrical vehicles (EVs), large-scale BESS, residential, and non-residential BESS are analyzed and tested to check its effectiveness in recovering the trembled frequency response. The authors in [8] explained the application and challenges of different AGC schemes in wind energy-based power plants, which affect the AGC regulations due to its low inertial and unsatisfactory primary performance. Further, the survey explored variable speed wind turbines and suitable frequency control techniques that can support the frequency contingencies. In [9], the authors focused on various intelligent AGC schemes and their implementation in renewable and distributed generation energy systems.

The frequency regulation of microgrids in autonomous mode is very critical as the generating sources are mostly renewable-based and therefore highly intermittent. Various control strategies employed for frequency regulation in a standalone microgrid are critically reviewed in [10]. Furthermore, different non-conventional frequency control methods, such as droop control, model predictive control, fuzzy logic control, H∞ base control, sliding mode, and demand-side control are analyzed and their implementation procedure in the microgrids are presented. The authors in [11] surveyed various robust frequency regulation strategies in a multi-area interconnected power system concerning technical and economical constraints along with the effect of RESs to attain a new type of collaboration between system performance, security, and dynamic robustness. The vertically integrated power system is now shifted towards the deregulated power system. Consequently, the analysis of the frequency control schemes in deregulated power networks is becoming a significant part of the power system operation and reliability. The authors in [12] provided an in-depth review of various deregulated power systems, their market structures, and contract agreements. Further, various traditional frequency controllers, such as I, PI, PID, robust and intelligent controllers for single and multi-area restructured power systems are reviewed and compared by considering their advantages and drawbacks. Likewise, an exhaustive literature survey on the application of AGC in different market scenarios of a deregulated power system considering several conventional and renewable energy sources is presented in [13]. Here, different AGC regulators including classical, modern, and intelligent controllers are reviewed and compared in terms of their transient and dynamic responses during sudden load perturbation and in the presence of different system parametric variations. All aforementioned reviews are listed in Table 1. In Table 1, "✔" supports the inclusion of a feature, while "✗" stands for the function not being reviewed in the referred study. Even though the above-mentioned surveys presented a comprehensive work on AGC schemes, they lack in providing an aggregating study on:

- Modern AGC techniques in present and in future smart power systems that can incorporate renewable energy sources, different fast energy storage devices, HVDC, and FACTS devices.
- Intelligent and pattern recognition-based AGC techniques that can handle non-linearities, parametric variations, uncertain states in demand identifications, and dynamics of the different loads. Furthermore, different virtual inertial controllers (VIC), which can support and improve the inertial response of renewable energy-based AGC systems.
- AGC schemes in different configurations of microgrids including stand-alone single area and multiple area microgrids and support the integration of non-dispatchable and high intermittent distributed generation sources.

- AGC techniques in smart grids, which incorporate and improve different features including demand-side response, data forgery attacks, and two-way communications.
- Efficient AGC models in a deregulated framework that can enhance the economic efficiency and stability of the restructured power market.
- AGC techniques in conjunction with the Economic Dispatch (ED) factor to improve its economic efficiency.
- Industrial practices of different AGC models around the world to explore and analyze different issues related to its practical implementation in the field.

Table 1. Summary of some generic state-of-the-art work on AGC.

Ref.	AGC-O&OF	AGC-CO	AGC-C& MCM	AGC-I&SCCM	AGC-ESS	AGC- FACTS	AGC-HVDCS	AGC-LREGS	AGC-MG & DG	AGC-SG	AGC-DPS	AGC-ED	AGC-WWP
[1]	✗	✔	✔	✔	✔	✔	✔	✗	✗	✗	✔	✔	✗
[2]	✗	✔	✔	✔	✔	✔	✔	✗	✗	✗	✔	✗	✗
[3]	✗	✗	✔	✗	✔	✗	✔	✗	✗	✗	✔	✗	✗
[4]	✗	✔	✔	✔	✔	✔	✔	✗	✗	✗	✗	✗	✗
[5]	✔	✔	✔	✗	✗	✔	✔	✗	✔	✔	✗	✗	✗
[6]	✗	✗	✔	✔	✔	✔	✗	✗	✗	✗	✔	✔	✗
[7]	✗	✗	✔	✔	✔	✗	✔	✔	✔	✗	✗	✗	✗
[8]	✔	✗	✔	✗	✗	✗	✗	✔	✔	✗	✗	✗	✗
[9]	✔	✔	✔	✔	✗	✗	✔	✔	✔	✗	✔	✗	✗
[10]	✗	✗	✗	✗	✗	✗	✗	✔	✔	✗	✗	✗	✗
[11]	✗	✗	✗	✗	✔	✗	✔	✔	✗	✗	✗	✗	✗
[12]	✗	✗	✔	✗	✔	✗	✔	✗	✗	✗	✔	✗	✗
[13]	✗	✔	✔	✗	✗	✗	✔	✔	✔	✔	✔	✗	✗
OR	✔	✔	✔	✔	✔	✔	✔	✔	✔	✔	✔	✔	✔

List of Abbreviations: AGC-O&OF: AGC Operation and Objective Functions; AGC-CO: AGC controller organizations; AGC-C&MCM: AGC Classical & modern control methods; AGC-I&SCCM: AGC Intelligent & soft computing control methods; AGC-ESS: AGC and Energy storage system; AGC-HVDCS: AGC and HVDC systems; AGC-LRES: AGC and large-scale renewable energy generation systems; AGC-MG: AGC and Microgrid; AGC-SG: AGC and Smart grid; AGC-DPS: AGC and deregulated power system; AGC- ED: AGC and Economic dispatch; AGC-WWP: AGC worldwide practices; OR: our review.

Considering the aforementioned issues, the main contributions of this review paper are:

- The developmental history of AGC models in traditional and renewable energy power systems is explored, which considers various constraints in performing the frequency control analysis. These constraints include generation rate constraint (GRC) and governor dead band (GDB) nonlinearities, parametric variations, inertial response, time delay problems, observability of state variables, and other stability issues.
- The general concept of AGC models in a multi-area interconnected power system is explored and different objective functions, which are based on several criteria and used to eliminate the area control errors, are presented from the literature.
- A state-of-the-art study of AGC schemes, focused on classical and modern control theories, is presented for current and future smart power systems. Furthermore, various intelligent AGC schemes based on fuzzy logic and artificial neural networks are explored and various soft computing control algorithms are comprehensively analyzed. All these control methods are critically compared using the tabular method showing their merits and demerits.
- The article addresses several frequency management systems integrating small and large renewable energy sources into the power grid for frequency regulation purposes. Further,

a comprehensive literature review on AGC strategies incorporating various energy storage systems (ESSs), HVDC interconnections, and FACTS devices is provided.
- A detailed overview of the AGC schemes in various microgrid configurations is presented and summarized for comparison in a tabulated form. Further, AGC approaches integrating various aspects of the smart grid are illustrated.
- The concept of a deregulated power system is addressed and the application and challenges associated with AGC implementation in different contract scenarios are presented.
- Different AGC schemes in conjunction with economic dispatch are reviewed from the literature and a detailed overview of worldwide AGC practices is provided to explore the industrial applications of AGC.

The article is organized as follows: Section 2 describes the history of the AGC schemes in different sets of power system models. The basic functioning of the AGC in multi-area power systems and the descriptions of the various objective functions are given in Section 3. Section 4 deals with the various configurations of the power system models associated with the AGC problems. The different control strategies used in the AGC operation are explained in Section 5. All the classical and modern control methods are discussed in Section 6. Likewise, Section 7 illustrates intelligent and soft computing control methods. AGC's incorporating ESS, HVDCs, and Fact's devices are presented in Section 8. Section 9 provides a comprehensive overview of different AGC schemes in renewable energy-based generation systems. AGC techniques in microgrid and smart grid are examined and overviewed in Section 10, while AGC schemes in deregulated power systems are analyzed in Section 11. AGC's secondary objective is studied in Section 12. Section 13 includes the worldwide AGC practices and Section 14 presents the future scope of this paper. Finally, the paper concludes with major findings. Figure 1 depicts the pictorial description of our paper organization into sections. The acronyms used in this manuscript are listed in Appendix A.

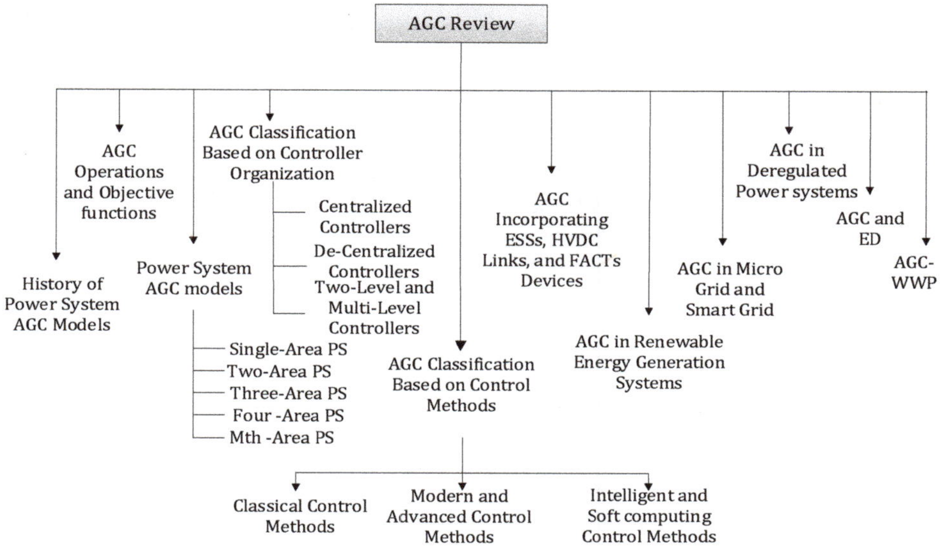

Figure 1. Structure of the survey.

2. History of the Power System AGC Models

An electrical power system is a large-scale complex system with underlying non-linear behavior. Any mismatch between the load and generation deviates the system frequency from its pre-defined level, which is initially counteracted by the retained kinetic energy, and then the frequency is balanced through activating the primary reserves, however, it is not restored to its nominal level. The system frequency is regulated to its pre-defined level by AGC/LFC, which is mainly responsible to keep the balance between generation and load demands in real-time. AGC uses ACE as the control signal and regulates it to zero in the occurrence of any oscillations, due to the deviations in the system frequency and scheduled tie-line interchanges. In the course of AGC operation, various constraints were found and, as a result, different changes were made to address these constraints. This section outlines the history of the AGC model during its development stages. The ACE equation, integrating frequency deviations and inadvertent interchanges in the tie-lines, was initially derived by Cohn [14] in which the time error correction factor was added to enhance the massive power transfer control technique. Various guidelines on the dynamic modeling of AGC for power system models including steam, hydro, and nuclear power plants [15] were then studied and small signal stability analysis was carried out for the AGC model by analyzing the system response characteristics. From the analysis, it was found that implementing the AGC strategy following the linearized model for a non-linear power system does not guarantee system stability. Keeping this in view, different destabilizing effects on the AGC system due to the GDB and GRC non-linearities were then studied [16] and it was revealed that non-linearities continuously produce oscillations in the transient response of ACE. The invisibility of different system state variables and parameters estimation in an integrated power system was a major challenge for the power control system. Modern and adaptive control theories were introduced [17] to the AGC system for the estimations of varying system parameters and unknown state variables using local inputs and outputs.

Centralized control strategies were initially adopted for the AGC regulators [18], however, the computational burden, communication delays, and hefty storage requirements precluded its application on a large scale. The disadvantages of the centralized control strategies were overcome via the implementation of a decentralized control system [19], where a better dynamic response was achieved for a complex integrated power system by reducing the controller complexities and raising its potential for effective implementation in large scale power systems. Digital approaches [20] were developed for the AGC framework to keep the scale small, reduce its size, and make it more adaptable. In the digital control method, the ACE was collected in discrete mode by sampling the frequency variations and the tie-line power flow, which was then transmitted via telemetry. Throughout the development, the structure of the power system network was changing due to the dynamics of the different load, incorporation of various equipment's, such as energy storage devices, HVDCs, and FACTS devices, and the integration of renewable energy sources into the conventional grid. This has further complicated the power system control problems, which required intelligent methods to have independence from the network models and do not need extensive knowledge of the problem. Intelligent controllers and computing-based algorithms [9,21] were introduced in this regard for the AGC to cope with such problems and increase network stability. Various impacts of renewables integration on the frequency control system were also studied [22]. Research on renewables integration proved that it has a non-zero impact on the system frequency, which becomes more significant in large-scale penetrations, where mounted inverters do not have a spinning mass and eventually minimizes system inertia. VICs were developed [23] to enhance the system inertial response and reduce the stability issues in the renewable energy-based AGC systems.

In the last decade, the shift from a vertically integrated electricity sector to a deregulated and dynamic energy market further modified the mode of power system operation. Holding these transformations in mind, a thorough study of AGC simulation and optimization after deregulation of the power system was conducted in [24]. In a restructured power network, there are Generation Companies (GENCOs), Transmission companies (TRANSCOs), and Distribution companies (DISCOs) with their policies. GENCOs sell their power at competitive rates to DISCOs. DISCOs in each area have the right to enter into a contract with GENCOs for power transactions, which are commonly known as bilateral transactions. Other types of transactions are poolco transactions, in which GENCOs of an area are bound to have a contract with DISCOs of the same area. All these deals are carried out under the supervision of ISOs. The AGC problem in such an environment was needed to accommodate all types of power contracts like unilateral, bilateral, and contacts with violations. Later on, a generalized AGC model was proposed [25] for a deregulated power structure to consider the possible outcome of bilateral agreements in the evolving generalized models. Augmented participation matrix (AGPM) was introduced, which defines the GENCO participation matrix agreed in consultation with DISCOS. In the AGPM matrix, the total number of rows and columns represents the strength of GENCO and DISCOS in the combined network. The suggested model helped visualize contracts and utilized new data signals in the strengths of the standard AGC systems.

3. AGC Operation and Objective Functions

The objective of determining the efficient, reliable, and stable power in an integrated power network is achieved via the implementation of AGC, which continuously monitors the load changes and alters the output of generators accordingly. In the AGC service, two variables need to be monitored consistently, namely tie-line interchanges and frequency variations. These two variables are combined to form one direct equation called ACE. To better understand the AGC operation, this section describes the basic functioning of AGC in a multi-area integrated power system and reviews some of the general-objective functions used in the literature to improve the AGC capacity. To analyze the AGC operation in an integrated multi-area power grid, consider a K-bus electric power system, where $\mathfrak{K} := \{1, \ldots, K\}$ having Z number of areas indexed by $\mathcal{A} := \{1, \ldots, Z\}$. Let $\mathfrak{L}$ denotes the set of lines, where $\mathfrak{L} := \{\mathcal{L}_1 \ldots, \mathcal{L}_N\}$. For simplicity, each line of the power system is represented by the order pair $\mathcal{L} := (k, k')$, where $k, k' \in \mathfrak{K}$. The network admittance matrix comprising the electrical parameters and information is represented by $Y := G + jB$, where $j := \sqrt{-1}$. The direction of the power flow is from k *to* k', whenever $P_\ell \geq 0$ and vice versa. There are N numbers of synchronous generators in each area indexed by $G := \{1 \ldots, \text{N}\}$ and represented from G_{z_i}, where G_z represents the set of all generators in area z and i represent the specific generator in that particular area. Figure 2 presents the schematic diagram of a multi-area interconnected power grid. The general equation for the actual power exchange between area z and z', when $P_\ell \geq 0$ is given as:

$$P_{zz'} = \sum_{\substack{i \in \mathfrak{B}_{zz'} \\ j \in \mathfrak{B}_{z'z}}} \mathbb{V}_i \mathbb{V}_j (G_{ij} cos(\vartheta_i - \vartheta_j) + B_{ij} sin(\vartheta_i - \vartheta_j) \tag{1}$$

where, $\mathbb{V}_I$ is the voltage magnitude of bus i, $G_{ij} + j\, B_{ij}$ is the (i j) entry of the system admittance matrix, and $\mathfrak{B}_{zz'}$ is the set of nodes in area z with tie lines to nodes in area z'. The actual frequency of the area z is given as:

$$f_z = \sum_{i \in G_z} \gamma_i \left(f_n + \frac{1}{2\pi} \frac{d\theta_i}{dt} \right) \tag{2}$$

where $\gamma_i = \frac{H_i}{\sum_{i \in G_z} H_i}$, in which H_i represent the $i-$th generator inertial constant. ACE equation for area z, denoted by ACE_z is given as:

$$ACE_z = \sum_{z' \epsilon \mathcal{A}_z} \beta_z \Delta f_z + (P_{zz'}^{Sch} - P_{zz'}^{Act}) \tag{3}$$

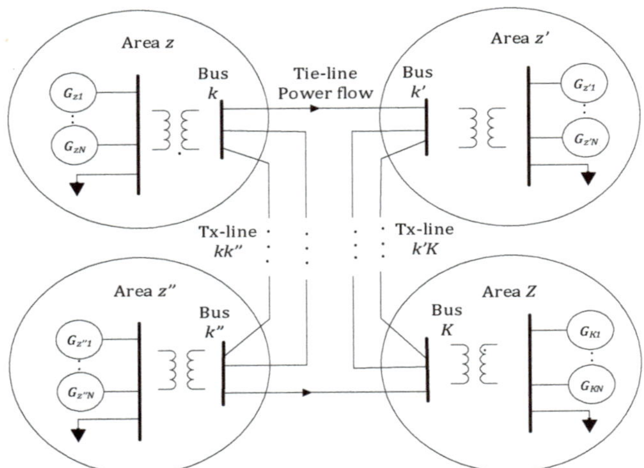

Figure 2. Schematic diagram of multiple areas interconnected power system.

In Equation (3), $P_{zz'}^{Act}$ and $P_{zz'}^{Sch}$ are the actual and scheduled power from area z to z' respectively. The variance between these two parameters is called tie-line error and is represented by ΔP_{tie}. β_z is the frequency bias constant for area z, and is given as $\beta_z = D_z + \frac{1}{R_z}$. Here, D_z is the power system damping, R_z is the governor droop and Δf denotes the frequency deviation from the nominal value. In a situation of contradiction between load and generation, ACE has a value other than zero and so the AGC system's primary objective is to restore the ACE to zero in each area of the power system network. To achieve this objective, the AGC regulator, simultaneously, changes the reference point ($\Delta P_{ref,i}$) of each generator taking part in the AGC operations.

A suitable objective function with the required specifications and constraints is selected to tune the parameters of the AGC regulators. Based on various time-domain requirements, such as peak overshoot and undershoot, settling time, and steady-state error, an appropriate objective function is selected. In the literature, different objective functions have been used with the AGC regulators to incorporate the two variables of AGC, namely frequency variance and tie-line interchanges. The most widely used objective functions are integral absolute error (IAE), squared error integral (ISE), time integral multiplied by absolute error (ITAE), and time integral multiplied by squared error (ITSE). Performance analysis of the aforementioned objective functions in AGC of single and multi-area integrated power systems is evaluated in [26]. The objective functions are used to optimize the gains of the PID-based AGC regulator with one percent load changes in a single area. The analysis revealed that the performance of the objective functions varies with the size of the power system. Moreover, the analysis proved that ITSE based objective function assured the minimum peak undershoot for the PID-based AGC system, compared to other criteria. The dynamic performance of the I, PI, PID, integral double derivative (IDD), and fuzzy-IDD (F-IDD) controllers are compared using the ISE criterion in formulating the objective functions in [27]. Sensitivity analysis is conducted for the F-IDD controller to check its robustness for a diverse range of loading conditions. Further, the proposed controllers provided larger values of speed regulation parameters that helped in the realization of the easier and cheaper governor. Based on the aforementioned literature, the commonly used objective functions incorporating ACEs for Z number of areas are as follows:

$$P_1 = IAE = \sum_{z=1}^{Z} \int_0^t \left(\partial_z |\Delta f_z(t)| + \sum_{\substack{z'=1 \\ z \neq z'}}^{Z} \partial_{zz'} |\Delta P_{tiez-z'}(t)| \right) \times dt \tag{4}$$

$$P_2 = ISE = \sum_{z=1}^{Z} \int_0^t \left(\partial_z |\Delta f_z(t)| + \sum_{\substack{z'=1 \\ z \neq z'}}^{Z} \partial_{zz'} |\Delta P_{tiez-z'}(t)| \right)^2 \times dt \tag{5}$$

$$P_3 = ITAE = \sum_{z=1}^{Z} \int_0^t \left(\partial_z |\Delta f_z(t)| + \sum_{\substack{z'=1 \\ z \neq z'}}^{Z} \partial_{zz'} |\Delta P_{tiez-z'}(t)| \right) \times t \times dt \tag{6}$$

$$P_4 = ITSE = \sum_{z=1}^{Z} \int_0^t \left(\partial_z |\Delta f_z(t)| + \sum_{\substack{z'=1 \\ z \neq z'}}^{Z} \partial_{zz'} |\Delta P_{tiez-z'}(t)| \right)^2 \times t \times dt \tag{7}$$

where ∂_z and $\partial_{zz'}$ represent the weight of the frequency and tie-line interchanges errors, respectively. Relying on the importance of the frequency variations and the tie-line interchanges in a particular location, an appropriate weight is assigned to that vicinity, which is comparably greater than the weight of the other control regions. Besides the single-objective optimization problems, researchers also targeted multi-objective optimization problems, which can be solved using the multi-objective optimization set of rules with clustering-based selection [28]. Here, the worst solution is decided in various non-dominated solutions discovered through the multi-objective optimization algorithm through the maximum value of each objective. The analysis proved that in terms of various performance indices, the controller primarily based at the multi-objective problems are more advanced than the controllers with a single-objective optimization problem. The multi-objective functions considering the damping of the frequency oscillation and settling time for both the frequency deviation and tie-line errors is given as:

$$\begin{aligned} P_5 = \omega_1 & \sum_{z=1}^{Z} \int_0^t \left(\partial_z \Delta f_z(t) + \sum_{\substack{z'=1 \\ z \neq z'}}^{Z} \partial_{zz'} \Delta P_{tiez-z'}(t) \right) \times t \times dt + \omega_2 \frac{1}{min(\{(1-\zeta_i),\ i=1\ldots n\})} \\ & + \omega_3 \left(\sum_{z=1}^{Z} \int_0^t \left(T_s \Delta f_z(t) + \sum_{\substack{z'=1 \\ z \neq z'}}^{Z} T_s \Delta P_{tiez-z'}(t) \right) \right) \end{aligned} \tag{8}$$

The weight (w) of each term in an objective function is set by the significance of each term in the combined objective function [29]. The objective functions are then optimized properly to achieve the required target. Several classical (deterministic) and recent (nondeterministic) heuristic optimization techniques proposed in the literature are discussed in a later section.

4. Power System AGC Models

For the power system configuration, most of the research relies on linearized models of single-area and multi-area power systems. This section explains the different methods of frequency control in major power system models.

4.1. Single Area Power Systems

AGC models in single area power systems are proposed from the very early stages. Several research articles are published on the implementation of AGC in single area power systems [30,31]. The general concept of a single area power system with a variable structure controller was initially presented in the study [30]. The variable structure technique was utilized to achieve a robust and invariant control system. In current power systems consisting of single areas, vulnerabilities are increased due to the sudden upsurge in the load demands and volatile nature of generation sources, resulting in an impact of communication delay on LFCs. The authors in [31] have proposed an analytic-graphical-based PI-AGC regulator for a single area power system to deal with the destabilizing performance due to the communication delays. The basis of the idea is to extract the stability region in parameters space within the defined gains and phase margin. Analysis of the suggested control technique has shown that it provides a faster response to the disturbance rejection as compared to the other conventional controllers.

4.2. Two Area Power Systems

The AGC regulator for a two-area interconnected power grid has been analyzed in the literature [32–35]. A network where there are two areas of unequal power distribution with diverse electricity generation sources and a PID controller is suggested in [32] for the investigation of the AGC regulator. The controller of this system is tuned using three optimization methods, such as particle swarm optimization (PSO), Bacteria foraging Algorithm (BFA), and Improved PSO. To make processing faster, the IPSO process is modified and integrated into a constraint treatment method called a heavy search space compression technique, allowing an increase in the speed of optimization process. The capacity of the proposed IPSO dual area power grid is measured using 1% step loading in each control area. The two-area deregulated power grid integrating the double fed induction generator and capacitive energy storage systems to accomplish the frequency control of the AGC system for the thermal, gas, and hydroelectricity generation units are discussed in the literature [33]. The proposed control scheme allows the kinetic energy of the turbine to provide the specified inertial support between the onset and the end of the extreme load perturbations. Moreover, the desired transient output is achieved by the coordinated action of both CES/DFIG to arrest the initial dip in the region frequency and reduce the tie-line power deviations. In [34], the authors have nominated AGC controller for dual area power grid inculcating interline power flow controller (IPFC) in series with Redox Flow Batteries (RFB) to regulate the frequency oscillation arising in the power grid due to a variety of available apparatus with large capacities and the slow response. A BFO algorithm is employed and used to organize and optimize the control model parameters to enhance its efficiency.

4.3. Three Area Power Systems

AGC schemes consisting of three areas are best described by the researchers in [36–38]. An AGC with three unequal areas having thermal sources with single reheat and GRCs in each area is presented in [36]. A two degree of freedom (2DOF) regulator with IDD is implemented in the study and compared its performance with the 2DOF- PI, PID, and DD regulator to check its superiority. The control parameters are tuned using the cuckoo search algorithm. Besides, the efficiency of several FACTS devices, such as Static synchronous series compensator (SSSC), Thyristor controlled series capacitor (TCSC), Thyristor controlled phase shifter (TCPS), and Unified power flow controller (UPFC) with the proposed 2DOF controller is checked and it has been found that the required dynamic response is achieved with IPFC. An AGC for a three-area power grid consisting of a solar thermal power plant (STPP) in one of the areas is described in [37] considering the effect of GRCs. Here, the performance of I, PI, and PID controller is checked and compared using the grey wolf optimization technique with and without the STPP. The analysis showed that the proposed GWO-based PID regulator delivers improved dynamic progress in term of overshoot, settling time, and oscillation damping. The authors in [38] applied the QOHS algorithm to the AGC-based PID regulator in a three-area energy system with thermal units and with appropriate GRCs for each area. The proposed algorithm is developed by combining the concept of quasi-oppositional with the existing harmony search technique. In addition, to compile quickly, the proposed algorithm focuses on the characteristics of both opposing point predictions and the point of its mirror.

4.4. Four Area Power Systems

The research work on AGC systems with four control areas is presented in [19,39,40]. A type-2 fuzzy PID-based AGC system [40] is used in a four-area integrated power grid with non-linear behavior and uncertainties. In case of system uncertainties, type-2 fuzzy sets with a grade of fuzzy membership are believed to fit well. The parameters of the aforementioned control scheme are optimized using a big bang–big crunch algorithm having the characteristic of fast convergence speed. The proposed control framework is contrasted with the results of most recent regulators and showed better execution in regard to settling time and stability under the occurrence of load variations. An AGC model based on the sliding mode control for a four-area power system, which considers hydropower plants, reheat thermal gas, and non-reheat thermal gas is presented in [39]. The analysis showed that a better system reaction to discard distortions and marinate the necessary quality of control is achieved by the proposed control technique. A Sugeno fuzzy-logic-based AGC system for a multi-units four area deregulated power system is proposed in the literature [19] to further strengthen the bilateral behavior and access the AGC performance. The parameters of the proposed controller are set-up using the quasi-oppositional harmony search algorithm.

4.5. M-th Area Power Systems

AGC model for multiple area power systems, in which several areas are considered with multiple sources is briefly explained in [41]. A GA-based PI controller is used with the ITSE objective function. Also, the robustness margin of the system and transient performance are optimized concurrently to attain the optimal controller parameters. In [42], a multi-area power system with M number of areas having reheat type steam turbine and gas turbine is proposed with appropriate GRC and GDB non-linearities. All the generators in a single area are simplified as one generating unit. Here, only the GAS turbine is included in the AGC operation, while steam turbine is available for generation of power. A fuzzy logic was utilized to develop an adaptive control technique and an algorithm for the parameter's upgradations.

5. AGC Classification Based on Controller Organizations

This section provides an overview of the different organizational structures of AGC in the power system network. They are mainly divided into three different categories, namely centralized, decentralized, and hierarchical control. The hierarchical control is subdivided into two and multi-level control concepts.

5.1. Centralized Controllers

The early part of the literature covers the centralized control concept for the power system AGC operations. In a centralized organization, a global controller operator takes information about all the states of the system and responds accordingly. The basis of the centralized control is the class of the disturbances. In [18], a centralized AGC regulator based on the mixed H_∞/H_2 control theory with pole placement technique is designed for an interconnected power system to eradicate the area frequency disturbances effects and enhance the transient response. Further, the pole placement techniques are utilized to place the poles in the required regions. The main disadvantage of the centralized control approach is the sharing of knowledge from the control region spread around the associated geographical regions. However, the computational burden and large storage requirement preclude the application of centralized control approach for the large interconnected power systems.

5.2. Decentralized Controllers

Contrary to the centralized control concept, the communication problem is effectively tackled in the decentralized controller organization developed for the power system control. This principle is focused on disintegrating the hefty power systems into several subsystems and having autonomous control over each subsystem. The AGC models using a decentralized approach in continuous and discrete power system models are discussed in the literature. The authors in [43] presented a decentralized optimal AGC in a three-area power system with appropriate non-linearities, which has resulted in a reduced frequency oscillation to an appropriate degree and has maintained the interchange power at minimal values. Furthermore, observers are being developed in each area of the power system to approximate the signal applied to each power plant. In a decentralized organization structure, the local controllers use only the local data and require no feedback from other areas. In such a situation, the reliability of the system is a concern as all local controllers operate concurrently. Therefore, an AGC regulator based on the exact plant model will not work well in such an environment. Considering this, an AGC scheme is designed for a multi-area power system using a Riccati-equation approach [44]. The control schemes consist of M number of AGC equations for M number of areas and a decoupling technique is used to separate the M number of Ricci-equations. The controller required feedback from the local area and is not dependent on the feedback from other areas. Moreover, the limits of the parameter uncertainties are also considered to enhance the robustness of the controller. The authors in [45] proposed a decentralized AGC system for a realistic Egyptian power system model incorporating both traditional generating units and hefty share of wind power plants. A Tustin method was developed and implemented in each area of the system, which reduces the implementation cost and enhanced the reliable performance of the controller under heavy load perturbation and wind power plant variations.

5.3. Two-Level and Multi-Level Controllers

To reduce the coordination costs, the feedback gains correlated with a certain state of the neighboring areas are not taken into account in the decentralized management approach. The type of cooperation among the areas renders the overall structure an unpredictable system. Two-level or multi-level control mechanisms are discussed in the literature to solve these limits. For an integrated power system, a two-level AGC regulator is suggested in [46]. Here, a multi-area interconnected power grid is disintegrated into many subsystems at the first stage and an optimization problem is solved in each region based on input from the local area and the other areas. At the second level, an iterative procedure is used to converge the local controllers to the overall optimal solution. The aforementioned approach reduced the computational time using the parallel processing approach and overcome other problems of the centralized control strategy in a hefty-scale interconnected power grid. A multi-agent-centered two-level coordinated control frame for the AGC of a three-area integrated power system is established in [47]. Here, to safeguard the frequency stability, the lower AGC agents maintain mutual power support among the interconnected areas, while the upper-level agent communicates with the lower-level agent using the unified approach to resolve the tension between ACE tapering and system frequency stabilization.

The multi-level AGC is best described by the researchers in [48,49]. A multi-level adaptive AGC based on a self-tuning regulator (STR) is used in [48]. In order to define the parameters on which the ACE for the given area is measured, the control scheme chooses each area as an individually regressive moveable sub-model and a recursive least square technique is utilized to select the parameters. The proposed model is an alternative to the integral-based AGC optimized power system model. The authors in [49] presented a detailed analysis on the implementation of multi-level control law for the AGC regulator in an interconnected power system. The centralized control is enforced at a group level, whereas a distributed control is implemented at a coordination level. When there are fewer groups than nodes at the coordination level, it is more efficient to reduce the expense for a system than the purely distributed control schemes. In addition to that, at the group level, the nodes are less when compared to the total number of nodes which enables the group to minimize the communication and computation overheads. Stability research is conducted on the proposed control systems utilizing the Lyapunov scheme, which has demonstrated greater stability at large sample rates. A three-level optimal control concept for AGC of an integrated power system and interconnected through asynchronous tie-lines are best described in [50]. The proposed control framework is divided into three tiers by the multi-level device concept: the first-level power control, the second-level local controls, and the third-level coordinating regulator. Local controllers manage the sub-systems, while the supervisor is used to boost co-operation factors and assign optimum settings to the DC capacity, which controls the tieline flow through converter controls.

6. AGC Classifications Based on the Control Methods

Since the advent of AGC, different forms of controls were created and introduced. With time, these methods of control were modified and as a result, new methods have emerged. This section overviews classical, modern, and advanced control methods, while intelligent and soft computing control techniques are discussed in the next section. Table 2 lists the summary of control techniques with the main features and drawbacks.

Table 2. Summary of control techniques with main advantages and disadvantages.

Sr. No	Control Method	Main Advantages	Drawbacks
1	Classical Control Methods	• Simple and easily implementable • Quickly provide transient and stability information's • Low initial cost and plan structure • Feedback controller	• Poor dynamic performance • Only valid for LTI and SISO systems • No very accurate and has poor quality • Low resistance to sensor and actuator faults
2	Optimal and Sub-optimal Control Methods	• Work in the MIMO system • Can handle non-linearities and delays • Can handle constraints on input and output.	• Requires observers • Complex structures and requires special configuration for a problem • Requires large storage space
3	Adaptive Control Methods	• Worked cell for a non-linear system with uncertain parameters and slow time response [51] • Offer a quick approach to change parameters in response to changes in process dynamics • Suitable for a system with limited conditions	• The complication in a system with large time delays • Risk of failure of estimation module • Mostly not consider the transient response • Not practical for a system having a large dimension

Table 2. *Cont.*

Sr. No	Control Method	Main Advantages	Drawbacks
4	Variable Structure Control Method	• Low sensitivity to parameters uncertainties • Handles non-linearities • Worked for a large order system • Finite-time convergence • Versatile control features	• Complex structure • State equation requires an observer • Chattering issue due to imperfect implementation
5	Robust Control Methods	• Used in multivariable systems • Provides the best functionality for a system with uncertainties and external disruptions • Past knowledge about uncertain inputs is not required.	• Not practical for a system having large dimensions and extensive parameter variations
6	Model Predictive control methods	• Can handle constraints explicitly • Optimizes the current time slot, while keeping future time slots in account • Works effectively in a system of multi-variable with bounds • Good capability of shifting the peak load	• Complex algorithm, which takes a longer time to execute than the ordinary one
7	Digital Control Methods	• Precise and reliable method [20] • The proposed controller has a smaller size • Adaptable and less noisy	• Appropriate selection of the digital controller is only feasible if the program requirements and the digital controller features are well defined.
8	Fuzzy logic-based control methods	• Instinctive design. • The rules demonstrate control action • Controllers may be set up through practice and the use of novel rules. • A specific model is not required.	• Problem with comprehensive rules and their durability. • Use of trial-and-error for optimization • Large number of tuning parameters • Stability is not certain.
9	ANN-based Control Methods	• Works well for a system with nonlinearities and uncertainties • Ideally suitable for multivariable and complex systems • No exact model is required [21] • Can be applied as feed-forward control.	• A large number of parameters are required for the adjustments • Size and structure are important to be determined • Can mostly be used in the trained region
10	Neuro-fuzzy based Control Methods	• Combines properties of a neuro network and fuzzy logic • Ideal for small system power systems • Can perfectly tackle system uncertainties and non-linearities.	• Problem with comprehensive rules and their durability for bulky power systems. • Difficulties in the analysis of the control system • Weak constraints handling

6.1. Classical Control Methods

For the first time, the flywheel governor of the synchronous machine was used to preserve the balance of the power system, but over time, this method became insufficient and a secondary control scheme was established to restore system frequency to its current valuations. This method constituted the classical control method (CCM), which is based on converting a differential equation to a transfer function to get the required system response. Examples of classical controls are Nyquist, Root Locus, and Bode plots. CCMs have the main advantage of continuously providing fast stability and transient response information, thus helping to know the consequences of different system parameters until an acceptable design is achieved. AGC problem in a multi-variable system based on the pole placement technique is addressed in the literature [52]. The pole placement technique assists in placing all the roots of the system characteristic equation in a required location resulting in a constant gain regulator. It is worth mentioning that if all the state variables are not accessible in the aforementioned technique then state estimators may be used for its implementation. The authors in [53] provided a comparison of different classical controllers including I, PI, ID, PID, and ID used in two, three, and five area power systems considering reheat turbine power plants with appropriate non-linearities. Furthermore, the gains of the controllers are optimized using the most recent evolutionary algorithm to improve its dynamic response characteristics. The detailed analysis of the power systems with classical approaches is well studied and in reality, CCM-based AGC designs are utilized in several power systems. However, there are major pitfalls in these designs, such as exhibiting poor dynamic performance during the parametric variations and in the presence of the non-linearities, and are only valid for LTI and SISO systems.

6.2. Optimal and Suboptimal Control Methods

Modern control theory applications in the power system paved the way for the implementation of optimal and suboptimal control methods in the power system frequency control services. Optimal control methods (OCM) fit in the calculus of variations and deal with the closed-set constrained variation problems. Several research articles on the optimal and sub-optimal controls are reported in the literature [43,46,54,55]. A detailed analysis of optimal control methods in the AGC system of an interconnected power system is presented in study [54]. An optimal PI regulator is used, which utilizes a performance index minimization criterion and employing a full feedback control strategy. Transient analysis is used to determine the dynamic response of a system, taking into consideration the settling period, over and undershooting, and damping parameters. The sensitivity analysis is conducted to ensure that the optimum controller is capable of operating effectively in the presence of abrupt load shifts and system parameter variations. It is worth mentioning that the viability of OCMs is feasible only if state variables are adequate enough to restore the input control signal. This requirement can only be met if the controller possesses an appropriately designed observer to regenerate the unavailable state from the available state vectors. The authors in the study [43] proposed an innovative function observer for the optimal AGC in a non-linear complex power system. The proposed control technique estimates directly the applied signal to each power plant with the help of observers. The suggested observer tackles parametric uncertainties, control loops, fault due to sensors, and cyber-attacks on input and outputs. Furthermore, to reduce the complexity and increase the chances for physical implantation, each observer is decoupled from another observer power plant. For performance comparison, the control strategy is evaluated on three regional power grids. The implementation issues of the complex optimal-AGC system are countered by introducing sub-optimal and near-optimal control methods for the AGC designs. In a study [55], the researchers created a sub-optimal AGC configuration for a dual area interconnected power system consisting of a non-reheat thermal energy system in each field and added an AC/DC power link for interconnection. The complexities of solving non-linear matrix equations in an optimal-AGC system are reduced by designing the sub-optimal AGC, which is based on the constrained feedback control technique and requires the feedback of only available states. Furthermore, various designs of weighting matrix 'Q' of state cost are presented in that work. Moreover, the dynamic performance of the suggested control technique is also analyzed by achieving the time domain plots in implementing the optimal and sub-optimal AGC system under one percent load perturbations. A sub-optimal AGC system, which is considered as an alternative for the optimal AGC in the hydrothermal power system, is presented in study [17] that works better in stabilizing a feedback system having few state variables. The proposed model is based on the state-space concept and is extracted using the partial fraction method for hydro and steam turbines. Moreover, the suggested decentralized AGC is synthesized using the minimum error excitation technique to further improve its performance.

6.3. Adaptive, Self-Tuning, and Model Reference Control Methods

Since the operating point of the system continually fluctuates, the control effect will not be optimal for the system. To hold system output close to the nominal value, the system's operating point must be tracked and the parameters should be modified accordingly. Adaptive controllers are known for their exceptional monitoring and parameter adjustment performance to enhance system reliability and robustness. An adaptive-AGC is considered based on the guided and indirect versatile fuzzy control methods [42] for a power system network with several areas. Fuzzy logic is utilized to update the algorithm and an H∞ tracking standard is applied to eradicate the estimated errors and other disturbances effects. The authors in study [56] expounded different AGC regulators in an interconnected hefty power system incorporating various FACTS devices. An adaptive fuzzy PID-based AGC regulator is designed and incorporated in a deregulated power system to improve the dynamic efficiency of the regulator. The gains and the scaling factors of the proposed control technique are tuned using the wild goat algorithm, which is a recently developed nature-inspired algorithm. The suggested control is compared with its counterpart by dispatching the power in different contract scenarios, including poolco, bilateral, and during contract volition. Furthermore, the efficacy of the proposed controller is assured under extreme load disturbances and parametric variations.

Self-tuning adaptive control methods (SACMs) are extended to the AGC regulators in the interconnected power systems, which are mostly used in the non-linear processes [48]. A self-tuning algorithm-based AGC for a thermal source-based multi-area integrated power network is suggested in the literature [51], in which every control area is shown by reducing order stochastic Auto regressive-Moving average model (AR-MAX) and the parameters are estimated through the extended least square technique. A general cost function minimization-based self-tuning algorithm is employed to compute the corrective control for installed generators. The control algorithm is further coupled with a load computing system to take into account the non-zero component of load diversion. Analysis of the proposed control system shown that the regulator demonstrates preferable performance in the presence of appropriate non-linearities (GDB and GRC). A self-tuning regulator-based multi-level adaptive AGC scheme is proposed in the study [48]. The algorithm in the proposed model considers each area as an auto-aggressive average moving model and utilizes a recursive least-squares technique for parametric identifications. Model reference ACMs are commonly used in the AGC regulators that use a reference model to define the desired closed-loop performance of the system. A model reference-based decentralized output-feedback AGC is analyzed in the literature [57] for three areas of integrated power systems with indefinite parameters. An adaptive observer is specifically designed that assisted in estimating the system parameters and state variables using local inputs and outputs. These estimates are then amalgamated with the reference model states to construct the proposed controller for each area that shows improved output in the company of parametric variations and non-linearities in the system. The aforementioned ACMs show encouraging performance but still face some issues, like the possibility of failure of the parameter estimation module, which can seriously put the power system at high risk. Moreover, the ACMs emphasize in bringing the steady-state error to zero and is deprived of mostly seeing the transient performance.

6.4. Variable Structure and Sliding Mode Control Methods

The variable structure control method (VSCM) was developed in the 1950s in a USSR research lab. The most eminent feature of this method is that any changes can be made in the structure of the control system during the transient process. Furthermore, the time at which these changes are made, the resultant structure is not determined by a fixed program, but in conformity with the system's current state. The versatility of VSCM recommends its implementation for power system control problems. Detailed analysis on the implementation of a variable structure AGC system in a single-area power grid is presented in study [30]. The basic concept of the variable structure control system is explained and implemented to improve the dynamic performance of the integral-based AGC regulator to eradicate the area frequency oscillations. A VSCM-based AGC design for dual and three areas integrated power grid is proposed in [58]. Two models are developed, i.e., optimal and pole placement variable structure AGC regulators. The switching decision of the proposed controllers is taken based on local data and those state variables that are accessible. The results revealed better dynamic response in terms of insensitivities to the uncertainties of system parameters. Sliding mode controllers are the kind of variable structure that drives and constrains the system to lie within a range that encompasses switching functions. In such type of a system, the dynamic behavior is tailored by the switching function resulting in the robustness of the control system. A discrete sliding mode controller (SMC) is utilized in the AGC framework of a power system with four areas interconnected [39]. The proposed control system is based on the entire-state-feedback system that takes the advantage of its applicability in both thermal and hydro power-based control areas keeping in view their non-minimum phase behaviors. The frequency deviation, swapping, generated power and exterior disturbance are measured using a rapid sample-based state estimated method. Simulation results showed that a better system reaction to reject distortions and marinate the necessary quality of control was achieved by a proposed control technique. In the literature [59] a multi-area, interconnected energy system with different power units is provided with an SMC-based AGC output feedback regulator. The controller is tuned using the TLBO algorithm and correlated with the output feedback SMC regulators based on DE, PSO, and genetic algorithm (GA) optimization techniques in a two-region power system with solar, water, and gas power plant in region 1, and thermal, hydro, and nuclear power plants in area 2. The proposed control system is optimized by using the TLBO algorithm. The results of the proposed control system are also checked by incorporating the effect of HVDC link. An SMC is investigated in [60] for the AGC system with a chattering lessening characteristic in a multi-area non-linear integrated power system. The above approach uses a GA algorithm to detect input gains following formulations of SMC designs as an optimization function. To minimize SMC chattering and increase complex performance, the optimization process also recommends double objective functions. VSCM-based frequency controllers have shown better dynamic response due to the unique property of proper selection of the plant parameters. However, due to the complex structure and the associated implementation problems, VSCM is not much appreciated for the AGC systems. Also, controllers centered on the linearized model state equations may require an observer to approximate unavailable state variables that would entail additional data telemetry cost.

6.5. Robust Control Methods

Robust control methods (RCMs) are used in the controller designs of the interconnected power systems to function accurately in case of any uncertainties or disturbances found in the network. RCMs are static and in contrast to ACMs adapt themselves to the measure of variations. Instead, the RCM-based controller continues work by assuming certain variables as unknown. Considering the importance of this method, several studies have been conducted, where RMCs have been implemented in the AGC system [25,44,61,62]. In [61], a systematic study is conducted on the robust AGC regulator implementation in an integrated power grid. A PID controller is used for the AGC system, which is tuned using the fine approach of fuzzy logic under load perturbations. The output of the suggested control techniques resulted in a very robust performance due to an extensibility feature of fuzzy logic, which furnishes an established framework to have wide observability on the control space. In [44], a Riccati-equation approach is utilized to develop an AGC for a multi-area power system with appropriate parametric uncertainties. The overall controller has N equations for N number of areas and N interrelated equations are generated initially, which then decoupled through a decoupling technique. Furthermore, the Riccati equation is made able to incorporate the bounds of parametric uncertainties for improving the robustness of the proposed regulator. A single AGC equation is then obtained from the solution of the respective decoupled Riccati equation requiring no-feedback from other areas. The proposed control system is analyzed after implementing it on a three-area power system, which demonstrates the reliability of the system when subjected to non-linearities. A robust multi-stage fuzzy PID regulator for the AGC framework is designed in [62] in a deregulated scenario of a bilateral contract scheme. The control system is fine-tuned to generate optimal performance based on the online information base and fuzzy inference, for which the membership functions are tuned using the improved genetic algorithm to increase the convergence speed. The proposed controller is tested under severe loading conditions and for a diverse range of parametric variations in the existence of various non-linearities. In all cases, the performance of the proposed control scheme was sufficient in mitigating the disturbing results. Parallel to all these advantages of RCM, massive system order, uncertainly linked subsystems, complex parametric variations, and excessive power network organizational layout hinder RCM's direct implementation in power system controls.

6.6. Model Predictive Control (MPC) Methods

Model predictive controls (MPC) are the most advanced control method used in the AGC problems to deal effectively with different constraints of the network. The general knowledge of MPC-based AGC regulator of a dual-area integrated power network with hydro and thermal energy systems is presented in [63]. The proposed model incorporated appropriate GRCs, GDBs, and time delays that occurred due to the thermodynamic process of the governor-turbine and communication channels. A bat-inspired algorithm is utilized to optimize the parameters of the suggested AGC system. Furthermore, the superiority of the MPC-based AGC system is shown by comparing its result with GA optimized PI controller over different load perturbations and parametric uncertainties. An MPC scheme is utilized in AGC framework of the Irish transmission network with significant penetration of wind power [64]. Such penetration of wind power resulted in decreased system inertia, which can have a direct effect on the operator's efficiency in regulating the frequency of the system. Various implementation problems and computational aspects are presented and the outcomes of the proposed control schemes are equated with traditional PI controllers for a conclusion. An adaptive MPC-based AGC for a dual area integrated power network with thermal and photovoltaic sources seeing the non-linear characteristic of governor and turbine is proposed in the literature [65]. First, the dynamic features are approximated using a discrete state-space model, then the prediction model is estimated using the model parameters and the control signal. The control signal is deployed using the control value of the weighted total of the residual errors and control input error. The superiority of the proposed control model is shown by comparing the results with those of the firefly algorithm, and GA-based PID regulators under extreme load disruptions and parametric uncertainties. An AGC system based on MPC is proposed for multiple terminal HVDC grids, which is a new technology in the development of global large-scale power grids [66]. These grids can efficiently regulate the frequency of AC systems. The advantages of MPC versus PI are shown for the enhancement of both frequency regulation and DC grid control power despite the delays and DC voltage constraints.

6.7. Digital Control Methods (DCMs)

Digital controllers are more accurate, smaller in size, having lower noise, and more adaptable. Considering these important features, researchers carried out extensive work on the application of DCMs in power system functions. A detailed study on the general usage of DCMs in AGC power systems was conducted in the literature [20], to achieve a discrete model of ACE by sampling the tie-line power flow and frequency variations. The outcome was then transmitted via telemetry. It was found that, unlike continuous mode, the discreet mode control vector must remain constant over sampling instants. Moreover, the dynamic criteria for the performance evaluation of the proposed controller were also presented. A new type of SMC-based digital AGC regulator using a full feedback control strategy to handle both thermal and hydropower plants is suggested in the study [39]. To enable the full feedback control strategy, a state estimation method is adopted, which is based on the fast sampling of measured outputs. The proposed control system is tuned using the GA for better disturbance rejection and to maintain the quality of the proposed control system. A digital model of PID-based AGC for the Egyptian power system (EPS) inculcating the communication delays is presented in the study [45]. The proposed model includes both traditional power generating units of Egypt like reheat and non-reheat as well as hydropower systems with non-linear properties and wind-generating units. To achieve the aim of an overall closed-loop system, the digital regulator-based Tustin method is developed for each substation of EPS. The analysis of the proposed controller showed better performance than the analog controller under heavy loading conditions, high wind penetrations, and communication delays. Furthermore, the proposed system reduces the implementation cost as it provides reliable performance at a huge sampling time. A discrete AGC system for a Spanish power system interconnected with French, Portuguese, and Moroccan power systems is proposed in one study [67].

7. Intelligent and Soft Computing Control Methods

Practically, non-linear systems like power systems are usually approximated by the use of reduced-order models, probably linear (related to the particular characteristics of the plant). Therefore, non-linear systems are accurate under some operating conditions in most situations. Otherwise, the control system would not be adequate or would require to adopt the modified system parameters. Moreover, the nature of the modern power system holds uncertainties, nonlinearities, and complexities, for which the traditional AGC system does not provide a sufficient solution. Intelligent control methods and soft computing-based control strategies are very effective in solving the above-mentioned problems. This section provides a comprehensive review of Intelligent and Soft Computing AGC approaches.

7.1. Intelligent Control Methods

7.1.1. Fuzzy Logic Control (FLC)

FLC theory was introduced in 1965 as an intelligent concept that advances the traditional control theories. FLC is used as an important tool for mathematical approaches to solve power system problems. Hence, significant literature is reported on FLC applications in power system control to enhances its reliability and robustness. A design procedure and numerical validation of a robust fuzzy logic-based tuning method for AGC regulators in a multi-area power grid are presented [61]. Here, a triangular membership function is used to convert the crisp value into a linguistic variable and a Mamdani inference mechanism is utilized for the proposed controller. It is worth to mention that the proposed controller is only activated encountering faulty situations, while in the steady-state only the conventional PI controller performs the activities. The AGC based on direct and indirect adaptive fuzzy logic for a multi-area energy system is suggested in study [42], in which different approximation capabilities of the fuzzy technique are explored to estimate a suitable regulator law and algorithm for parameters upgradations. The performance criterion of H∞ has been used to eradicate the area errors and external disturbances. Correlation of the proposed control framework with the traditional PID and type-2 fuzzy controller has shown better results in providing stability of the overall closed loop. The research work on the PI and PID controllers centered on fuzzy logic concepts has also been discussed in the literature. In [62], the authors introduced a multi-stage fuzzy AGC regulator in a three-area restructured power grid using a bilateral arrangement. In the proposed control system, the control signal is tuned online containing a minimal number of references, and has only two basic laws. In addition, membership functions are designed using the modified GA to ensure a global optimum value and to reduce the speed of algorithm convergence. The analysis of the proposed control scheme delivered better performance under the large load disturbance and variations in the parameters. A self-tune PID regulator is proposed in study [68] for the AGC system in a dual area integrated power grid. A bunch of control rules is produced for the regulator from which the control signal is determined using an information-based and fuzzy inference. In addition, a variable is utilized for the proposed regulator, which adjusts the I/O scaling factors of derivative and integral coefficient in the proposed PID type fuzzy rationale system.

The parallel combination of fuzzy logic and PID proved a useful control strategy in non-linear power systems. The resultant controller delivers better dynamic performance when states are not near to the equilibrium and parallel to that it also holds a lower control signals profile. However, this approach is only efficient in effectively selecting the tuning parameters and optimizing them for optimal solutions. Therefore, opting for the optimization algorithm for parameters tuning is of significant importance. The authors in [69] has applied a fuzzy PID regulator for the AGC framework in two region power system with hydro and nuclear energy sources. The overall parameters are tuned by employing a gray wolf optimizer algorithm. The GRC of 3 percent for thermal power plant and 270 percent up and 360 percent down for hydropower plant has been considered. The performance parameters are measured using the TCPS in series with a tie-line. The analysis showed that the power full gray wolf optimizer algorithm can efficiently damp the oscillations in both area and tie-lines. In addition, the robustness of the proposed controller has been tested under different loading conditions. An online keen procedure dependent on the blend of self-adaptive bat algorithm and fuzzy strategy to ideally tune the PI parameters of an AGC framework in a four-territory incorporated force framework is introduced in the literature [70]. The simultaneous optimization of controller parameters and input/output membership function increases the robustness and security of the complete system against outside disturbances. The supremacy of the anticipated online control system is checked by comparing its result with the latest fuzzy PID control techniques.

7.1.2. Artificial Neural Network (ANN) Control

The prime interest of researchers in artificial neural networks is its best approximation of arbitrary non-linear functions and its usage in parallel processing and multivariable systems. Further, a neural network mimics the human brain and can learn in a complex multi-layer network based on which it responds intelligently. Various types of topologies exist for the ANN in deep learning algorithms that are applied to both supervised and unsupervised learning approaches. Besides, the newly evolved reinforcement learning algorithms for ANN are catching the ground in real-world applications. Throughout the development of AGC schemes, ANN schemes are suggested and used by the researchers [21,25,71]. A comparative study of various savvy control procedures for the AGC of a dual area integrated power network considering three kinds of turbines, i.e., reheat, non-reheat, and the hydraulic turbine is proposed in the literature [71]. The kind of control techniques incorporates ANN, fuzzy logic, and regular PID-based controllers. A multi-layered perceptron is utilized as an example in ANN, where numerous input-neurons work in correspondence to frame the layers, which are then used to build up the entire network. The network is trained through the Levenberg-Marquardt back propagation algorithm and considered both transient and steady-state performance of the plant output. To address the disadvantage of requiring an immense amount of training time and a large number of neurons in a simple neural network, a generalized neural network is practiced in [21]. The current downside is removed in the generalized configuration of the neuron. The suggested control scheme was tested in various system scenarios and with different types of loads, showing a quicker response due to its frequency shift rates to predict load variations.

An emotional reinforcement learning control (ERL) scheme for AGC of a dual area interconnected grid and the China Southern Power System is suggested and applied in the literature [72]. The agent in the proposed scheme considers two portions, namely a mechanical logical part and a humanistic emotional part. The suggested controller creates various control techniques depending upon the requirement of the operating scenarios by highly integrated artificial emotional functions including quadratics, exponential, and linear functions, and with the RL elements such as learning rate, reward function, and actions. Moreover, a deep forest RL algorithm is designed in [73] as a preventive strategy for the dual and triple areas AGC problems. The proposed method consists of two components of deep forest and multiple subsidiaries. The deep forest is responsible for forecasting the next systematic condition, while several subsidiaries are used to lean on the characteristics of the power structure. The superiority of the proposed control is tested on ten other traditional controls in two and three regional power grids and China's southern power grid. The proposed control showed the best efficiency in emergency scenarios and reduced the dimensionality challenge. A reinforcement learning (RL) scheme is also used to train the ANNs used in the AGC systems. Parallel to this, the hybrid structure of the ANN controller and other intelligent controllers are also reported in the literature. A combination of the ANN and a conventional control methodology is presented in [74] for the synthesis of AGC in the interconnected power system. Further, a non-linear ANN-based AGC model is suggested, which is centered on the μ-synthesis theory that deals with the uncertainties in the power system. Moreover, the flexibility of the ANN has been increased over time using the flexible sigmoid functions. The resultant network is called a flexible neural network (FNN) used for the design of the AGC system. An artificial FNN based AGC for a three-area integrated power grid network under a restructured environment is proposed in the literature [75] to attain the minimum acceptable regulation in the area frequency and to eradicate the effect of the disturbance during the heavy loading conditions and line disturbances. Here, the authors have used a sigmoid unit function to shape the neural network as a flexible unit.

7.2. Soft Computing Control Methods

7.2.1. Genetic Algorithm (GA)

GA is a meta-heuristic optimization technique that handles strongly non-linear and noisy cost functions. GA uses the individual fitness value and can be effectively converted to a near-optimal solution in complex engineering applications. Many literature studies report the use of GA for tuning the control parameters of the conventional AGC regulator system. A GA is utilized in the AGC system of an interconnected power system with a sliding model controller incorporating the non-linearities of GRC and integral control limiter [60]. The GA is utilized to find gains of the feedback and the switching vector of the regulator. A hybrid Taguchi-genetic algorithm (HTGA) is proposed in [76] for the AGC of a dual area integrated power grid with multiple sources (thermal, hydro, and wind power sources) based on the doubly-fed induction generator (DFIG). To increase the performance of the genetic algorithm in respect to a large standard deviation of the fitness value, a Taguchi technique is utilized, which is centered on the adapted statistical method and systematic cognitive capability. The supremacy of the suggested modified HTG algorithm is proved by comparing its optimum gains for the AGC regulator with the conventional GA-based AGC system for the dynamic stability of the power system.

7.2.2. Particle Swarm Optimization (PSO)

PSO is another population-based evolutionary modeling approach inspired by social behavior analysis rather than suitable survival. The implementation of the PSO in the AGC method shows that the PSO technique produces the optimum gains required in the shortest possible time. The implementation of the PSO and improved PSO algorithm in the AGC framework of two unequal areas interconnected power systems is presented [32]. The improved PSO methodology uses a dynamic search squeezing approach to speed up the optimization process. The efficiency of the suggested regulator is assessed by the objective function of the ITAE and with a 1% step load change in either control field. Further, the results revealed that the suggested IPSO-based regulator increased the dynamic efficiency of the network by considering AC/DC lines instead of AC tie-lines. A PSO-based optimal-AGC system is designed in [45] for an Egyptian power system with conventional and renewable energy sources by considering the communication delay. To safeguard the stability of the complete closed-loop system, an optimal regulator is designed for each subsystem, which uses the Tustin approach. The contrast of the suggested control system with other conventional controllers proved that it has more robust performance and delivers unswerving performance at large sampling time.

7.2.3. Firefly Algorithm (FA)

The firefly algorithm (FA) is a metaheuristic algorithm that is recently established and is inspired by fireflies' attractive and flashing behavior. The use of an FA in the power system is suggested in [65,77]. The authors in [77] presented a PID-based AGC system for a five-region interconnected power framework considering the reheat thermal energy station in every zone. Each area has its appropriate governor, reheater unit with turbine, generator, and regulator unit for speed. To properly tune the parameters of the proposed PID-based AGC regulator, the nature-inspired firefly algorithm is used. Moreover, the predominance of the suggested approach is tried by contrasting its outcome with

GA-tuned PID-AGC and PSO-tuned PID-AGC framework. The proposed FFA-based AGC framework achieved better execution in regard to settling time, top overshoot, and undershoot. Besides, FFA likewise showed better execution with an enormous number of iterations.

7.2.4. Artificial Bee Colony (ABC) Algorithm

The ABC algorithm is a swarm-based algorithm that depicts a honey bee swarm's clever hunting behavior. In the literature it has been frequently used by researchers to solve AGC problems [78,79]. In [78], a detailed analysis of the ABC algorithm, when applied to the AGC of the interrelated thermal energy system is provided. The algorithm is utilized to change PI/PID regulator boundaries in the AGC system. Further, the strength of the proposed regulator is checked by uncovering the AGC framework to a wide scope of varieties in the load and generation boundaries. Moreover, the proposed controller performance is compared with the recently developed heuristic algorithm by determining their transient performance. An ABC-based AGC model for a stand-alone microgrid having hybrid renewable energy sources is considered in [79].

7.2.5. Differential Evolution (DE) Algorithm

DE is a population-centered process that provides the solution to noisy, non-linear, non-continuous, and multidimensional problems. DE uses the iteration process to provide better results concerning a defined standard. A study [80] demonstrated the behavior of DE-based 2DOF-PID-AGC regulator for a dual area integrated power network with a thermal power plant. The proposed controller is designed using the ITAE objective function and the robustness of the regulator is checked under the diverse loading condition of 1%, 5%, and 10% SLP, applied in either area, which has shown better performance in term of percent overshoots and settling time. A hybrid DE and PSO technology are used in [81] to configure the PID-based AGC system parameters. The proposed controller employed a hybrid form of DE and PSO technique to regulate the parameters. The dominancy of the suggested control process is shown over the independent usage of the DE and PSO technique. Also, the control algorithm is observed in different system parameters and it was noted that even under extreme loading conditions, the optimum gain of the proposed control did not need any modification.

7.2.6. Bacterial Foraging Optimization (BFO) Algorithm

The newly designed optimization algorithm BFO is focused upon *Escherichia coli*'s searching behavior and used by researchers in the power system control problems for its efficacy in the results. Researchers have described the implementation of AGC based on the BFOA algorithm in [34]. The authors in [34] have utilized a BFOA optimized Integral regulator centered AGC system for a dual area power network to tune its parameters under different load perturbations. The proposed control system based on the BOFA algorithm has been tested with IPFC units to advance the dynamic performance of the overall control. In [53], various classical AGC regulators including I, PI, ID, PID, and ID utilized in different area thermal energy-based power system are proposed, which are tuned using the BFO algorithm. The authors in [27] presented and compared I, PI, PID, IDD, and F-IDD regulators for the AGC system of hydrothermal energy-based power plants using the ISE in formulating the objective functions. A sensitivity study indicates the robustness of the F-IDD regulator under a variety of loading situations.

7.2.7. Bat-Inspired Algorithm

The bat-inspired algorithm relies on the behavior echolocation of microbats with different emission and loudness pulse rates. The issues involved with the usage of this algorithm in AGC are investigated in the literature [63,70]. A BIA optimized MPC-based AGC regulator in a dual area power network having non-linearities of GDP and GRCs is presented in [63]. The proposed control system showed superior results in terms of overshoots/undershoots and settling time by comparing its results with a GA-optimized PI-based AGC regulator. An online intelligent control scheme for the AGC is suggested in a multi-area interrelated power grid using a hybrid form of a bat algorithm and fuzzy logic [70]. The hybrid control technique incorporates the properties of both the fuzzy logic and bat algorithm to optimize the typical PI regulator parameters that ensured the robustness and stability of the proposed control system.

7.2.8. Quasi Oppositional Harmony Search (QOHS) Algorithm

A QOSH algorithm added to the AGC of the power system network having two, three, and five areas interconnected with each other is presented in [38]. In the first step, QOSH is used to change PID controller parameters, while in the second phase, a PID is optimized with the proposed algorithm. Further, the suggested algorithm is also examined in the occurrence of nonlinearities and parametric uncertainties. In the end, a comparative comparison was rendered with other traditional control algorithms to authenticate the efficacy of the proposed process. The application of the QOSH algorithm to the SFL centered AGC of a multi-area restructured power grid is analyzed [19]. A total of four areas having eight GENCOS and eight DISCOS is considered as a completed plant model to examine the performance of AGC in the occurrence of intensified bilateral behaviors. The key benefit of QOSH is a good dynamic response over different parametric variations.

7.2.9. Teaching Learning-Based Optimization (TLBO)

TLBO is a recently developed method that emulates the teaching and the learning method of the classroom and does not require the parameters associated with specific algorithm controls. It helps to find a global solution and provides an analysis of AGC design and performance. The AGC system for dual area power grids with a thermal energy source and based on the output feedback SMC regulator is presented in the literature [59]. The parameters of the proposed regulator are improved utilizing the recently created TLBO algorithm. In addition, the implementation of the suggested regulator is also expanded to a multi-area power system incorporating the HVDC link. A detailed overview of the execution of a fuzzy PID controller in two unequal power grids is given in the literature [82]. The architecture problem is developed as an optimization problem and the parameters of the proposed controller are obtained using the TLBO algorithm. In addition, a comparison of the proposed approach with some of the recent optimization strategies is shown, which is comparatively stronger for maximizing dynamic performance in terms of settling time and peak overshoots and undershoots.

7.2.10. Cuckoo Search Algorithm (CSA)

CSA is an effective algorithm created by Yang and Deb that solves problems of global optimization. CSA is influenced by some cuckoo species' compulsory parasitism as they put their eggs in the nests of other birds. In [36], a two-degree of freedom regulator for a multi-area AGC framework is proposed considering reheating turbine-based thermal sources. Various FACTS devices such as SSSC, TCSC, TCPS, IPFC are used and compared using the CSA-based 2DOF-IDD controllers for the AGC system. Further, sensitivity analysis showed that the parameters obtained for a CSA-based 2DOF-IDD-based AGC framework in the company of IPFC are robust and need not be re-set in the presence of large load perturbations. The authors in [83] presented a CSA-based AGC system for a dual area power grid with hydro, thermal, and wind power plant based on DFIG. The parameters of the DFIG and AGC are tuned using the CSA algorithm. Furthermore, the effect of the different load conditions on frequency regulation is studied in different environments.

7.2.11. Grey Wolf Optimizer (GWO) Algorithm

GWOA is a modern algorithm influenced by the social behavior of grey wolves, exploring this capacity to solve real-life problems. An AGC regulator based on I, PI, and PID for a three-area power framework with thermal energy sources is provided in [37] considering the necessary GRC constraints. The proposed controllers are calibrated using a GWO algorithm and sensitivity analysis is conducted, which showed that the obtained parameters are safe and are not expected to be modified due to significant changes in the system conditions. A detailed implementation of the GWO algorithm for the implementation of fuzzy PID regulator in the AGC of two-area power grid with appropriate non-linearities of GRCs is presented in [69]. The use of TCPS in tie-line with the proposed GWO based AGC system showed the best performance under different loading conditions and system and parametric variations.

7.2.12. Other Computing Control Methods

The other soft computing algorithms for the AGC problems in the power system are wind-driven optimization (WDO) algorithm [29], big bang-big crunch algorithm [40], hybrid local unimodal sampling technique (LUS) [84], ant colony optimization algorithm [26], wild goat algorithm (WGA) [56], grey wolf optimizer algorithm [69], self-adaptive modified bat algorithm [70], quasi-oppositional selfish-herd optimization algorithm [85], and whale optimization technique [86]. In [29], a detailed investigation is carried out on the implementation of the WDO-based AGC regulator in the integrated power system. Further, the analysis on WDO-based AGC system considering different objective functions is carried out and robustness of the suggested control technique in situations of varying system limits is confirmed. In [40], the authors used a big bang-big crunch algorithm to optimize the parameters of type-2 fuzzy PID-based AGC system in a deregulated power system to further improve the convergence speed. A hybrid LUS-TLBO algorithm for tuning the fuzzy-PID regulator is proposed [84] for two unequal areas integrated power systems. For a load agitation in area 1, the dynamic output of the suggested regulator and traditional PID regulator is implemented and compared. Further, the strength of the aforementioned regulator for an AC tie-line is also carried out by varying the parameters from −50% to 50%. A quasi-oppositional selfish-herd optimization algorithm is proposed in the literature for the frequency control management system of an isolated micro grid, incorporating bio-gas, micro hydro, solar thermal power unit, and bio diesel generator unit. Moreover, suitable demand response strategy is devised for the proposed micro grid in isolated and interconnected mode. The system response is studied in term of its adaptability in four different scenarios of source variations and three scenarios of demand response variations. The authors in [86] presented a whale optimization technique for a Fractional order PID with filer of AGC in a dual area power network under restructured environment. Area 1 had a thermal generating unit and DGs including WTS, DSTS, AE and fuel cell, while area 2 had gas and thermal power plants. The obtained parameters for proposed AGC system using WOA showed greater robustness under a wide range of parameters. In tables 3 and 4, the detailed comparative analysis of AGC systems in two, three, and five area power system is provided using different controller approaches.

Table 3. Performance evaluation of AGC in two-area power systems with conventional sources.

Ref.	Power System Configuration	Controller Approach	Operating Scenarios	Peak Overshoot			Settling Time		
[80]	Two area non-reheat thermal power system	DE based 2DOF-PID regulator	Controller comparisons with other types:	Δf_1	Δf_2	ΔP_{tie1-2}	Δf_1	Δf_2	ΔP_{tie1-2}
			2DOF-PID	0.0144	0.00598	0.00711	11.1	7.2	13.8
			PID	0.0251	0.0172	0.0083	24.8	24.1	23.3
[26]	Two area non-reheat thermal power system	ACO based PID regulator	Comparisons of objective functions:						
			ISE	1.1×10^{-6}	0.0002	0.0001	29.62	37.20	50.41
			ITSE	0.0016	0.0001	0.0005	25.69	25.43	45.54
			IAE	3.66×10^{-6}	0.0000	0.0001	26.29	32.12	37.64
			ITAE	0.001	0.0000	0.0005	23.69	32.12	33.63
[59]	Two area reheat thermal, hydro, gas and nuclear power plant	TLBO based AGC system with output feedback SMC	Controller comparisons with other types:						
			SMC with output feedback with TLBO	0.001	2.242×10^{-4}	2.883×10^{-5}	1.3	1.46	1.05
			SMC with output feedback with DE	0.0018	6.389×10^{-4}	7.680×10^{-5}	1.4	1.9	1.1
			SMC with output feedback with PSO	0.0016	4.301×10^{-4}	8.050×10^{-5}	1.52	1.54	1.24
[84]	Two area reheat thermal, hydro, gas and nuclear power plant	LUS-TLBO based Fuzzy PID controller	Performance evaluations:						
			Without AC-DC tie-lines	0.000551	0.000219	0.0000826	5.26	2.96	2.36
			With AC-DC tie-lines	0.000280	0.000208	0.0001353	1.85	4.14	2.55

Table 4. Performance evaluation of AGC in three and five area power systems under different conditions.

Power System Configurations	Controller Approach	Operating Scenarios	Peak Over Shoot						Settling Time					
Three area power system with hydro-thermal sources	BFA based IDD and FDID regulator	Controller comparisons	Δf_1	Δf_2	Δf_3	ΔP_{tie1-2}	ΔP_{tie1-3}	ΔP_{tie2-3}	Δf_1	Δf_2	Δf_3	ΔP_{tie1-2}	ΔP_{tie1-3}	ΔP_{tie2-3}
		BFA-IDD	0.02142	0.01335	0.01081	0.00113		0.00117	60.26	64.81	65.47	61.00		73.63
		BFA-FIDD	0.01554	0.01071	0.00742	0.00059		0.00069	49.75	52.76	55.21	51.37		58.75
Three area non-reheat thermal power system with non-linearities	CSA based 2DOF-ID controller	Use of FACTS devices												
		SSSC	0.0013	0.0019	0.0012	0.0002	0	0	33.9	24.28	30.61	51.12	40.42	29.12
		TCSC	0.00004	0.01071	0.00742	0.0002	0	0	33.06	34.46	33.28	51.13	40.20	34.35
		TCPS	0.0012	0.0013	0.0004	0	0	0	31.7	33.36	29.21	50.97	40.23	34.32
		IPFC	0	0	0	0	0	0	28.53	23.37	26.03	34.56	40.07	29.12
Three area non-reheat thermal power system with non-linearities	GWO based PID controller	Use of ESs												
		With-STTP	0.007231	0.006335	0.006087	0.001814	0.001761	0.0008671	34.81	35.56	32.35	42.84	41.67	38.82
		Without-STTP	0.01432	0.01213	0.006886	0.001026	0.0007518	0.001709	24.47	23.35	21.81	21.26	24.11	32.97
Five- area reheat thermal power system	FFA based PID controller	Controller comparisons	ACE_1	ACE_2	ACE_3	ACE_4	ACE_5		ACE_1	ACE_2	ACE_3	ACE_4	ACE_5	
		GA-PID	0.00071	0.00063	0.0006	0.00089	0.0006		17.84	26.63	17.19	24.33	17.19	
		PSO-PID	0.00042	0.00065	0.00042	0.00078	0.00042		16.33	22.7	16.54	23.23	16.33	
		FFA-PID	0.00038	0.00075	0.00038	0.00085	0.00038		13.53	21.8	14.55	22.77	13.79	

8. AGC Incorporating ESSs, FACTs Devices and HVDC Link

This section briefly describes the applications of various energy storage systems, FACTs devices, and the integration of HVDC links into the AGC network to further enhance its capabilities. This section also discusses the implications of these devices in the modern era.

8.1. AGC Incorporating Energy Storage Systems (ESS)

The significance of ESSs can be estimated by how their utilization in AGC productively diminishes frequency varieties in the power system networks. Keeping in view such importance, researchers have investigated the utilization of ESSs in AGC models for interconnected power systems. The authors in [33] introduced the AGC framework for a deregulated two-region interconnected power system having different thermal and hydro gas-producing plants and incorporating capacitive energy storage systems (CES) and DFIG for the frequency regulation purposes under different load perturbations. The performance of the proposed control framework is tried in various contract scenarios, which showed that the integration of CES units with TCPS into the proposed control system provided better transient performance in arresting the frequency dips and the deviation in the tie-lines. Redox flow batteries usually do not age out by the frequent charging and discharging. Further, it is equivalent to the SMES and results in a very quick response characteristic. Keeping this in view, the authors in [34] have presented an Integral-based AGC regulator for dual area deregulated power system, where different devices with fast consumption rates and different capacities may cause and sustain frequency oscillations in the system. To tackle this problem, the RFB system is mounted in successions with IPEC and a tie-line for investigation. Simulation results have revealed the better dynamic performance of the proposed control system in the wake of sudden load perturbations. In [87], the authors have explored the usage of the double-layer capacitor (DLC) with the collaboration of fuel cells in LFC of a standalone hybrid renewable-based microgrid. The dynamic efficiency of the result revealed that DLC provides a better solution for damping the oscillation in an isolated distributed generation system. The study in [88] provides an in-depth analysis on the implementation of supercapacitor and Li-battery into AGC of traditional multi-area power grid to improve its dynamic efficiency against the frequency oscillations and other external disturbances. Two theories of state feedback and matrix inequality are combined to design the proposed controller, which estimates the power required from the batteries to be injected into the system. Comparisons of the proposed control system with a conventional PID-based AGC system proved its performance in stabilizing the whole system and increasing the robustness of the overall controller. The authors in [86] presented a deregulated two un-equal region power systems. Area 1 has a distributed generation system and thermal units, while area two has only thermal generation sources. Further, the DG system consists of WTS, DSTS, AC, FC, DEG, and battery energy storage systems. The AGC based on the filter-PID cascaded with a fractional order PID (PIDN-FOPID) regulator is proposed for the suggested power system and is tuned using the whale optimization algorithm. The matchless quality of the proposed regulator is checked by contrasting it with I, PI, and PIDN based AGC framework, showing that the former one outperformed the other controllers in reducing the settling time and oscillation during the variable power flow from both the wind and solar power sources. Also, the dynamic performance has been further improved with the utilization of the BES system.

The use of VICs in the AGC framework is another exciting approach to boost the inertial response and stability of renewable energy-based power system networks, where mounted inverters do not have a spinning mass and eventually minimize system inertia. In [23], an additional VI-based MPC is incorporated into LFC system of a microgrid structure composed of wind and thermal power plants. All the required inherent requirements and specific constraints relating to the operation of thermal power plants are engaged into account in understanding the accurate microgrid model. The control of the inertial block is refined by conveying energy storage frameworks, where the inertial help is given to the microgrid system similarly to coordinated rotors, and in this way, the transient response and unwavering quality of the microgrid framework are improved. VIC for an integrated power system with high reconciliation of environment friendly power sources is also adopted in the study [89]. The proposed VIC loop uses second-order characteristics that assist in enabling the damping and inertia emulations in the integrated power system and as a result, increases frequency stability and resilience. The inclusion of the HVDC connection to the interconnected power grids has further raised the demand of the VIC regulators. In [90], a virtual synchronous control-based VIC is studied in the HVDC connection of the interconnected power grid. To mirror the impact of virtual inertia parameters, a progression of affectability investigations is conducted to decide the correct determination of basic control parameters for dependability of the system. The prevalence of the proposed control framework is tried on two and three region power system networks.

8.2. AGC Incorporating FACTS Devices

From the past several decades, FACTS devices are reliably helping in using the transmission lines at their full limit as opposed to adding new and additional lines. The utilization of FACTS devices in AGC regulators of the power framework is accounted for in the literature [36,69]. Comparative performance of Static synchronous series compensator (SSSC), Thyristor controlled series capacitor (TCSC), Thyristor controlled phase shifter (TCPS), and IPFC in the presence of 2DOF-IDD based PID-AGC system for a three un-equal areas power grid is offered in the literature [36]. Unlike other facts devices, the existence of IPFC in all three areas provides improved efficiency in the system under differing load conditions, and as a result, provided better system output. In [69], the TCPS phase angle is regulated by a fuzzy-PID regulator in a dual-area power grid with hydro and thermal energy sources to increase the overall dynamic efficiency of the system. The analysis revealed that the overall frequency oscillation can be efficiently damped out using the proposed control scheme. The problem of disturbances among the interconnection and the control of tie-line power using the phase angle property of the TCPS is tackled in [91]. Furthermore, the tuning of the parameters using the PSO technique for the PID controlled AGC regulator in a bilateral contracted scenario after the large variation in load demands is investigated. The major findings of this research work revealed that the use of TCPS in tie-line traffic is very efficacious in suppressing the frequency deviations and oscillation in the tie-line traffic. A new robust decentralized AGC regulator for the interconnected power system is designed in [92] using the SSSCs cascaded with the tie-lines. The proposed decentralized AGC system converts the installed SSSCs into the MIMO system, thereby making each LFC of the SSSC system, to be designed independently, effective against the oscillations. Moreover, the stability margin of the suggested control is further increased in systems with uncertain states and having parametric variations by the use of a multiplicative stability margin. The comparative performance of TCSC-based AGC system with SSSC and TCPS as damping controllers in two equal-area integrated power systems is presented in [93]. The Taylor expansion is used to model the TCSC impact for low load turbulences in the tie-line control. The parameters of this control are set by an improved PSO algorithm using an ITSE objective function. The performance analysis showed that the TCSC-based AGC regulator effectively dampens the frequency and tie-line oscillations.

8.3. AGC with HVDC Link

With the potential to effectively transfer cumbersome power over long distances in the interconnected power systems, the usage of the HVDC connection in the power system control service is highly significant [90]. Important research on the frequency regulation of interrelated power grids via the HVDC link is presented [32,59,66,90,94]. An HVDC design technique for the AGC system in a multi-area interrelated power grid is demonstrated in the study [94]. In this process, the modeling of the T_{dc}, which is the time obligatory for the HVDC link to established the DC current after the load perturbations is challenged, and based on that analysis, a new accurate model is suggested for the HVDC link. Also, the inertia-emulation-based technique for AGC is introduced to harness the power from the capacitors of the HVDC link. In [32], three meta-heuristic enhancement methodologies are utilized to change the parameters of the PID-controlled AGC system in a double territory interconnected power grid. Various sources, for example, hydro, thermal diesel, and wind power plants are utilized in every region considering the AC/DC tie-line impact. In each control area, the output of the controller is checked for a 1% load shift. The results revealed that better dynamic performance of the proposed control mechanism can be achieved by using the AC/DC tie-lines instead of only AC tie-lines.

9. AGC in Renewable Energy Generation Systems

Currently, unparalleled developments are happening in the power markets that are not only limited to problems of deregulation and the implementation of modern regulatory strategies but mostly due to the amalgamation of RESs into the existing grid that significantly impact the system's inertia. However, it is difficult to dilute the value of renewables as green energy is a reliable clean energy option that comes from nature. Moreover, renewable energies are both the least costly and less expensive than other electricity generation sources. Research on renewable integration has shown that it influences the frequency and power volatility of the grid, which after large penetrations become more significant. Further, massive penetration requires network augmentation due to the risk of tie-line overloading. The AGC system in this regard requires more intelligence and flexibility to balance the fluctuating power and regulate the frequency deviations due to the integration of renewable energy sources. Therefore, substantial research is witnessed in the literature, in which the AGC performance criteria, capabilities, and technologies are revived in the presence of RERs.

Solar power is the most common and easiest to use in all renewable energy resources. The impact of power resulting from the photovoltaic generation on the load frequency controller is analyzed in [95]. The analysis showed that the integration of a photovoltaic source of 10% into the grid would approximately require a 2.5% increase in the capacity of AGC, compared to the conventional AGC regulators. The AGC technique for a power system with multi-areas connected and having photovoltaic generation sources is presented in [65], incorporating the characteristic of GDB and GRC non-linearities. A discrete time state-space model established the characteristics of the system, and then a predictive model is developed using the state vector and applying the control signal roll optimization technique. The suggested control model is compared with the new heuristic models grounded on FA, GA, and PSO algorithms, which showed the superiority of the later techniques. The ability of the AGC for the performance intermittency of the PV module is determined in [96] that is dependent on the isolation data taken from different points. The investigation showed that AGC capacity depends on the intermittency speed and if it is assumed that a certain power plant will respond to the PV fluctuation, then the required AGC capacity would be large than that of the PV system. Further, the AGC capacity largely depends on the observed insulation at a single point.

An AGC scheme of integrating wind power sources into the grid is described in the literature. Wind power can be used for control purposes and also as a single electricity generating unit. The authors in [83] introduced a study on the incorporation of DFIG-based wind power plants in the AGC of two territory power frameworks for the frequency regulation ability of the system. Two situations of an ordinary power plant are introduced, in which DFIG based wind power plant is integrated, first the two zones have nuclear energy stations and in second one nuclear energy station in zone one supplanted by hydropower plant. A CSA is utilized to concurrently optimize the parameters of the AGC and DFIG to additionally improve the frequency control ability of the system. Moreover, the analysis showed that utilization of wind power in a hydro-thermal power system is better when the DFIG-based wind power plant is installed in the hydropower area as compared to the thermal area. Integrating large-scale wind power into the system is a new test for AGC due to the unpredictable nature of wind energy. Generally, the active power control of WT is accomplished by pitch angle control (PAC) or rotor speed control (RSC), which give an exceptionally sluggish reaction and restricted controlling range. To improve the regulating performance, an altered control technique is proposed in the literature [97] to simultaneously enact both RSC and PAC and accomplish full leeway of the rotor kinetic energy to assimilate or deliver the required power. Besides, the AGCs based on intelligent control schemes such as NN, FL, and GA are also proposed in the literature to make feasible the large wind power integrations into the conventional power system for frequency control capabilities. In this regard, a decentralized fuzzy logic controlled AGC system is proposed in the literature [98] for an integrated power system having a considerable penetration from the wind power sources. The controller parameters are tuned using the PSO technique for increasing the dynamic efficiency of the regulator. The results revealed that the proposed AGC system provided a better response relative to the conventional AGC system following the abrupt load fluctuations and varying levels of wind power penetration. The AGC scheme, based on the PO-2-DOF-PID control technique in the power grid for the incorporation of large quantities of wind power is suggested in [99] to minimize variability and boost the system inertial reaction. The regulation is evaluated for three region power systems with high and low wind power production in each case and has a satisfactory solution to the adverse effects induced by the inclusion of wind power in the AGC. Table 5 provides the comparison of AGC schemes with and without RESs from recent literature.

Table 5. Summary of AGC Schemes with/without RES from recent literature.

Ref.	Type of Power System	Areas	Energy Generation Sources	Additional Devices	Controller Approach
[32]	Traditional	2	Thermal, hydro, wind-Diesel	AC/DC link	IPSO based PID
[33]	Deregulated	2	Wind, hydro, thermal, gas	TCPS, CES	ISE based I
[34]	Deregulated	2	Thermal	IPFC + RFB	BFO based I and FL
[36]	Traditional	3	Thermal	TCPS, UPFC	CSA based 2DOF-IDD
[37]	Traditional	3	Solar thermal, thermal	–	GWO based I, PI, PID
[38]	Traditional	1,3,5	Thermal	–	QOHS based PID/IDD
[76]	Traditional	2	Thermal, wind power	–	HTGA based PID
[40]	Traditional	4	Thermal, hydro	–	BB-BC based IT2FPID
[93]	Traditional	3	Thermal, hydro, gas	TCSC, TCPS, SSSC	IPSO based I
[99]	Traditional	3	Wind-thermal	–	PO-2-DOF-PID
[100]	Deregulated	2	Hydro, thermal, wind	TCPS, SMES	CRPSO based I
[101]	Traditional	2	Wind, thermal, hydro	TCPS, SMES	MWO based Fuzzy-PIDF

Abbreviations: UPFC: Unified Power Flow Controller; RFB: Redox Flow Batteries; TCSC: Thyristor controlled series capacitor; CRPSO: Craziness based-PSO; OGSA: Opposition based gravitational search algorithm; FOPID: Fractional Order-PID; FOFPID: Fractional Order Fuzzy-PID; CES: capacitive energy storage; CPSS: Conventional power system stabilizer; IPSO: improved particle swarm optimization; MWO: Multi-Verse Optimizer, MWO: Imperial competitive algorithm.

10. AGC in Microgrids and Smart Grids

10.1. AGC in Microgrids

A microgrid (MG) is a fairly recent phenomenon in the contemporary electricity sector, comprising small power systems with the ability to work in isolation from the main grid and is known for its resilience and reliability. MG is capable of handling traditional and non-conventional generation units like micro-turbine generators (MTG), diesel engine generators (DEG), fuel cells with aqua-electrolyzer, wind turbine generators (WTG), solar photovoltaic generators, solar thermal power generators, and energy storage units like, flywheel storage system, ultra-capacitor, and SMES. MG supports these units as a single entity and offers essential control capabilities. This section overviews the frequency management services in a single area stand-alone MG system and multiple areas stand-alone MG. Figure 3 shows the transition of a single area standalone microgrid to four area standalone microgrid systems, while Table 6 provides a comparative analysis of AGC schemes in both types of MGs network.

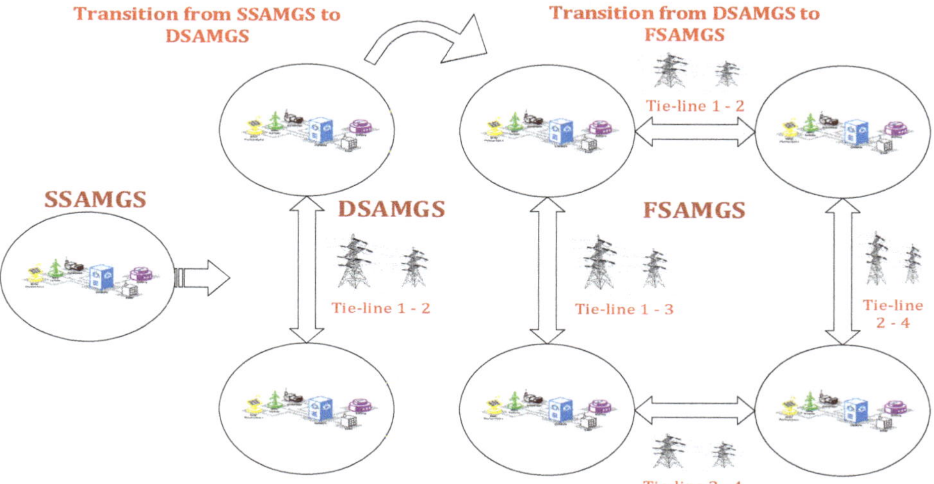

Figure 3. Transition from Single Area Microgrid to Dual and Four Areas Interconnected Stand-Alone Microgrid System, SSAMGS: Single Area Stand-Alone Microgrid System, DSAMGS: Dual Areas Stand-Alone Microgrid System, FSAMGS: Four Areas Stand-Alone Microgrid System.

Table 6. Comparison of AGC schemes in microgrids networks.

Ref.	Type of Power System	Areas	Energy Generation Sources	Additional Devices	Controller Approach
[79]	SSAMGS	1	DEG, WTG	FESS, FC	ABC based fuzzy-PID
[85]	SSAMGS & MSAMGS	1,2	Wind, MH, and BG,	-	NQOSO based PID
[87]	SSAMGS	1	WTG, PV	DLC, FC	GA based PID
[102]	SSAMGS	1	DEG, WT, PV,	FC, BESS, FESS	ABC based TSMC
[103]	SSAMGS	1	PV, WTG, DEG,	AE, FC, BESS	PI control
[104]	SSAMGS	1	WTG, DEG, PV	BESS, FESS	LADR control
[105]	SSAMGS	1	Wind, PV, DEG	BESS, SMES	PSO based ANN
[106]	MSAMGS	2	WTG and PV	SMES, BES	SSO based PID
[107]	MSAMGS	2	MT, PV DEG	FC, BESS	I-SSO based type-II fuzzy PID
[108]	MSAMGS	3	WTG, PV, PTC	ESS	MBA based 2DOF-PID

Abbreviations: NQOSO: novel quasi-oppositional selfish-herd optimization algorithm, SSO: Social-spider optimizer; I-SSO: improved-salp swarm optimization; DSTS: Dish-Stirling solar thermal systems, MBA: mine blast algorithm, FISCA: Fuzzy improved sine cosine algorithm, AWEC: Archimedes wave energy conversion, VIC: Virtual Inertial Controller: LADR: Linear Active Disturbance Rejection.

10.1.1. Single Area Stand-Alone MG Systems (SSAMGS)

MG in stand-alone mode is mostly used in areas that are inaccessible to the central grid. Here the purpose of the AGC is to control the frequency and hold the value of each generation unit within the optimum range. However, due to the stochastic nature of RESs, microgrids are highly exposed to unstable oscillations in frequency. In this respect, the authors in [102] proposed an intelligent terminal SMC-based AGC system for an islanded microgrid having diverse renewable energy sources. The problem of conversion of different states of the system in a finite time is handled using the non-linear sliding surface and by the usage of fractional power term in the surface. The parameters of the suggested controller are tuned using the ABC algorithm, which showed an excellent performance in the microgrid. To encourage the reliable participation of wind power and photovoltaic power into the LFC of a stand-alone microgrid system, it is necessary to study the accurate prediction model of solar and wind sources. In this respect, the authors in [103] have analyzed a long short-term memory (LSTM) recurrent neural network (RNN) to accurately forecast wind and solar power. The forecasted power from both solar and wind are utilized in the LFC model of a SSAMGS to analyze the behavior of LFC and the response of these sources under sudden load perturbation. The study in [79] presents an artificial bee colony-based AGC model for a SSAMGS system, incorporating DEG, FC, FESS, and WTG as reserves for the active power injection. The analysis showed that the proposed control system has effectively dealt with different problems of isolated microgrids such as low inertia, and unpredictable nature of RES. Furthermore, simulation results have shown that the above-mentioned RES has effectively handled the execrations in frequency during the load perturbations. A new LFC regulator employing fuel cell and capacitor banks to control the frequency of the single area stand-alone MG is explained in [87]. The MG incorporates a wide range of energy sources. The proposed control is tested under different loading conditions utilizing real weather data. Moreover, to effectively reduce the frequency oscillations in the stand alone microgrid, the authors in [104,105] proposed a linear active disturbance rejection control technology for LFC in a SSAMGS. The concept of extended state observer has been used to approximate the extended states and compensate estimation of disturbances. Furthermore, the analysis of the various demand coefficient influence on LFC has also been carried out to check the response speed of the suggested controller. Notwithstanding the far and wide utilization of AGC controllers in SSAMGS, it faces a few difficulties, for example, dealing with the low moment of inertia and the discontinuous nature of the RESs.

10.1.2. Multiple Areas Stand-Alone MG Systems (MSAMGS)

To meet the consistently increasing daily load demands and to supply reliable power, the idea of the multiple areas stand-alone microgrid system (MSAMGS) is another promising solution. Villages far away from the main grid can develop MSAMGS to meet their daily load demands, but as mentioned earlier microgrids with multiple areas are still susceptible to the frequency oscillations due to many factors such as a low moment of inertia and the volatile nature of renewable energy sources. In this respect, the struggle has been carried out by the researchers to sustain the power quality of the microgrids with multiple areas. A dual area interconnected MG system consisting of wind and PV source is presented in [106] with real-time-based experimental setups. The performance of the proposed Social-spider optimizer (SSO)-based PID control scheme is tested with different energy storage systems in the presence of diverse solar irradiance and wind speeds. In this work, the outpower power from solar and wind have not been used for the frequency regulation due to the unique property of tracking maximum power. The output of dynamic response showed that the proposed controller is effectively participating in damping the frequency oscillations. In [107], a robust type-II fuzzy-PID AGC system for an islanded dual area MG system is proposed, where MT, DEG, and FC are mainly responsible for the power generation and balancing. The gain values of the proposed controller are obtained using the improved slap swam optimization (I-SSO) algorithm. Analysis of the suggested controller is conducted by contrasting its performance with recently developed meta-heuristic algorithm-based techniques, which has proved that the proposed regulator exhibits superior performance in multi-area standalone AC-MG system. The idea of extending the two areas stand-alone MG system to three areas MG system is proposed in [108]. Highly vulnerable sources like WTGs, parabolic trough collectors, and PV systems are used with the backup source of the diesel generator and energy storage system to make the whole system reliable. The frequency deviation is controlled utilizing the proposed MBA-based 2DOF-PID controller, which gives better synchronization among the distinctive fuel sources to keep up the required power quality. Indeed, MSAMGS is a smart idea for satisfying the power demands of remote regions, yet at the same time, certain requirements forestall the execution of MSAMGS on a large scale. A sudden fault in any area may adversely affect the operation of the whole power system. Also, the principal system cost is very high and due to the presence of RESs, the implementation of AGC in this system in the deregulated environment is difficult.

10.2. AGC in Smart Grids

Designing an AGC system for the smart grid (SG) is another prominent issue that attracted the attention of the researchers. The smart management of EVs power provides us the best opportunity to use it as a vehicle-to-grid (V2G) source for AGC regulators in an integrated power system. The authors in [109] addressed various challenges related to the integration of EVs such as uncertainty in capacity and time delay problems due to charging and discharging. A static output feedback-based H2/H∞ regulator is designed for the AGC system to provide better reference tracking to both EVs and other conventional sources for ensuring the system frequency and tie-line interchange at their minimal value. The performance of the suggested regulator is tested on single and three area power networks, which showed that the proposed controller and EVs respond intelligently in different scenarios such as load perturbation, parameters variations, dead band effects, and time delay problems. Dynamic demand response is the main aspect of the SG and is suggested for AGC models in [110]. Here the problem of controlling the domestic demands directly from the smart meters is investigated. An alternative LFC scheme is designed that measures the load frequency using the smart meters. An experimental setup is arranged to test and demonstrate the proposed LFC scheme, which revealed that for a smart meter to have a frequency response characteristic must have a speed of 3 to 200 milliseconds in measuring the system frequency. In [111], the authors presented a detailed analysis of the effect of DR control intelligence on communication delays in the LFC of a single energy system with a thermal and wind power station. In addition, line matrix linear and linear-quadratic controller (LMI-LQR) have been introduced as a link between secondary control loop and DR loop to suppress frequency oscillations and other external disturbances. The results demonstrated the ability of DR control loop for LFC performance in the SG perspective. An active power balance control strategy that stressed the integration of flexible loads and wind energy in AGC is suggested in [112]. The study revealed that the incorporation of flexible loads and wind power into the AGC system can eradicate the fluctuations via provisioning of active power support to the real-time imbalances. The betterment of the connectivity networks in a multi-area interrelated power grid for AGC problems is enlightened in [113]. Here the network mediated effects like time delay, packet loss, bandwidth, quantization, and change in communication topology are considered to examine the system efficiency. The decentralized controller and linear matrix inequality-based linear quadratic regulator are applied to reduce the dynamic performance (mean square error of states variables) of the power system as communication topology shifts. The reliability of the LFC schemes is highly vulnerable to cyber-attacks, which can amend the critical variables data of LFC system and make received data un-reliable leading to system frequency oscillations and even collapse. The authors in [114] proposed a detection scheme for the detection of cyber-attacks on actual data using the dual-source information of compromised variables. Initially, a variable observer is designed to establish a relationship among the compromised variables and know security variables. In the second phase, a Siamese network (SN) is developed to observe the relationship between characteristics of the data measured and the data observed. Finally, a detection scheme is used to detect and analyze the similarity of the dual-source data.

11. AGC in Deregulated Power Systems

Compared to the conventional power system, the deregulated power structure is divided into several different entities, namely generation (GENCOs), transmission (TRANSCOs), and distribution companies (DISCOs), and independent system operators (ISOs). These are the market players in an uncontrolled power system that controls loading and production required in a highly competitive control environment. Each DISCO has a transaction contract with GENCOs and transactions between them are managed by an ISO [12,13]. These types of transactions are divided into three types, i.e., unilateral contracts [100], bilateral contracts [75], and infringement contracts [55]. If DISCOs are contracted by the GENCOs in the same area, then it is classified as a pool-co/unilateral based transactions. If DISCOs made contract for power transaction from the GENCOs in other area, then it is known as a bilateral transaction. If DISCO break the contract and request more power than the stated value, it is known as a contract infringement operation. The contract of each GENCO with DISCO is simulated by DISCO Participation Matrix (DPM) [24]. The general form of a DPM for a two-area power system with two GENCOs and DISOCs is shown below:

$$\text{DPM} = \begin{bmatrix} cpf_{11} & cpf_{12} & cpf_{13} & cpf_{14} \\ cpf_{21} & cpf_{22} & cpf_{23} & cpf_{24} \\ cpf_{31} & cpf_{32} & cpf_{33} & cpf_{34} \\ cpf_{41} & cpf_{42} & cpf_{43} & cpf_{44} \end{bmatrix} \tag{9}$$

The GENCO and DISCO numbers are reflected respectively in different rows and columns in the DPM matrix. Each entry of the DPM, which is the participation ratio of each GENCO to the overall DESCO demand, is represented by the contract participation factor (*cpf*). The diagonal and off-diagonal components in DMP matrix represent local and non-local contributions. Here, the GENCO is supposed to provide load demanded by DISCOs of the same and the other areas. The general form of the scheduled tie power flow between areas is as given by [86]:

$$\Delta P_{tie1-2(scheludeid)} = \begin{pmatrix} \text{DISCO demand in Area 2} \\ \text{from GENCO in Area 1} \end{pmatrix} - \begin{pmatrix} \text{DISCO demand in Area 1} \\ \text{from GENCO in Area 2} \end{pmatrix} \tag{10}$$

$$\Delta P_{tie1-2(actual)} = \frac{2\pi T_{12}}{s}(\Delta f_1(s) - \Delta f_2(s)) \tag{11}$$

$$\Delta P_{tie1-2(error)} = \Delta P_{tie1-2(Actual)} - \Delta P_{tie1-2(Schedule)} \tag{12}$$

From these equations, the ACE can be calculated using the equation given below:

$$\text{ACE}_\text{i} = B_i \Delta f_i + \Delta P_{tie1-2(Error)} \tag{13}$$

The main difference in the operation of vertical structure and restructured power systems is also depicted in Figure 4. Modern energy sector analysis shows that common regulations are the most beneficial resources in a deregulated power system. However, various uncertainties and complications existing in the real power system become more severe with the effect of deregulation. Therefore, the conventional and classical control schemes for the current AGC system may not be suitable and will require a more efficient control scheme to handle the complex structure of the deregulated power system. In recent years, several research papers are published pertaining to the frequency control schemes focused on optimal and existing intelligent control approaches in the deregulated power system [19,33,35,62,115,116]. A thorough analysis on the basic operation of AGC in a multiarea power system under a deregulated scenario is presented in [19]. Eight GENCOs and DISCOs are considered as the plant model to intensify the bilateral behavior and to analyze the dynamic performance of the proposed AGC regulator. The conversion of the electricity market into a restructured market under the ISO management is primely carried out to get an independent and unforce electricity system. The authors in [115] presented a deregulated AGC regulator for a quadruple area power network incorporating reheat thermal energy system, diesel generator, gas, and hydropower system. To damp the abrupt oscillation of the frequency, the optimization of PID regulator with a low pass filter is converted to a single objective-function optimization problem considering parametric uncertainties of the system and eigenvalues. To further, improve the performance, the regulator parameters are optimized using a modified virus colony search (MVCS) algorithm. The proposed control system is tested in a large-scale deregulated power system under all possible contracts. Table 7 provides the detailed comparative of different ACG schemes in deregulated power systems.

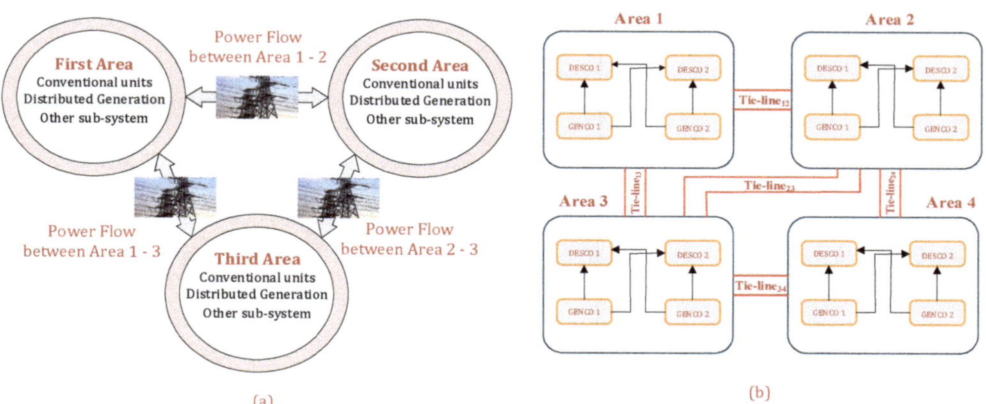

Figure 4. Schematic diagrams for (**a**) three area (conventional) and (**b**) deregulated interconnected power systems.

Table 7. Performance evaluation of AGC in Two area deregulated power system under different conditions.

Ref.	Power System Configuration	Controller Approach	Operating Scenarios	Over Shoots			Settling Times		
				Δf_1	Δf_2	ΔP_{tie1-2}	Δf_1	Δf_2	ΔP_{tie1-2}
[33]	Two area power system with thermal, hydro and gas power plants in coordination with CES/DFIG and TCPS	Integral controller	Contract scenarios:						
			Unilateral contract	0.03279	0.029	0.0038	50.22	47.31	90.2
			Bilateral contract	0.0756	0.07932	0.01114	50.05	49.01	65.56
			Contract with violation	0.07567	0.079	0.01123	94.3	87.03	116.1
[35]	Two area multiple units reheat thermal and, gas power plant with GRCs and GDBs	MSCA-FPID based AGC regulator	Contract scenarios:						
			Unilateral contract	0.000517	0.000471	0.000311	2.073	6.461	4.98
			Bilateral contract	0.00291	0.004555	0	6.273	4.555	1.837
			Contract with violation	0.004255	0.003285	0	3.731	2.398	0.642
[55]	Two area multiple units non-reheat thermal system with HVDC	WGA based SA-FPID Controller	Contract scenarios:						
			Unilateral contract	0.000135	0.0000662	0.0000329	0.57	5.07	3.94
			Bilateral contract	0.0000813	0.0000813	0.0001680	2.29	2.29	5.07
			Contract with violation	0.0003905	0.0003529	0.0001680	2.96	3.47	5.68
[34]	Two area multiple units reheat thermal power system with IPFC and RFB	BFO based Integral Controller	Controller comparisons:						
			Integral controller	0.321	0.224	0.081	16.69	15.48	14.46
			Integral controller with IPFC	0.204	0.112	0.059	6.27	5.47	7.46
			Integral controller with IPFC and RFBs	0.148	0.082	0.042	4.98	4.72	6.12

A detailed study in [33] is conducted to examine the efficiency of RESs in a deregulated power system. The contribution of frequency controllers in two-area deregulated interconnected energy grid consisting of multiple energy sources such as doubly-fed induction generation (DFIG) and BESS is studied. Furthermore, to enhance the system's dynamic response, in collaboration with TCPS, both DFIG and CES units are used in control areas. The proposed controller's gains are set using the ISE methodology and output is evaluated in different contract scenarios including the violation of the contract. Analysis proved that the inertial support of the wind turbine during the load perturbation is phenomenal, which controlled the frequency oscillation and tie-line power traffic efficiently. The effects of the AC/DC lines on the AGC of an integrated multi-area system with hydro and thermal energy sources in a deregulated environment are studied in [116]. The DPM definition is addressed and a study of the eigenvalue is carried out to assess the effect of AC/DC lines on the proposed control scheme. In the proposed restructured non-linear power grid, a PI regulator is considered to analyze all potential bilateral contracts. In comparison with the controller performance of the recently created CRAZYPSO and BFOA- based PI control units, the efficacy of the proposed control is presented. The authors in [35] implemented an intelligent fuzzy-PID based AGC regulator in a dual area restructured power system using the modified sign and cosine algorithm (MSCA). An ITAS objective function is used to define the required controller gains. Furthermore, three different scenarios are considered for investigations. To analyze, the proposed regulator, settling time, overshoot and undershoot are considered as transient parameters. The superiority of MSCA based FPID AGC system is shown by comparing its outputs with the conventional PID and FPID regulators. Moreover, the robustness of the suggested regulator is showed by changing some of the important system parameters during the operations. Apart from these, there is an emergent need for a communications infrastructure to sustain the increasing variety of auxiliary resources for the efficient operation of AGC schemes after restructuring the power sector. The communication network specifications in an integrated power grid for third-party AGC services are explained in [117] and proposed models of data communication with an emphasis on queuing theory. Moreover, certain challenges are still there in the deregulated power system, such as the selection of the proper DPM and suitable non-linear control method, bandwidth limitation, and computational burden.

12. AGC and Economic Dispatch (ED)

The primary purpose of the power system operation is the continuous provisioning of interruptible power supply at the lowest possible utility rate. This task must be performed while adhering to the prescribed equilibrium of active power between the generation and the load. AGC in conjunction with LFC is reasonable to perform such activities in which LFC controls the production of the generator at all times, while ED tunes the participation factor after every several minutes to reduce the total generation cost of the system. However, the ED process is distinguished based on the time, exhibited by ED in execution. Many literature studies are available on AGC, which performs both the LFC and ED operation simultaneously [49,118–120]. For example, the authors in [118] provided a detailed analysis on the decentralized control system for an interconnected power system that incorporated both AGC and ED objectives and operated in a well-specified manner at a specific time. To keep the decentralized structure and make fewer modifications, two approaches are utilized, i.e., reverse engineering and forward engineering. In the first approach, the conventional AGC is studied to interpret it as a partial primal-dual gradient algorithm for solving an optimization problem. In the second approach, the optimization problem is studied to inculcate the ED factor with AGC to guarantee resource-consumer balance and economic efficiency. In the final stage, a distributed generation control technique is then proposed for AGC system. Furthermore, the engineered optimization problem shared the same optima as ED, and thereby resulting in the control technique, which inculcates ED factor automatically into AGC system. A hierarchical control scheme is suggested in the literature [49], for large-scale interrelated power grid, which utilizes a centralized control scheme at the group level and a distributed control mechanism at the coordination level. In the proposed control structure, the number of groups at the coordination level is less than the number of nodes, which minimizes the generation cost along with the reduction in oscillations of frequency and tie-line interchanges. It is worth mentioning that the optimality of ED is more likely to be at risk when a significant amount of RESs is incorporated into the generation mix, since more generation reserves are expected to be activated for frequency regulation. Furthermore, the thermal limits of the tie-line power flow are often exceeded in the presence of high penetration of RESs, which may lead to network failure. The authors in [119] proposed an AGC regulator in a distributed framework of a multi-area interconnected power system, which minimizes the real time operational cost of the power system in conjunction with the objective of reducing the area control error in the presence of violation of tie-line flow thermal constraints. The optimization formulation is established in a combined manner with the constraints, and is based on the load flow steady state conditions. An approximation of a Lagrange dual of $\sigma-$logarithmic barrier function is utilized to develop a distributed AGC structure and to obtain the controller gains as a function of fuel cost. Analysis of the proposed control structure revealed that the control law reduces the network operation cost in the presence of highly variable RESs, without risking the thermal overloads. Reference [120] provides a detailed analysis on the implementation of an AGC regulator in a 40 bus four-area non-linear interconnected power system with diverse energy generation sources that cannot be aggregated into a single generation source. An algorithm is proposed to optimally tune the individual generator parameters, accounting for the dynamics of each generator, while aggregating the generator output to match the target area output. The proposed control technique has outperformed other traditional economical AGC techniques in not only optimal allocation of control signal among the interconnected areas but also dispense the signal to each generator in every area to settle them to an optimal dispatch point. A two-stage control concept is introduced [121] that inculcates both the ED and AGC in a different fashion. In the first stage, the state estimators' results initiate the OPF, while in the second stage, the ED is executed in conjunction with the AGC in an optimum manner. The proposed control technique is tested on three area power systems with the IEEE-118 bus system and revealed no problem in the algorithm convergence. A modified AGC approach is presented in [122] that supported ED in conjunction with frequency regulation and tie-line control problems. The frequency regulation is a decentralized control structure, while the ED is a centralized control operated on a slower scale than the former one.

13. Worldwide AGC Practices

The basic purpose of TSOs/ISOs is to procure the secondary regulatory services from ancillary markets to maintain the system frequency within the described limits by ensuring a minute-to-minute equilibrium between load and generation. In various parts of the world, AGC is being used to enable these ancillary services to ensure the long-term and short-term security of the power supply. Denmark's National Transmission System Operator (Energinet. dk) has established a secondary regulating strategy for building larger markets of automated secondary reserves in collaboration with the German and Nordic Transmission System Operators [123]. In western Denmark, LFC is currently deployed to release primary reserves and regulate the frequency to 50 Hz. It is also used to control unintended imbalances to accomplish the desired interchange of TenneT GmbH with Western Denmark. Therefore, Energinet.dk is purchasing approximately 90 MW of LFC per month based on the maximum utilization in that month. In Germany, a decentralized AGC is implemented as there are four regions, which are operated by different TSOs. The deployment time and the control cycle time of AGC, implemented in the balancing region, as per the NERC recommendation are 5 min and 1–2 s, respectively, which are much lower than the Union for the Co-ordination of Transmission of Electricity (UCTE) recommendation of 5 s [124].

In Continental Europe (CE), the individual control actions among the synchronous areas are implemented in a decentralized fashion, which consists of AGC block and AGC areas. Further, some of the synchronous areas in a CE are interconnected and have an AGC block with a centralized control structure regulated by UCTE. Inside the interconnected power system, there is an independent AGC of each area having its control as long it does not affect the interconnected power system. Among the interconnected European power systems, the Spanish power system is one the area [67], which consists of different zones controlled by a hierarchal AGC system. The Spanish power system is interconnected with the French, Portuguese, and Moroccan power systems. In normal operation, Spanish AGC regulates its interchange power with the French system to a scheduled interchange biased by an UCTE-assigned interconnection. AGC is operated in the Spanish power system based on the results of a secondary reserve (that associated with secondary regulation) market. This market determines the amount of reserve assigned to each zone in the system and therefore the participation factor of each zone in the regulating requirements.

Similarly, there is a pluralistic AGC installed in North America, which has eight independent regions controlled by approximately 100 balancing authorities [125]. The Pennsylvania–New Jersey–Maryland Interconnection (PJM) market is previewed by RFC and SERC regions while the New-England market is previewed by the NPCC region. As per the NERC guidelines, for deployment time, in both the PJM and New England markets, AGC should be enforced within five minutes and for at least one hour. In Prince Edward Island and Saskatchewan Research Council's Cowessess, Regina, a project under the Program of Energy Research and Development (PERD) is executed to integrate the wind power and energy storage devices into the AGC system for providing secondary regulation services. Different scenarios are performed using the 10 MW Wind R&D Park in North Cape and the Saskatchewan Research Council's Cowessess [126], which has 800 kW Type IV wind turbine and 744 kWh battery storage connected through a 400 kW inverter. The performance is measured from the matrix of performance score of the National Research Council (NRC), Canada, and the Pennsylvania Jersey Maryland (PJM) system operator. Results in the former case shown that the wind turbines performed well in providing secondary frequency regulation. However, to provide the AGC service, it involves power curtailment. In the latter case, the use of IEC Type IV16 turbine technology has made it able to get 100 percent active power flow through the power electronics. Results have shown that the IEC Type IV wind turbine is typically capable of supplying secondary frequency control with a reasonably limited and stable error.

14. Future Scope of Work

AGC is a significant control issue in the design and operation of an interconnected power system framework, which is becoming more important as a result of growing power system size, structural changes, enormous entrance of environmentally friendly power sources, vulnerabilities and parametric varieties, different specialized and ecological limitations, and other complexities of the power system. The AGC system needs improved insight and flexibility to guarantee that, after extreme unsettling influences, it can sustain a generation-load equilibrium. The future AGC models need to have such intelligence that can handle multi-objective optimization problems with a high degree of diversification in policies, control methodologies, and have a different kind of load and generation profiles. The basis of such intelligence should be advanced information technology, flexible algorithms, and a stable networking system. In this struggle, the AGC area is enriched with valuable research contributions from time to time. The AGC design is improved significantly to handle the uncertainties, non-linearities, parametric variations, integration of different systems, for example, energy storage systems, HVDC link, wind and photovoltaic sources, and other non-conventional fuel sources. Distinctive analog and computerized control strategies for single/various areas and traditional/deregulated power systems are developed and tested. The most developed control strategies in the AGC configuration to adapt to a non-linear power system or having deficient information about the system is the development of intelligent control techniques like fuzzy logic, neural networks, as well as evolutionary and heuristic optimization techniques. In light of some completely contemplated literature, the spectrum of the future for the AGC to be further explored is given below:

- Explore AI techniques to train the AGC algorithm for activation of optimum reserves to secure the operation of the power system with large-scale integration of RESs.
- Explore and include more constraint coefficients like transmission line congestions into the objective functions to make the system more efficient in the practical scenario.
- Explore adaptive and robust control methods for the AGC to effectively handle system parametric variations.
- Explore various control techniques for AGC to perfectly predict the load and forecast the weather in large- and small-scale renewable energy-based power systems.
- An in-depth study on state estimation for AGC in real-time is required to effectively deal with the packet loss problems in the communication process.
- Susceptibilities of various AGC schemes to cyber-attacks should need to be explored further.

15. Conclusions

AGC is an integral part of the electric power system that ensures sufficient, effective, and consistent power delivery. In a deregulated power system, its importance increases as it plays a major role in power-sharing and improving energy market conditions. The key goals of AGC are to regulate the system frequency and tie-line power, while continuously monitoring the load demands in the existence of various uncertainties, system non-linearities, and in different multi-variable power system conditions, which make it a multi-objective optimization problem. This article discusses AGC's latest advancement in traditional, deregulated, and renewable energy power systems. The paper discoursed single and multi-area power system networks. A detailed history of the development of power system AGC models is provided. Efforts are being made to cover AGC basic operation in a multi-area power system and analyzed some of the most commonly used single and multiple objective functions incorporating the ACE equations. Different classical and modern frequency control methods including optimal and suboptimal control, model predictive control, variable structure control, adaptive, robust, and digital control methods are comprehensively discussed. Furthermore, the applications of various intelligent and soft computing control mechanisms to the AGC system such as fuzzy logic, artificial intelligence, heuristics, and metaheuristic algorithms are surveyed and their merits and demerits are highlighted through tabular analysis. AGC incorporating different Energy storage and FACTS devices, and interconnection with the HVDC link is discussed in detail. The behavior of AGC is studied in the power system models incorporating various RESs for frequency regulation purposes. Moreover, the applicability of AGC in various configurations of microgrids and smart grids are investigated and a tabular summary is provided. The applications and challenges of AGC in a restructured power system is overviewed deeply. The collective control structure of AGC and ED is analyzed for ensuring the economic efficiency of the control structure. Moreover, to explore the industrial usage of the AGC in different parts of the world, a summary of worldwide AGC practices and experiences is provided in a detailed manner. Finally, the paper concluded with an emphasis on future works in the field of AGC. It is envisioned that this work will be a beneficial source for the researcher working in the field of automatic generation control.

Author Contributions: Conceptualization, K.U. and A.B.; methodology, Z.U.; software, K.U.; validation, K.U., A.B. and Z.U.; formal analysis, S.A.; investigation, Z.U.; resources, H.H.; data curation, A.B.; writing—original draft preparation, K.U. and Z.U.; writing—review and editing, S.A. and H.H.; visualization, K.U.; supervision, A.B.; project administration, S.A. and H.H.; funding acquisition, S.A. and H.H; All authors have read and agreed to the published version of the manuscript.

Funding: This research received no external funding.

Institutional Review Board Statement: Not applicable.

Informed Consent Statement: Not applicable. **Data Availability Statement:** This study did not report any data. **Acknowledgments:** The writers give special thanks to M. Naeem Arbab, whose helpful advice and comments significantly improved the content of the manuscript. In addition, the writers are grateful for the assistance of UET Peshawar for providing research facilities to complete the manuscript.

Conflicts of Interest: The authors declare no conflict of interests.

Appendix A

Table A1. Nomenclature.

Acronym	Definition	Acronym	Definition
AGC	Automatic generation control	FES	Flywheel energy storage
TSO	Transmission system operator	ISO	Independent system operator
AE	Aqua-electrolyzer	LTI	Linear time-invariant
SMES	Super magnetic energy storage system	RFC	Reliability first corporation (us)
RESs	Renewable energy sources	GRC	Generation rate constrains
BESS	Battery energy storage systems	GDB	Governor dead band
UPFC	Unified power flow controller	DISCOs	Distribution companies
MLCS	Modified load control scheme	GENCOs	Generation companies
DGs	Distributed generation sources	SERC	Southeastern electric reliability council (us)
FLC	Fuzzy logic control	TRANSCOs	Transmission companies
MTSA	Multiple tabu search algorithm	DPM	Disco participation matrix
IPFC	Interline power flow controller	ITAE	Integral time multiplied by absolute error
LMI	Linear matrix inequalities	AGPM	Augmented participation matrix
IAE	Integral of absolute error	EVs	Electrical vehicles
LTI	Linear time-invariant system	NPCC	Northeast power coordinating council (na)
GNN	Generalized neural network	SISO	Single input-single output

References

1. Kumar, P.; Kothari, D. Recent Philosophies of Automatic Generation Control Strategies in Power Systems. *IEEE Trans. Power Syst.* **2005**, *20*, 346–357.
2. Shayeghi, H.; Shayanfar, H.; Jalili, A. Load frequency control strategies: A state-of-the-art survey for the researcher. *Energy Convers. Manag.* **2009**, *50*, 344–353. [CrossRef]
3. Shankar, R.; Pradhan, S.; Chatterjee, K.; Mandal, R. A comprehensive state of the art literature survey on LFC mechanism for power system. *Renew. Sustain. Energy Rev.* **2017**, *76*, 1185–1207. [CrossRef]
4. Pandey, S.; Mohanty, S.; Kishor, N. A literature survey on load frequency control for conventional and distribution generation power systems. *Renew. Sust. Energy. Rev.* **2013**, *25*, 318–334. [CrossRef]
5. Alhelou, H.; Hamedani-Golshan, M.-E.; Zamani, R.; Heydarian-Forushani, E.; Siano, P. Challenges and Opportunities of Load Frequency Control in Conventional, Modern and Future Smart Power Systems: A Comprehensive Review. *Energies* **2018**, *11*, 2497. [CrossRef]
6. Zurfi, A.; Zhang, J. Exploitation of Battery Energy Storage in Load Frequency Control -A Literature Survey. *Am. J. Eng. Appl. Sci.* **2016**, *9*, 1173–1188. [CrossRef]
7. Obaid, Z.A.; Cipcigan, L.M.; Abrahim, L.; Muhssin, M.T. Frequency control of future power systems: Reviewing and evaluating challenges and new control methods. *J. Mod. Power Syst. Clean Energy* **2019**, *7*, 9–25. [CrossRef]
8. Wu, Z.; Gao, W.; Gao, T.; Yan, W.; Zhang, H.; Yan, S.; Wang, X. State-of-the-art review on frequency response of wind power plants in power systems. *J. Mod. Power Syst. Clean Energy* **2018**, *6*, 1–16. [CrossRef]
9. Bevrani, H.; Hiyama, T. *Intelligent Automatic Generation Control*, 1st ed.; CRC Press: Boca Raton, FL, USA, 2017.

10. Veronica, A.J.; Kumar, N.S. Control strategies for frequency regulation in microgrids: A review. *Wind. Eng.* **2019**. [CrossRef]
11. Bevrani, H. Robust Power System Frequency Control. Springer: New York, NY, USA, January 2009.
12. Pappachen, A.; Fathima, A.P. Critical research areas on load frequency control issues in a deregulated power system: A state-of-the-art-of-review. *Renew. Sustain. Energy Rev.* **2017**, *72*, 163–177. [CrossRef]
13. Brinda, M.D.; Suresh, A.; Rashmi, M. A literature survey on LFC in a deregulated electricity environment. *World Rev. Sci. Technol. Sustain. Dev.* **2018**, *14*, 1. [CrossRef]
14. Cohn, N. Techniques for Improving the Control of Bulk Power Transfers on Interconnected Systems. *IEEE Trans. Power Appar. Syst.* **1971**, *PAS-90*, 2409–2419. [CrossRef]
15. Report, I. Dynamic Models for Steam and Hydro Turbines in Power System Studies. *IEEE Trans. Power Appar. Syst.* **1973**, *PAS-92*, 1904–1915. [CrossRef]
16. Tripathy, S.C.; Bhatti, T.S.; Jha, C.S.; Malik, O.P.; Hope, G.S. Sampled data automatic generation control analysis with reheat steam turbines and governor dead-band effects. *IEEE Trans. Power Appar. Syst.* **1984**, *PAS-103*, 1045–1051. [CrossRef]
17. Malik, O.; Hope, G.; Tripathy, S.; Mital, N. Decentralized suboptimal load-frequency control of a hydro-thermal power system using the state variable model. *Elecrtr. Power Syst. Res.* **1985**, *8*, 237–247. [CrossRef]
18. Bensenouci, A.; Ghany, A. Mixed H∞/H2 with pole-placement design of robust LMI-based output feedback controllers for multi-area load frequency control. In Proceedings of the EUROCON 2007—The International Conference on "Computer as a Tool", Warsaw, Poland, 9–12 September 2007.
19. Shiva, C.K.; Mukherjee, V. Automatic generation control of multi-unit multi-area deregulated power system using a novel quasi-oppositional harmony search algorithm. *IET Gener. Transm. Distrib.* **2015**, *9*, 2398–2408. [CrossRef]
20. Ross, C.; Green, T. Dynamic Performance Evaluation of a Computer Controlled Elecrtr.ric Power System. *IEEE Trans. Power Appar. Syst.* **1972**, *PAS-91*, 1158–1165. [CrossRef]
21. Chaturvedi, D.K.; Satsangi, P.S.; Kalra, P.K. Load frequency control: A generalised neural network approach. *Int. J. Elecrtr. Rical Power Energy Syst.* **1999**, *21*, 405–415. [CrossRef]
22. Sanjeevikumar, P.; Sarojini, R.K.; Palanisamy, K.; Padmanaban, S.; Holm-Nielsen, J.B. Large Scale Renewable Energy Integration: Issues and Solutions. *Energies* **2019**, *12*, 1996.
23. Kerdphol, T.; Rahman, F.S.; Mitani, Y.; Hongesombut, K.; Küfeoğlu, S. Virtual Inertia Control-Based Model Predictive Control for Microgrid Frequency Stabilization Considering High Renewable Energy Integration. *Sustainability* **2017**, *9*, 773. [CrossRef]
24. Donde, V.; Pai, M.; Hiskens, I. Simulation and optimization in an AGC system after deregulation. *IEEE Trans. Power Syst.* **2001**, *16*, 481–489. [CrossRef]
25. Shayeghi, H.; Shayanfar, H.; Malik, O. Robust decentralized neural networks based LFC in a deregulated power system. *Elecrtr. Power Syst. Res.* **2007**, *77*, 241–251. [CrossRef]
26. Jagatheesan, K.; Anand, B.; Dey, K.N.; Ashour, A.S.; Satapathy, S.C. Performance evaluation of objective functions in automatic generation control of thermal power system using ant colony optimization technique-designed proportional–integral–derivative controller. *Elecrtr. Eng.* **2008**, *100*, 895–911. [CrossRef]
27. Saikia, L.C.; Sinha, N.; Nanda, J. Maiden application of bacterial foraging based fuzzy IDD controller in AGC of a multi-area hydrothermal system. *Int. J. Elecrtr. Power. Energy Syst.* **2013**, *45*, 98–106. [CrossRef]
28. Fini, M.H.; Yousefi, G.R.; Alhelou, H.H. Comparative study on the performance of many-objective and single-objective optimisation algorithms in tuning load frequency controllers of multi-area power systems. *IET Gener. Trans. Distrib.* **2016**, *10*, 2915–2923. [CrossRef]
29. Alhelou, H.; Golshan, M.; Fini, M. Wind Driven Optimization Algorithm Application to Load Frequency Control in Interconnected Power Systems Considering GRC and GDB Nonlinearities. *Elecrtr. Power Compon. Syst.* **2018**, *46*, 11–12.
30. Sivaramakrlshnan, A.Y.; Hariharan, M.V.; Srisailam, M.C. Design of variable-structure load-frequency controller using pole assignment technique. *Int. J. Control.* **1984**, *40*, 487–498. [CrossRef]
31. Saxena, S.; Hote, Y.V. PI Controller Based Load Frequency Control Approach for Single-Area Power System Having Communication Delay. *IFAC-PapersOnLine* **2018**, *51*, 622–626. [CrossRef]
32. Barisal, A.; Mishra, S. Improved PSO based automatic generation control of multi-source nonlinear power systems interconnected by AC/DC links. *Cogent Eng.* **2018**, *5*, 1422228. [CrossRef]

33. Dhundhara, S.; Verma, Y.P. Evaluation of CES and DFIG unit in AGC of realistic multisource deregulated power system. *Int. Trans. Elecrtr. Energy Syst.* **2017**, *27*, e2304. [CrossRef]
34. Chidembaram, I.; Paramasvam, B. Optimized load-frequency simulation in restructured power system with redox flow batteries and interline power flow controller. *Int. J. Elecrtr. Power. Energy Syst.* **2013**, *50*, 9–24. [CrossRef]
35. Nayak, N.; Mishra, S.; Sharma, D.; Sahu, B.K. Application of modified sine cosine algorithm to optimally design PID/fuzzy-PID controllers to deal with AGC issues in deregulated power system. *IET Gener. Transm. Distrib.* **2019**, *13*, 2474–2487. [CrossRef]
36. Dash, P.; Saikia, L.C.; Sinha, N. Comparison of performances of several FACTS devices using Cuckoo search algorithm optimized 2DOF controllers in multi-area AGC. *Int. J. Elecrtr. Power Energy Syst.* **2015**, *65*, 316–324. [CrossRef]
37. Sharma, Y.; Saikia, L.C. Automatic generation control of a multi-area ST—Thermal power system using Grey Wolf Optimizer algorithm based classical controllers. *Int. J. Elecrtr. Power Energy Syst.* **2015**, *73*, 853–862. [CrossRef]
38. Shiva, C.K.; Mukherjee, V. A novel quasi-oppositional harmony search algorithm for automatic generation control of power system. *Appl. Soft Comput.* **2015**, *35*, 749–765. [CrossRef]
39. Vrdoljak, K.; Perić, N.; Petrović, I. Sliding mode-based load-frequency control in power systems. *Elecrtr. Power Syst. Res.* **2010**, *80*, 514–527. [CrossRef]
40. Yesil, E. Interval type-2 fuzzy PID load frequency controller using Big Bang–Big Crunch optimization. *Appl. Soft Comput.* **2014**, *15*, 100–112. [CrossRef]
41. Toulabi, M.; Shiroei, M.; Ranjbar, A. Robust analysis and design of power system load frequency control using the Kharitonov's theorem. *Int. J. Elecrtr. Power. Energy Syst.* **2014**, *55*, 51–58. [CrossRef]
42. Yousef, H.A.; Al-Kharusi, K.; Albadi, M.H.; Hosseinzadeh, N. Load Frequency Control of a Multi-Area Power System: An Adaptive Fuzzy Logic Approach. *IEEE Trans. Power Syst.* **2014**, *29*, 1822–1830. [CrossRef]
43. Alhelou, H.H.; Golshan, M.E.H.; Hatziargyriou, N.D. A Decentralized Functional Observer Based Optimal LFC Considering Unknown Inputs, Uncertainties, and Cyber-Attacks. *IEEE Trans. Power Syst.* **2019**, *34*, 4408–4417. [CrossRef]
44. Lim, K.; Wang, Y.; Zhou, R. Robust decentralised load–frequency control of multi-area power systems. *IEE Proc. Gener. Transm. Distrib.* **1996**, *143*, 377. [CrossRef]
45. Magdy, G.; Shabib, G.; Elbaset, A.A.; Kerdphol, T.; Qudaih, Y.; Bevrani, H.; Mitani, Y. Tustin's technique based digital decentralized load frequency control in a realistic multi power system considering wind farms and communications delays. *Ain Shams Eng. J.* **2019**, *10*, 327–341. [CrossRef]
46. Rahmani, M.; Sadati, N. Two-level optimal load–frequency control for multi-area power systems. *Int. J. Elecrtr. Power Energy Syst.* **2013**, *53*, 540–547. [CrossRef]
47. Yang, S.; Huang, C.; Yu, Y.; Yue, N.; Xie, J. Load Frequency Control of Interconnected Power System via Multi-Agent System Method. *Elecrtr. Power Compon. Syst.* **2017**, *45*, 1–13. [CrossRef]
48. Rubaai, A.; Udo, V. Self-tuning load frequency control: Multilevel adaptive approach. *IEEE Proc. Gener. Trans Distrib.* **1994**, *141*, 285–290. [CrossRef]
49. Xi, K.; Lin, H.X.; Shen, C.; Van Schuppen, J.H. Multilevel Power-Imbalance Allocation Control for Secondary Frequency Control of Power Systems. *IEEE Trans. Autom. Control.* **2020**, *65*, 2913–2928. [CrossRef]
50. Bengiamin, N.; Chan, W. 3-level load-frequency control of power systems interconnected by asynchronous tie lines. *Proc. Inst. Elecrtr. Eng.* **1979**, *126*, 1198. [CrossRef]
51. Lee, K.; Yee, H.; Teo, C. Self-tuning algorithm for automatic generation control in an interconnected power system. *Elecrtr. Power Syst. Res.* **1991**, *20*, 157–165. [CrossRef]
52. Memon, A.; Memon, L.; Shaikh, A. Pole placement design for load frequency control (LFC) of an isolated power system. *J. Eng. Appl. Sci.* **2002**, *21*, 11–16.
53. Saikia, L.C.; Nanda, J.; Mishra, S. Performance comparison of several classical controllers in AGC for multi-area interconnected thermal system. *Int. J. Elecrtr. Power. Energy Syst.* **2001**, *33*, 394–401. [CrossRef]
54. Arya, Y.; Kumar, N. Optimal control strategy-based AGC of Elecrtrical power systems: A comparative performance analysis. *Optim. Cont. Appl. Methods.* **2017**, *38*, 982–992. [CrossRef]
55. Hasan, N.; Ibraheem; Kumar, P.; Nizamuddin. Sub-optimal automatic generation control of interconnected power system using constrained feedback control strategy. *Int. J. Elecrtr. Power Energy Syst.* **2012**, *43*, 295–303. [CrossRef]

56. Sahoo, S.; Jena, N.K.; Dei, G.; Sahu, B.K. Self-adaptive fuzzy-PID controller for AGC study in deregulated Power System. *Indonesian. J. Elecrtr. Eng Inform.* **2019**, *7*, 650–663.
57. Kazemi, M.H.; Karrari, M.; Menhaj, M.B. Decentralized robust adaptive-output feedback controller for power system load frequency control. *Elecrtr. Eng.* **2002**, *84*, 75–83. [CrossRef]
58. Al-Hamouz, Z.; Abdel-Magid, Y. Variable structure load frequency controllers for multiarea power systems. *Int. J. Elecrtr. Power Energy Syst.* **1993**, *15*, 293–300. [CrossRef]
59. Mohanty, B. TLBO optimized sliding mode controller for multi-area multi-source nonlinear interconnected AGC system. *Int. J. Elecrtr. Power Energy Syst.* **2015**, *73*, 872–881. [CrossRef]
60. Hamouz, Z.; Duwaish, H.; Musabi, N. Optimal design of a sliding mode AGC controller: Application to a nonlinear interconnected model. *Elecrtr. Power Syst. Res.* **2011**, *81*, 1403–1409. [CrossRef]
61. Khezri, R.; Golshannavaz, S.; Shokoohi, S.; Bevrani, H. Fuzzy logic based fine-tuning approach for robust load frequency control in a multi-area power system. *Electr. Power Comp. Syst.* **2016**, *44*, 2073–2083. [CrossRef]
62. Shayeghi, H.; Jalili, A.; Shayanfar, H. Robust modified GA based multi-stage fuzzy LFC. *Energy Convers. Manag.* **2007**, *48*, 1656–1670. [CrossRef]
63. Elsisi, M.; Soliman, M.; Aboelela, M.; Mansour, W. Bat inspired algorithm based optimal design of model predictive load frequency control. *Int. J. Electr. Power Energy Syst.* **2016**, *83*, 426–433. [CrossRef]
64. Mc Namara, P.; Milano, F. Efficient implementation of MPC-based AGC for real-world systems with low inertia. *Electr. Power Syst. Res.* **2018**, *158*, 315–323. [CrossRef]
65. Zeng, G.-Q.; Xie, X.-Q.; Chen, M.-R. An Adaptive Model Predictive Load Frequency Control Method for Multi-Area Interconnected Power Systems with Photovoltaic Generations. *Energies* **2017**, *10*, 1840. [CrossRef]
66. McNamara, P.; Milano, F. Model Predictive Control-Based AGC for Multi-Terminal HVDC-Connected AC grids. *IEEE Trans. Power Syst.* **2017**, *33*, 1036–1048. [CrossRef]
67. Egido, I.; Fernandez-Bernal, F.; Rouco, L. The Spanish AGC System: Description and Analysis. *IEEE Trans. Power Syst.* **2008**, *24*, 271–278. [CrossRef]
68. Yesil, E.; Guzelkaya, M.; Eksin, I. Self tuning fuzzy PID type load and frequency controller. *Energy Convers. Manag.* **2004**, *45*, 377–390. [CrossRef]
69. Lal, D.K.; Barisal, A.K.; Tripathy, M. Grey Wolf Optimizer Algorithm Based Fuzzy PID Controller for AGC of Multi-area Power System with TCPS. *Procedia Comput. Sci.* **2016**, *92*, 99–105. [CrossRef]
70. Khooban, M.H.; Niknam, T. A new intelligent online fuzzy tuning approach for multi-area load frequency control: Self Adaptive Modified Bat Algorithm. *Int. J. Elecrtr. Power Energy Syst.* **2015**, *71*, 254–261. [CrossRef]
71. Azeer, S.; Ramjug-Ballgobin, R.; Hassen, S.S. Intelligent Controllers for Load Frequency Control of Two-Area Power System. *IFAC-Papers On-Line* **2017**, *50*, 301–306. [CrossRef]
72. Yin, L.; Yu, T.; Zhou, L.; Huang, L.; Zhang, X.; Zheng, B. Artificial emotional reinforcement learning for automatic generation control of large-scale interconnected power grids. *IET Gener. Trans. Distrib.* **2017**, *11*, 2305–2313. [CrossRef]
73. Yin, L.; Zhao, L.; Yu, T.; Zhang, X. Deep forest reinforcement learning for preventive strategy considering automatic generation control in large-scale interconnected power systems. *Appl. Sci.* **2018**, *8*, 2185. [CrossRef]
74. Shayeghi, H.; Shayanfar, H. Application of ANN technique based on μ-synthesis to load frequency control of interconnected power system. *Int. J. Elecrtr. Power Energy Syst.* **2006**, *28*, 503–511. [CrossRef]
75. Bevrani, H.; Hiyama, T.; Mitani, Y.; Tsuji, K.; Teshehlab, M. Load-frequency regulation under a bilateral LFC scheme using flexible neural networks. *Eng. Intell. Syst.* **2006**, *14*, 109–117.
76. Hasan, N.; Nasirudin, I.; Farooq, S. Hybrid Taguchi Genetic Algorithm-Based AGC Controller for Multisource Interconnected Power System. *Elecrtr. Power Syst. Res.* **2019**, *47*, 101–112. [CrossRef]
77. Jagatheesan, K.; Anand, B.; Samanta, S.; Dey, N.; Ashour, A.S.; Balas, V.E. Design of a proportional-integral-derivative controller for an automatic generation control of multi-area power thermal systems using firefly algorithm. *IEEE/CAA J. Autom. Sin.* **2019**, *6*, 503–515. [CrossRef]
78. Kumar, B.V. Power System loss minimization by using UPFC placed at optimal location given by Artificial Bee Colony Algorithm. *Int. J. Res. Advent Technol.* **2019**, *7*, 127–132. [CrossRef]
79. Abazari, A.; Monsef, H.; Wu, B. Coordination strategies of distributed energy resources including FESS, DEG, FC and WTG in load frequency control (LFC) scheme of hybrid isolated micro-grid. *Int. J. Electr. Power Energy Syst.* **2019**, *109*, 535–547. [CrossRef]

80. Mishra, S.K.P.; Sekhar, S. Modelling of Differential Evolution Based Automatic Generation Control for Two Area Interconnected Power Systems. *J. Inform. Math. Sci.* **2019**, *11*, 19–30.
81. Sahu, B.K.; Pati, S.; Panda, S. Hybrid differential evolution particle swarm optimisation optimised fuzzy proportional–integral derivative controller for automatic generation control of interconnected power system. *IET Gener. Transm. Distrib.* **2014**, *8*, 1789–1800. [CrossRef]
82. Sahu, B.; Pati, S.; Mohanty, P.; Panda, S. Teaching–learning based optimization algorithm based fuzzy-PID controller for automatic generation control of multi-area power system. *Int. J. Elecrtr. Power Energy Syst.* **2015**, *27*, 240–249. [CrossRef]
83. Chaine, S.; Tripathy, M. Performance of CSA optimized controllers of DFIGs and AGC to improve frequency regulation of a wind integrated hydrothermal power system. *Alex. Eng. J.* **2019**, *58*, 579–590. [CrossRef]
84. Sahu, B.K.; Pati, T.K.; Nayak, J.R.; Panda, S.; Kar, S.K. A novel hybrid LUS–TLBO optimized fuzzy-PID controller for load frequency control of multi-source power system. *Int. J. Electr. Power Energy Syst.* **2016**, *74*, 58–69. [CrossRef]
85. Barik, A.K.; Das, D.C. Proficient load-frequency regulation of demand response supported bio-renewable cogeneration based hybrid microgrids with quasi-oppositional selfish-herd optimisation. *IET Gener. Transm. Distrib.* **2019**, *13*, 2889–2898. [CrossRef]
86. Saha, A.; Saikia, L.C. Combined application of redox flow battery and DC link in restructured AGC system in the presence of WTS and DSTS in distributed generation unit. *IET Gener. Transm. Distrib.* **2018**, *12*, 2072–2085. [CrossRef]
87. Nayeripour, M.; Hoseintabar, M.; Niknam, T. Frequency deviation control by coordination control of FC and double-layer capacitor in an autonomous hybrid renewable energy power generation system. *Renew. Energy* **2011**, *36*, 1741–1746. [CrossRef]
88. Yan, W.; Sheng, L.; Xu, D.; Yang, W.; Liu, Q. H∞ Robust Load Frequency Control for Multi-Area Interconnected Power System with Hybrid Energy Storage System. *Appl. Sci.* **2018**, *8*, 1748. [CrossRef]
89. Kerdphol, T.; Rahman, F.S.; Mitani, Y. Virtual Inertia Control Application to Enhance Frequency Stability of Interconnected Power Systems with High Renewable Energy Penetration. *Energies* **2018**, *11*, 981. [CrossRef]
90. Rakhshani, E.; Remon, D.; Cantarellas, A.; Garcia, J.M.; Rodriguez, P. Modeling and sensitivity analyses of VSP based virtual inertia controller in HVDC links of interconnected power systems. *Electr. Power Syst. Res.* **2016**, *141*, 246–263. [CrossRef]
91. Shayeghi, H.; Shayanfar, H.; Jalili, A. LFC design of a deregulated power system with TCPS using PSO. *Int. J. Electr. Electron. Eng.* **2009**, *3*, 632–640.
92. Ngamroo, I.; Tippayachai, J.; Dechanupaprittha, S. Robust decentralised frequency stabilisers design of static synchronous series compensators by taking system uncertainties into consideration. *Int. J. Electr. Power Energy Syst.* **2006**, *28*, 513–524. [CrossRef]
93. Morsali, J.; Zare, K.; Hagh, M.T. Performance comparison of TCSC with TCPS and SSSC controllers in AGC of realistic interconnected multi-source power system. *Ain Shams Eng. J.* **2016**, *7*, 143–158. [CrossRef]
94. Pathak, N.; Verma, A.; Bhatti, T.S.; Nasiruddin, I. Modeling of HVDC Tie Links and Their Utilization in AGC/LFC Operations of Multiarea Power Systems. *IEEE Trans. Ind. Elecrtr.* **2019**, *66*, 2185–2197. [CrossRef]
95. Asano, H.; Yajima, K.; Kaya, Y. Influence of photovoltaic power generation on required capacity for load frequency control. *IEEE Trans. Energy Convers.* **1996**, *11*, 188–193. [CrossRef]
96. Yanagawa, S.; Kato, T.; Kai, W.; Tabata, A.; Yokomizu, Y.; Okamoto, T.; Suzuoki, Y. Evaluation of LFC capacity for output fluctuation of photovoltaic power generation systems based on multi-point observation of insolation. In Proceedings of the 2001 Power Engineering Society Summer Meeting. Conference Proceedings (Cat. No.01CH37262), Vancouver, BC, Canada, 15–19 July 2001.
97. Luo, H.; Hu, Z.; Zhang, H.; Chen, H. Coordinated Active Power Control Strategy for Deloaded Wind Turbines to Improve Regulation Performance in AGC. *IEEE Trans. Power Syst.* **2019**, *34*, 98–108. [CrossRef]
98. Bevrani, H.; Daneshmand, P.R. Fuzzy Logic-Based Load-Frequency Control Concerning High Penetration of Wind Turbines. *IEEE Syst. J.* **2011**, *6*, 173–180. [CrossRef]
99. Zhao, X.; Lin, Z.; Fu, B.; He, L.; Fang, N. Research on Automatic Generation Control with Wind Power Participation Based on Predictive Optimal 2-Degree-of-Freedom PID Strategy for Multi-area Interconnected Power System. *Energies* **2018**, *11*, 3325. [CrossRef]

100. Bhatt, P.; Ghshal, S.; Roy, R. Coordinated control of TCPS and SMES for frequency regulation of interconnected restructured power systems with dynamic participation from DFIG based wind farm. *Renew. Energy* **2012**, *40*, 40–50. [CrossRef]
101. Kumar, A.; Suhag, S. Effect of TCPS, SMES, and DFIG on load frequency control of a multi-area multi-source power system using multi-verse optimized fuzzy-PID controller with derivative filter. *J. Vib. Control.* **2018**, *24*, 5922–5937. [CrossRef]
102. Bagheri, A.; Jabbari, A.; Mobayen, S. An intelligent ABC-based terminal sliding mode controller for load-frequency control of islanded micro-grids. *Sustain. Cities Soc.* **2021**, *64*, 102544. [CrossRef]
103. Kumar, D.; Mathur, H.D.; Bhanot, S.; Bansal, R.C. Forecasting of solar and wind power using LSTM RNN for load frequency control in isolated microgrid. *Int. J. Model. Simul.* **2020**, 1–13. [CrossRef]
104. Liu, K.; He, J.; Luo, Z.; Shen, X.; Liu, X.; Lu, T. Secondary Frequency Control of Isolated Microgrid Based on LADRC. *IEEE Access* **2019**, *7*, 53454–53462. [CrossRef]
105. Safari, A.; Babaei, F.; Farrokhifar, M. A load frequency control using a PSO-based ANN for micro-grids in the presence of electric vehicles. *Int. J. Ambient Energy* **2019**, *42*, 688–700. [CrossRef]
106. Fergany, A.; Hameed, M. Efficient frequency controllers for autonomous two-area hybrid microgrid system using social-spider optimiser. *IET Gener. Trans. Distrib.* **2017**, *11*, 637–648. [CrossRef]
107. Sahu, P.C.; Mishra, S.; Prusty, R.C.; Panda, S. Improved-salp swarm optimized type-II fuzzy controller in load frequency control of multi area islanded AC microgrid. *Sustain. Energy Grids Netw.* **2018**, *16*, 380–392. [CrossRef]
108. Ranjan, S.; Das, D.; Latif, A.; Sinha, N. LFC for autonomous hybrid micro grid system of 3 unequal renewable areas using mine blast algorithm. *Int. J. Renew. Energy* **2018**, *8*, 1297–1308.
109. Khan, M.; Sun, H.; Xiang, Y.; Shi, D. Electric vehicles participation in load frequency control based on mixed H2/H∞. *Int. J. Electr. Power Energy Syst* **2021**, *125*, 106420. [CrossRef]
110. Samarakoon, K.; Ekanayake, J.; Jenkins, N. Investigation of Domestic Load Control to Provide Primary Frequency Response Using Smart Meters. *IEEE Trans. Smart Grid* **2012**, *3*, 282–292. [CrossRef]
111. Bharti, K.; Singh, V.P.; Singh, S.P. Impact of Intelligent Demand Response for Load Frequency Control in Smart Grid Perspective. *IETE J. Res.* **2020**, 1–12. [CrossRef]
112. Basit, A.; Ahmad, T.; Ali, A.Y.; Ullah, K.; Mufti, G.; Hansen, A.D. Flexible Modern Power System: Real-Time Power Balancing through Load and Wind Power. *Energies* **2019**, *12*, 1710. [CrossRef]
113. Singh, V.P.; Kishor, N.; Samuel, P. Load Frequency Control with Communication Topology Changes in Smart Grid. *IEEE Trans. Ind. Inform.* **2016**, *12*, 1943–1952. [CrossRef]
114. Bi, W.; Zhang, K.; Chen, C. Cyber Attack Detection Scheme for a Load Frequency Control System Based on Dual-Source Data of Compromised Variables. *Appl. Sci.* **2021**, *11*, 1584. [CrossRef]
115. Ghasemi-Marzbali, A. Multi-area multi-source automatic generation control in deregulated power system. *Energy* **2020**, *201*, 117667. [CrossRef]
116. Arya, Y.; Kumar, N. AGC of a multi-area multi-source hydrothermal power system interconnected via AC/DC parallel links under deregulated environment. *Int. J. Electr. Power Energy Syst.* **2016**, *75*, 127–138. [CrossRef]
117. Bevrani, H.; Mtani, Y.; Tsji, K.; Bevrani, H. Bilateral based robust load frequency control. *Energy Conv. Mang.* **2005**, *46*, 1129–1146. [CrossRef]
118. Li, N.; Zhao, C.; Chen, L. Connecting automatic generation control and economic dispatch from an optimization view. *IEEE Trans. Control Netw. Syst.* **2015**, *3*, 254–264. [CrossRef]
119. Patel, R.; Li, C.; Yu, X.; McGrath, B. Optimal Automatic Generation Control of an Interconnected Power System Under Network Constraints. *IEEE Trans. Ind. Electron.* **2018**, *65*, 7220–7228. [CrossRef]
120. Patel, R.B.; Li, C.; Meegahapola, L.G.; McGrath, B.; Yu, X. Enhancing Optimal Automatic Generation Control in a Multi-Area Power System with Diverse Energy Resources. *IEEE Trans. Power Syst.* **2019**, *34*, 3465–3475. [CrossRef]
121. Bacher, R.; Meeteren, H.P. Real-time optimal power flow in automatic generation control. *IEEE Trans. Power Syst.* **1988**, *3*, 1518–1529. [CrossRef]
122. Eidson, D.; Ilic, M. Advanced generation control with economic dispatch. In Proceedings of the 1995 34th IEEE Conference on Decision and Control, New Orleans, LA, USA, 13–15 December 1995; Volume 4, pp. 3450–3458.

123. Energinet.dk. *Energinet.dk's Ancillary Services Strategy*; Energinet.dk, Doc. No. 77566/11 v1, Case No. 10/5553; Energinet.dk: Fredericia, Denmark, 2011.
124. Balancing and Frequency Control—NERC. Available online: https://www.nerc.com/comm/OC/RSLandingPageDL/RelatedFiles/NERCBalancingandFrequencyControl040520111.pdf (accessed on 20 April 2021).
125. Pandurangan, V.; Zareipour, H.; Malik, O. Frequency regulation services: A comparative study of select North American and European reserve markets. In Proceedings of the 2012 North American Power Symposium (NAPS), Champaign, IL, USA, 9–11 September 2012.
126. Program of Energy Research and Development Research. Available online: https://weican.ca/docs/publications/PERDPublicSummaryReport.pdf (accessed on 20 April 2021).

Article

Towards Short Term Electricity Load Forecasting Using Improved Support Vector Machine and Extreme Learning Machine

Waqas Ahmad [1], Nasir Ayub [1], Tariq Ali [2], Muhammad Irfan [2], Muhammad Awais [3], Muhammad Shiraz [1] and Adam Glowacz [4,*]

1 Department of Computer Science, Federal Urdu University of Arts, Science and Technology, Islamabad 44000, Pakistan; saqaw@ymail.com (W.A.); nasir.ayubse@gmail.com (N.A.); drmuhammadshiraz@fuuastisb.edu.pk (M.S.)

2 College of Engineering, Electrical Engineering Department, Najran University, Najran 61441, Saudi Arabia; tariqhsp@gmail.com (T.A.); irfan16.uetian@gmail.com (M.I.)

3 School of Computing and Communications, Lancaster University, Bailrigg, Lancaster LA1 4YW, UK; m.awais11@lancaster.ac.uk

4 Department of Automatic Control and Robotics, Faculty of Electrical Engineering, Automatics, Computer Science and Biomedical Engineering, AGH University of Science and Technology, al. A. Mickiewicza 30, 30-059 Kraków, Poland

* Correspondence: adglow@agh.edu.pl

Received: 30 April 2020; Accepted: 2 June 2020; Published: 5 June 2020

Abstract: Forecasting the electricity load provides its future trends, consumption patterns and its usage. There is no proper strategy to monitor the energy consumption and generation; and high variation among them. Many strategies are used to overcome this problem. The correct selection of parameter values of a classifier is still an issue. Therefore, an optimization algorithm is applied with deep learning and machine learning techniques to select the optimized values for the classifier's hyperparameters. In this paper, a novel deep learning-based method is implemented for electricity load forecasting. A three-step model is also implemented, including feature selection using a hybrid feature selector (XGboost and decision tee), redundancy removal using feature extraction technique (Recursive Feature Elimination) and classification/forecasting using improved Support Vector Machine (SVM) and Extreme Learning Machine (ELM). The hyperparameters of ELM are tuned with a meta-heuristic algorithm, i.e., Genetic Algorithm (GA) and hyperparameters of SVM are tuned with the Grid Search Algorithm. The simulation results are shown in graphs and the values are shown in tabular form and they clearly show that our improved methods outperform State Of The Art (SOTA) methods in terms of accuracy and performance. The forecasting accuracy of Extreme Learning Machine based Genetic Algo (ELM-GA) and Support Vector Machine based Grid Search (SVM-GS) is 96.3% and 93.25%, respectively. The accuracy of our improved techniques, i.e., ELM-GA and SVM-GS is 10% and 7%, respectively, higher than the SOTA techniques.

Keywords: electricity load forecasting; smart grid; feature selection; Extreme Learning Machine; Genetic Algorithm; Support Vector Machine; Grid Search

1. Introduction

Load forecasting has a huge impact on routine electric functions including fuel resource management and for accurate decision making to stabilize the demand and supply of electricity. After revolutionary overhauling of the electricity market internationally, the importance of load forecasting has increased multifold and also encompassed other areas of significance, e.g., financial planning and energy trading, etc. Precision in load forecasting lays the foundation of a system to establish a spot pricing, which in turn benefits the system in acquiring minimum purchasing cost of electricity in the local and regional markets. Load forecasting has a great significance in demand-side management planning, storage planning to include its maintenance and schedule, integrating renewable energy sources and multiple other utilities involving smart grid implementations. Load forecasting also enables the electric consumers in creating a link between price and demand for electricity and get benefits by adjusting the usage pattern in line with cost advantages. Electric grids are defined as a network of electric power generators in which users are interconnected by transmission lines and well-synchronized through control centres. Moreover, the power grid usually refers to the transmission system of electric power. Similarly, Traditional Grids (TGs) are defined as the grids which connect power providing system to distribution network further. Whereas, a TG is operated on a centrally controlled system as shown in Figure 1.

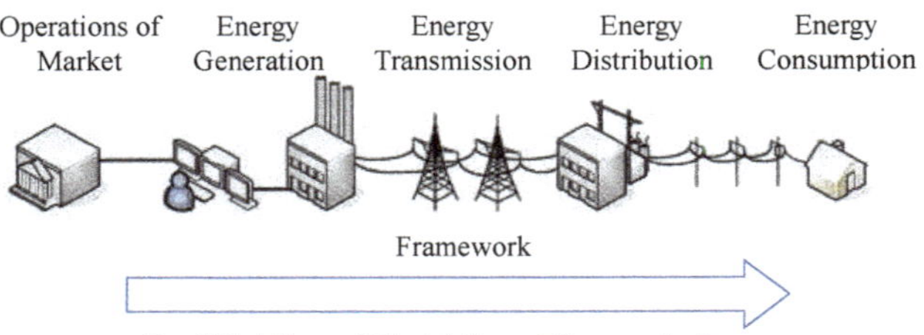

Figure 1. Traditional Grid.

Besides, Smart Grid (SG) is defined as an enhanced electric grid that operates in two-ways between a provider and consumer. It incorporates artificial intelligence, intercommunication between and smartly operating distributor network among them. SG brings a revolution by efficiently managing the power generation. Moreover, it distributes the electricity as per the requirement and consumption of the end-users. The latest technologies and types of equipment are also incorporated to establish SGs. Furthermore, it is a requirement in today's energy-starved world due to an acute shortage of energy in summers and other conditions like bad weather, etc. SG incorporates the latest technologies and techniques from power generation to power consumption as shown in Figure 2. In SG, the work mainly focuses on infrastructure, management system and protective structure [1].

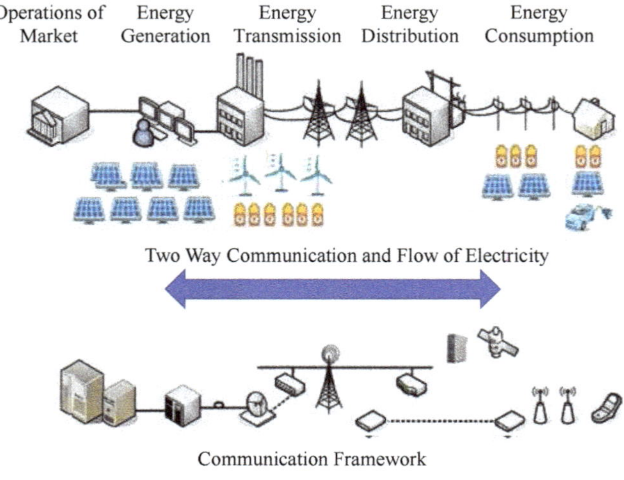

Figure 2. Smart Grid.

As per the aforementioned literature, the power grid is one of the most complex structures consisting of a myriad of transformers, substations, generator sets and long-distance power lines. The main purpose of the power grid is to supply power to end-users. It extends from a small local design to meet the daily needs of hundreds or millions of consumers through ultra-long high-voltage and low-voltage lines and makes an excellent interconnect structure. The basic components of this structure are shown in Figure 3, i.e., the power plant, transmission line and the distribution system.

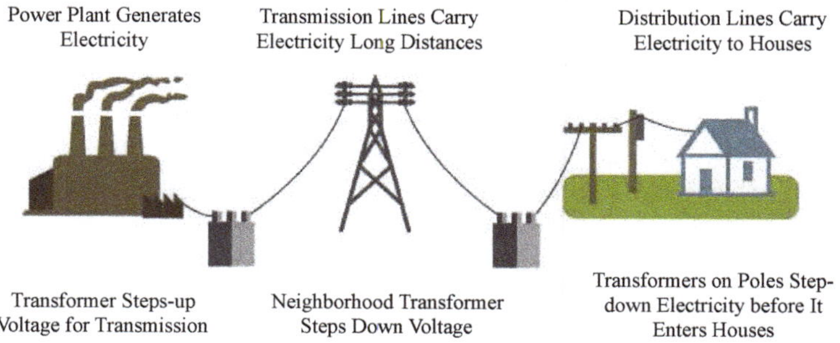

Figure 3. Electricity generation, transmission and distribution.

In a nutshell, the TG system is not well-equipped to fulfil the requirements of the end-users. Whereas, in SG, these grids are utilized in a manner to fulfil the need of the consumer. At this end, Smart Meters (SMs) are used to handle the appliances. Information is transferred between consumer and utility using these SMs, which benefit both the provider and the consumer [2]. The main techniques employed in SG are energy scheduling and management. Algorithms and subsystems exchange records and their statistics are used to further improve the energy generation and consumption in a closed-loop [3].

SM is an advanced energy meter that is used to get the consumption pattern information from the user as shown in Figure 4. So, there are two different types of SMs depending upon different communication technologies, i.e., power line carrier and radiofrequency [3–5].

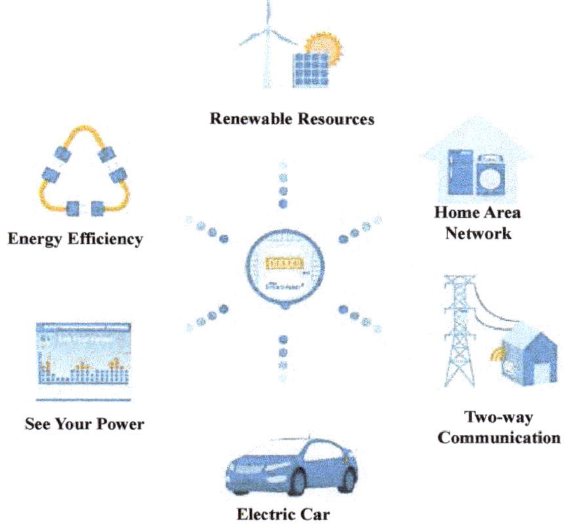

Figure 4. Smart Meter.

With the help of electricity load forecasting, the utility company can plan and have a better understanding of future consumption and load demand. The minimum risk for the utility company and understanding the pattern of the electricity help to fulfil the load demand. It helps the utility in understanding consumer demand and to define a proper period for the maintenance of electricity in residential areas [5–9]. Moreover, it helps the utility to avoid the generation of energy, in operation and supply chain management. The utility aims to seek for economical generation and maintains reliability in terms of an increase in load demand.

Transmission planning is also very important for the utility. It identifies the areas that need to be expanded in terms of the load to maintain reliability and growth rate. It also ensures that power can be delivered from generators [9–11]. Similarly, distribution planning is also a very important process that helps to meet the requirements, if required, i.e., determine location, size and time of installation of distribution equipment.

Nowadays, energy forecasting is being implemented through various models in the last few decades. However, increased industrialization and urbanization have led to an increased energy demand with already depleting energy resources. Hence, there is a strong need to significantly alter the existing models or to establish the latest forecasting models to incorporate the rising customer awareness and concerns and to make them an active participant in an SG environment [12].

There are few challenges which may be encountered during the development of the newer models, i.e., availability of greater volumes of electric power consumption data, hybrid grouping making profiling, correct identification of the leading factors contributing to electricity consumption, identifying major factors contributing to consumption, particularly in off-peak hours and quantifying and incorporating outside environmental parameters and incorporating in a climate system or other resources of energy, and scenario-based feedback collection, i.e., non-availability of demand and corresponding cost history data.

Therefore, many techniques are proposed to address these aforementioned issues; however, there still exist some challenges, i.e., fluctuation in energy generation and consumption to control the fluctuating behaviour between the energy consumption pattern and generation pattern, technique accuracy and tuning the hyperparameters for the prediction of electricity load data. To address these issues, a machine learning and deep learning-based model is proposed. The main contributions of this work are:

- a machine learning and deep learning-based model is proposed, i.e., Extreme Learning Machine based Genetic Algorithm (ELM-GA) and Support Vector Machine based Grid Search (SVM-GS),
- the hyperparameter values are tuned using an optimization algorithm to obtain maximum accuracy,
- DT, XGboost and RFE are used in the feature engineering process for removing the redundancy and cleaning the data,
- the GA and GS optimization algorithms are applied to the ELM and SVM to calculate the optimum hyperparameter values.

2. Related Work

Electricity load forecasting is important in terms of production and transmission of energy. The basic work of any system which deals with power or energy is to keep checking the system load at all possible times. This tracking is for different periods. Many factors can cause variations in the load. So, a proper watch must be done on power generation as these sources are not much flexible [12]. The fuel must be available in large quantities for power generation. The following are the three types of load forecasting mainly. The duration of one to years are considered in long term forecasting. Medium-term load forecasting which covers months to weeks, whereas short term forecasting is done on the scheduling of days or usually an hour of a day [13–16]. A similar day lookup approach method is also used for electricity load forecasting and is based upon historical data or previous data for any period, which has the same characteristics, for example, someday of a week, different dates, weather, humidity, etc. This previous data can be used to make trend analysis using regression [17]. The regression-based approach in this technique, linear regression is used to examine dependent variation to specify independent variation. First, consider the independent variable, because most changes occur in the independent variable. In load forecasting, the dependent variable is usually electricity demand and depends on electricity production. Independent variables are usually related to weather, such as humidity, temperature, and wind. The future value of the dependent variable can be estimated by this method [18–21]. The artificial neural network performs nonlinear modeling and adaptation. There is no need to presume a functional relationship between weather and load variables as they are not required in advance. The ANN can adapt the new data when it is exposed. ANN is currently being used in power system problems such as alarm processing, topological observability, fault diagnosis and security assessment, etc. [22]. Time series analysis is done on properly sequenced data which is at uniform intervals for the desired results. TSA is used to know the pattern of the data to predict or know future results based on the previous events that occurred. This method is usually applied to the short time span for the prediction of future events [23].

Expert systems are very intelligent. It can increase their knowledge and work when new data is given to the system. In expert systems, knowledge engineers are added to acquire knowledge and used to create new predictive models for load prediction [24]. Support Vector Machines (SVMs) are a powerful way to provide solutions to regression and classification problems. Linear functions are used in SVM to create linear decision boundaries. In the neural network, there is a problem in choosing the correct architecture, whereas, in SVM, difficulty occurs in selecting suitable kernel [25]. Fuzzy logic is a method that uses the same logic as Boolean logic. Boolean accepts input truth values as 0 and 1. In fuzzy logic, inputs are based on comparisons. In this technique, mathematical models are not used for mapping the value of the input and the result. Fuzzy logic is not affected by noise. When processing is done by this technique "defuzzification" is used to obtain an accurate output or result [26]. The time factor method is completed with short-term load forecasting since it is used every hour for a short time [27]. The weather factor technique uses the load pattern of different weather conditions as the load varies season to season. These loads can be of different appliances heavily used in a different season. Temperature is an important factor in terms of the effect created in this method [28]. The random disturbance is the method in which load demand depends on the sudden disturbance like the suddenness of Television when Minister of any country goes public or come in any program. During this time, the power system contains electrical machines that are being turned on or turned off, which has an impact on sudden changes in the loads. This change is called a random disturbance [29]. Cultural events also cause random disturbance on load. Economic factors have a big influence on LTLF and less effect on STLF. The economic factors have a clear effect on energy consumption. Different factors like industries' heavy appliances generally economic trends have an impact on load growth [30].

An intelligent modular approach is a technique, which is used on STLF. This method has a great impact on the day ahead load forecasting models such as fuel purchases and planning for energy. It aims to increase forecast accuracy without spending more execution time by using a hybrid artificial neural network [31]. To improve the accuracy of electricity load forecasting, Restricted Boltzmann Machine (RBM) and Rectified Linear Unit (ReLU) are implemented. For short term electricity load forecasting, a neural network with the ReLU activation function is implemented. RBM process and train the data, while ReLU performs electricity load forecasting [32]. Deep Long Short-Term Memory (DLSTM) with DNN is applied for the forecasting of electricity load and price. The combination of DLSTM with DNN improves the processing, which can easily be carried out on big data [33]. Deep Auto Encoders (DAEs) provide excellent results in obtaining or achieving accuracy and understanding data. DAE has improved performance in obtaining the accuracy of the results [34]. Gated Recurrent Units (GRU) method is applied to forecast the electricity price [35]. Enhanced SVM and enhanced CNN are implemented using the feature engineering model for classification [36]. Neural network techniques, i.e., the layers of Long Short-Term Memory (LSTM) is added in the layers of Convolutional

Neural Network (CNN) for forecasting the load [37–39]. CNN and LSTM perform better when they are combined rather than performing separately with several other models. The hybrid of CNN and LSTM gives better results [40]. Stack berg is a theoretical game used for the minimization of operational cost and scheduling appliances [41,42]. Feature engineering is used for selection and extraction [43]. There are many techniques used for feature engineering in electric loads [44–47]. Residential load management uses meta-heuristic and machine learning techniques to check energy costs and achieve maximum benefits or comfort during peak hours [48]. In [49], author tuned the hyperparameters of SVM and achieve better accuracy. Wind speed and power prediction is performed using machine learning algorithms in [50].

3. Problem Statement and Motivation

In machine learning, every technique has pros and cons. However, better performance and accuracy are the key concerns in forecasting the electricity load. At this end, a huge amount of data produces hindrance to achieve accuracy during forecasting. As a result, many techniques are designed and modified to address these issues within the limited period; however, there still exist some challenges, i.e., fluctuation in energy generation and consumption to control the fluctuating behavior between the energy consumption pattern and generation pattern [32], technique accuracy and tuning the hyperparameters for the prediction of electricity load data [26,49] and computational complexity during the unclear information, e.g., irrelevant and redundant features in the data, which increase the computation time of the training process and decrease the accuracy of electricity load forecasting. To address these issues, a machine learning and deep learning-based model is proposed. Moreover, the hyperparameter values are tuned using an optimization algorithm to obtain maximum accuracy. Furthermore, DT, XGboost and RFE are used in the feature engineering process for removing the redundancy and cleaning the data. In the end, the GA and GS optimization algorithms are applied to the ELM and SVM to calculate the optimum hyperparameter values.

4. Proposed Model

To address the mentioned problems, we have proposed a model in which firstly, all the features/data from the dataset are imported. Secondly, a hybrid feature selector; XGboost and DT are applied to select useful features. Thirdly, the more relevant features are extracted using RFE method. After working and preprocessing of data, ELM classifier with GA and SVM classifier with GS is applied for classification and forecasting load data as shown in Figure 5. Figure 6 shows the flow chart of the proposed model.

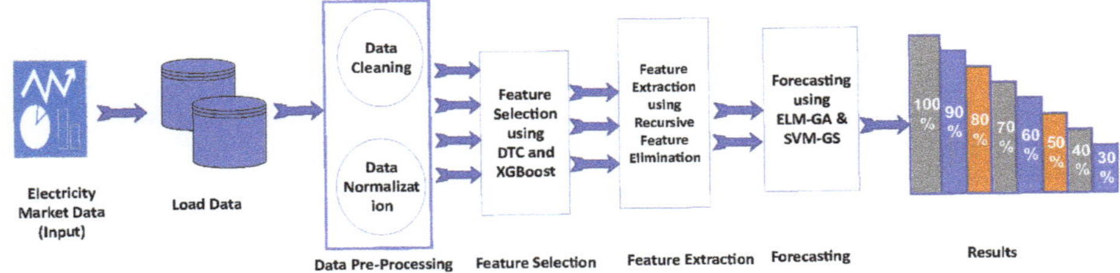

Figure 5. Proposed model.

4.1. Dataset

The daily electricity load data of three years, i.e., January 2017 to December 2019 are used in this paper, which is taken from Independent System Operator New England (ISO NE) (https://www.iso-ne.com). It supplies electricity to different cities in England. The dataset contains dependent and independent data, i.e., weather, temperature, humidity, etc. A column named "electricity load" is our target data. All the features other than the target features have an impact on the target data.

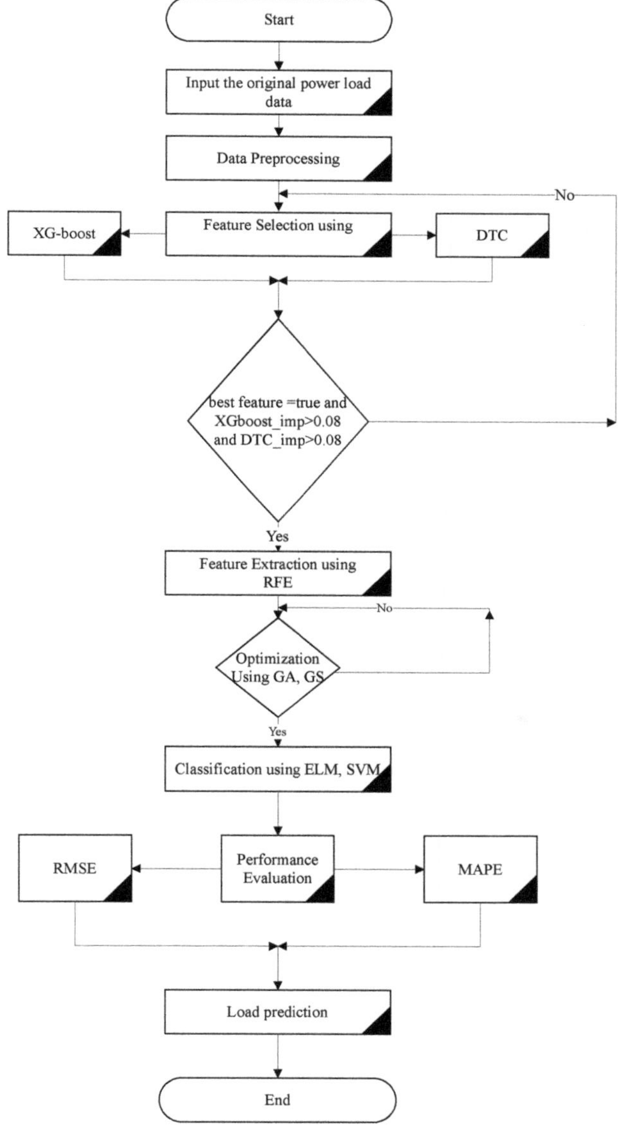

Figure 6. Flow Chart.

As the consumption pattern of the electricity load of the similar month in every year is approximately the same. Therefore, we have taken three years of data, i.e., 36 months. At this end, we have divided the dataset into sets, i.e., training and testing. Therefore, 90% of data is used as training while 10% of the data is used for testing because more the data is provided for training, more will be the learning rate of the model. Furthermore, the data of similar months of every year like January 2017, January 2018, and January 2019 are combined to perform a short-term load forecast for December 2019.

The month-wise arrangement of data in the dataset helps us in better training of our model to find out the load pattern of months. For weekly forecasting, all the data expect the first week of December 2019, i.e., 1 December 2019 to 7 December 2019 are taken as training. The training model is tested in the first week of December. Moreover, for the first five months of the year 2019 are also considered for training and testing. Likewise, all data except January 2019 are also taken for training and testing purposes. Moreover, the same scenario is followed for February 2019, March 2019, April 2019 and May 2019. Simulation and results show the efficacy of the proposed models. Figure 7 shows the data overview and feature names.

Date	DA_Demand	RT_Demand	DA_LMP	DA_EC	DA_CC	DA_MLC	RT_LMP	RT_EC	RT_CC	RT_MLC	Dry_Bulb	Dew_Point	System_Load	Reg_Capacity_Price	Reg_Service_Price
1/1/2017	16679	16391	63	62	0	1	45	45	0	0	30	9	16688	6	0
1/2/2017	16924	16649	57	56	0	1	22	22	0	0	31	15	16934	7	0
1/3/2017	15999	16624	53	52	0	0	58	58	0	0	27	25	16958	12	1
1/4/2017	14404	15765	41	40	0	0	77	76	0	1	41	40	16058	21	1
1/5/2017	18036	18557	83	82	0	1	63	62	0	1	19	-2	18867	8	3
1/6/2017	18504	19259	82	82	0	1	130	130	0	1	15	10	19554	28	0
1/7/2017	19409	19883	115	114	0	1	127	126	0	1	10	-11	20159	24	0
1/8/2017	19192	19813	107	106	0	1	90	89	0	1	19	0	20143	11	0
1/9/2017	17996	18155	85	85	0	1	60	60	0	1	23	8	18454	5	0
1/10/2017	17734	17620	102	101	0	1	81	80	0	1	17	-3	17899	8	0
1/11/2017	17029	17601	113	113	0	1	87	86	0	1	27	12	17874	11	2
1/12/2017	17489	17694	97	96	0	1	85	84	0	1	34	32	17993	17	0
1/13/2017	19463	19034	116	115	0	1	104	101	3	1	15	-2	19307	11	1
1/14/2017	18440	18901	97	96	0	1	93	93	0	1	21	15	19213	9	1
1/15/2017	17852	18212	107	106	0	1	99	98	0	1	26	17	18513	19	1
1/16/2017	18037	17913	114	112	0	1	88	87	0	1	20	-1	18219	23	1
1/17/2017	17640	17321	104	103	0	1	60	60	0	0	20	0	17574	9	0
1/18/2017	15825	15683	53	53	0	0	7	7	0	0	44	41	15929	13	0
1/19/2017	17233	17185	73	72	0	0	49	49	0	0	35	23	17443	17	0
1/20/2017	17716	17683	72	71	0	1	56	56	0	0	30	9	17917	17	0

Figure 7. Dataset Overview.

4.2. Feature Engineering

The statistical mechanics of the feature selection process of the proposed model is applied to the dataset and the most relevant features are selected from the dataset by calculating feature importance. We achieved accurate results by combining DT and XG boost as shown in Figure 8. By setting the feature selection threshold, select features whose importance is equal to or higher than the threshold while deleting other features. The features are selected according to the following formula:

$$f(s) = \begin{cases} reserveif, & IXG[f] + IDT[f] \geq \text{t}, \\ dropif, & \text{IXG[f]} + \text{IDT[f]} < \text{t}. \end{cases} \tag{1}$$

where, variable IXG [i] indicates the calculated feature importance by XGB, and IDT [i] by the DT method. The symbol t represents the threshold for selection of features, and the f symbol represents the feature.

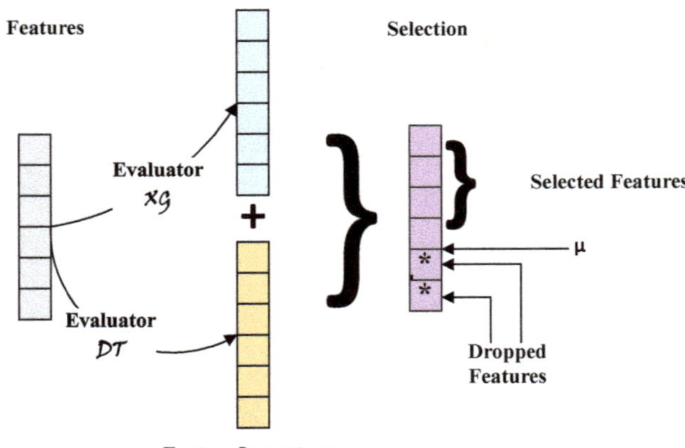

Figure 8. Hybrid feature selector.

Where evaluator XG and DT show the feature' importance calculated by XGboost and Decision Tree Technique. μ represents the threshold set for the selection of features.

After the feature selection step, the feature extraction is carried out by the RFE method. The feature extraction technique extracts non-redundant features and also those features that have a high impact on the target feature in the dataset. RFE, as its name suggests that it removes features recursively and builds a model with the help of extracted attributes and calculates or gives the accuracy of the model. RFE can work on the combination of different attributes that can help in predicting the target variable.

4.3. Classification and Forecasting

ELM with GA algorithm and SVM with the GS algorithm is used as a classifier in the proposed model. The ELM technique is a feed-forward NN used for classification, clustering, regression and compression. ELM can produce better generalization performance when tuned with GA. These models can learn faster than networks based on backpropagation. ELM-GA models can give better performance than SVM-GS. ELM is used for classification and regression problems. ELM is used to learn about a parameter of hidden nodes, input weights that are randomly used and are not tuned. ELM provides extremely fast learning with greater speed, better performance with less involvement of humans [48]. There are different variants of ELM, i.e., voting-based ELM, incremental ELM, error-minimized ELM, pruning ELM, and evolutionary ELM, etc. The main applications of ELM are pattern recognition, classification, image processing, regression, and forecasting, etc.

ELM is very effective as a new training algorithm for SLFN. It is effective and very efficient. Therefore, for a single hidden layer of ELM, we suppose that the i-th node is

$$h_i(\mathbf{x}) = G\left(\mathbf{a}_i, b_i, \mathbf{x}\right). \tag{2}$$

where, a_i and b_i represents the parameters of i-th node. The output function of ELM of L hidden layer is

$$f_L(\mathbf{x}) = \sum_{i=1}^{L} \beta_i h_i(\mathbf{x}). \tag{3}$$

where B_i is the output of i-th hidden node

$$\mathbf{h}(\mathbf{x}) = \left[G\left(h_i(\mathbf{x}) \ldots, h_L(\mathbf{x})\right)\right], \tag{4}$$

is the hidden layer of output mapping of ELM. Where N training sample and output matrix H is

$$\mathbf{H} = \begin{bmatrix} \mathbf{h}\left(\mathbf{x}_1\right) \\ \vdots \\ \mathbf{h}\left(\mathbf{x}_N\right) \end{bmatrix} = \begin{bmatrix} G\left(\mathbf{a}_1, b_1, \mathbf{x}_1\right) & \cdots & G\left(\mathbf{a}_L, b_L, \mathbf{x}_1\right) \\ \vdots & \vdots & \vdots \\ G\left(\mathbf{a}_1, b_1, \mathbf{x}_N\right) & \cdots & G\left(\mathbf{a}_L, b_L, \mathbf{x}_N\right) \end{bmatrix}, \tag{5}$$

and T is training data of the target matrix

$$\mathbf{T} = \begin{bmatrix} \mathbf{t}_1 \\ \vdots \\ \mathbf{t}_N \end{bmatrix}. \tag{6}$$

ELM is a type of NN in which the parameter, i.e., mapping of hidden layers is tuned GA (an optimization algorithm). Furthermore, its main object is:

$$Minimize: \|\boldsymbol{\beta}\|_p^{\sigma_1} + \mathbf{C}\|\mathbf{H}\boldsymbol{\beta} - \mathbf{T}\|_q^{\sigma_2}, \tag{7}$$

$$\text{where } \sigma_1 > 0, \sigma_2 > 0, p, q = 0, \frac{1}{2}, 1, 2, \cdots, +\infty.$$

Different combinations of the above parameters are used and their results are used in different algorithms for classification and regression.

The SVM technique draws a hyperplane between the data. Accurate classification of data can be made by the optimal values of the hyperparameters of SVM. The optimized value of the main three parameters of SVM, i.e., intensive loss function (ϵ), cost penalty (c), and kernel parameter (k) is still a serious issue. In our proposed model, the hyperparameters of SVM are tuned with the GS algorithm, which increases the performance of SVM.

5. Simulation Setup

To simulate our proposed model, we have used anaconda spyder (a desktop application) based on python libraries, i.e., Keras v2.3.0, NumPy v1.18.4, TensorFlow v2.0, and Scikit-learn v0.23. The system specifications are Intel core i3, 512 GB of storage and 8 GB RAM. Whereas, the electricity load dataset used in our proposed model is taken from https://www.iso-ne.com. Simulation results show the efficacy of the proposed models, respectively.

(1) Feature Selection based on Hybrid Feature Selector

The hybrid feature selector (XG-Boost and DT) is applied to the dataset, which calculates the importance of features in numeric format. The high feature importance value is calculated via feature selection technique, which shows the dependency of the input features (high or low) on target value, i.e., in the current scenario this is named "system load" in the respective dataset. Features with high importance are elected as the best feature and features with low importance value are considered as unimportant features. Hereafter, the unimportant features are removed from the dataset. The removal of unimportant features is according to the threshold, i.e., 0.08 in the given scenario. Figure 9 shows the features' importance calculated by feature selection techniques.

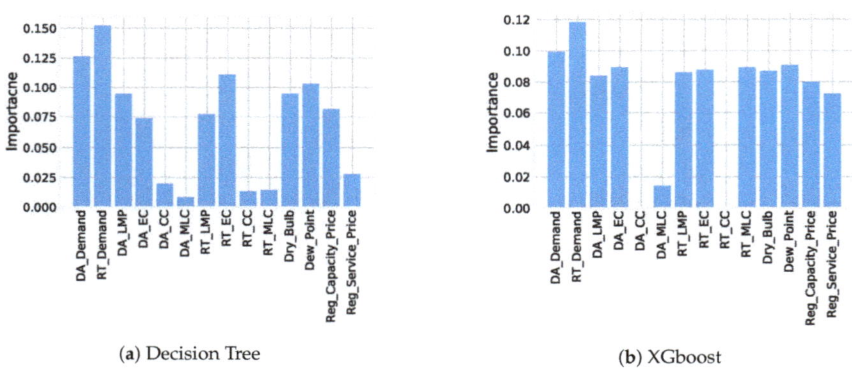

(a) Decision Tree (b) XGboost

Figure 9. Feature importance.

(2) Feature Extraction using RFE

Table 1 shows the target features and other influential features of the dataset. After feature selection, RFE is used as a feature extractor, which results in dimension reduction. The features with the "TRUE" dimensions are not redundant and have a high impact on the target feature, which are selected as an input to the classifier. While the features with the "False" dimensions are redundant and rejected.

Table 1. Features Overview and Dimensions calculated by Redundacny removal using Feature Extraction (RFE).

Target Feature	Features	Short Name	Dimension
System Load	Day-Ahead Cleared Demand	DA_Demand	TRUE
	Regulation Market Service clearing price	Reg_Capacity_Price	TRUE
	Real-Time Demand	RT_Demand	TRUE
	The dewpoint temperature	Dew_Point	FALSE
	Day-Ahead Locational Marginal Price	DA_LMP	FALSE
	The dry-bulb temperature	Dry_Bulb	FALSE
	Energy Component of Day-Ahead	DA_EC	FALSE
	Marginal Loss Component of Real-Time	RT_MLC	FALSE
	Congestion Component of Day-Ahead	DA_CC	FALSE
	Congestion Component of Real-Time	RT_CC	FALSE
	Marginal Loss Component of Day-Ahead	DA_MLC	FALSE
	Energy Component of Real-Time	RT_EC	TRUE
	Real-Time Locational Marginal Price	RT_LMP	TRUE
	Regulation Market Capacity clearing	Reg_Service_Price	FALSE

(3) ELM-GA and SVM-GS performance and Comparison with SOTA Algorithms

Figure 10 shows the normal load of duration 1 January 2019 to 31 December 2019. The consumption pattern of electricity load is shown in Figure 10, which is according to the dataset provided by ISO-NE. Whereas, the variation between the days of the month is shown in Figure 10. It is due to the arrangement of data in months.

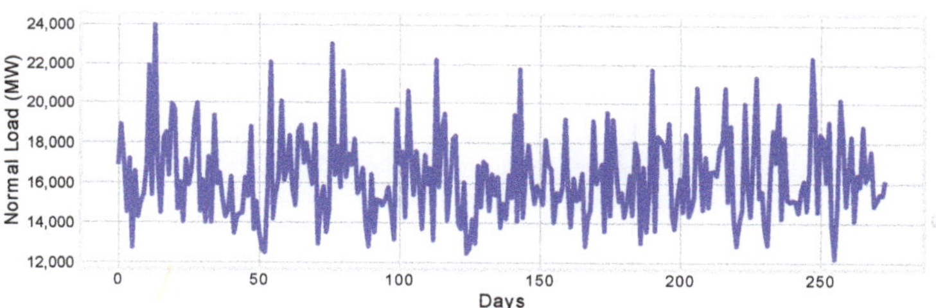

Figure 10. Normal electricity load according to the dataset.

In Figure 11, the results of the forecasted load of the first week of December, i.e., 1 December to 7 December 2019 is shown. Results validate that one of the proposed techniques, i.e., ELM-GA performs better than SVM-GS in terms of load forecasting accuracy.

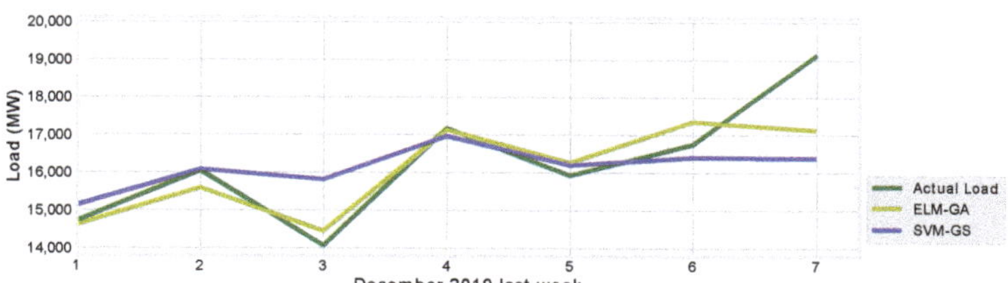

Figure 11. One week prediction (1 December 2019 to 7 December 2019).

The comparison of forecasted results of the proposed methods, i.e., ELM-GA and SVM-GS with an actual load of respective months are shown in Figure 12. Simulation results show that the performance of the proposed technique ELM-GA is nearly the same as the actual data, while the graph of proposed method SVM-GS shows a little deviation as compared to the actual values. However, SVM-GS has better performance than SOTA approaches. In Figure 12, all data except January 2019 is taken as training data. After training the model, it is tested on January 2019. The same cases are revised for February 2019, March 2019, April 2019, May 2019 and June 2019, respectively.

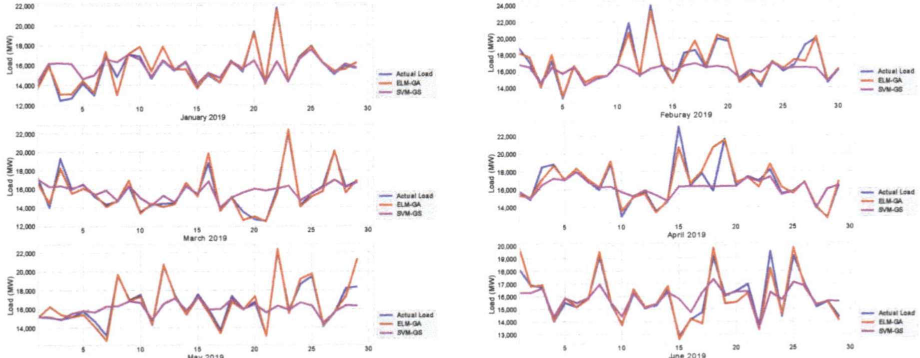

Figure 12. Comparison of forecasting results of different months with actual month load data.

Figure 13 shows the comparative analysis of the proposed models against SOTA techniques. We have used the optimization techniques to tune the parameters of the classifier and to get the prediction accuracy of the classifier that is close to the actual values given in the dataset.

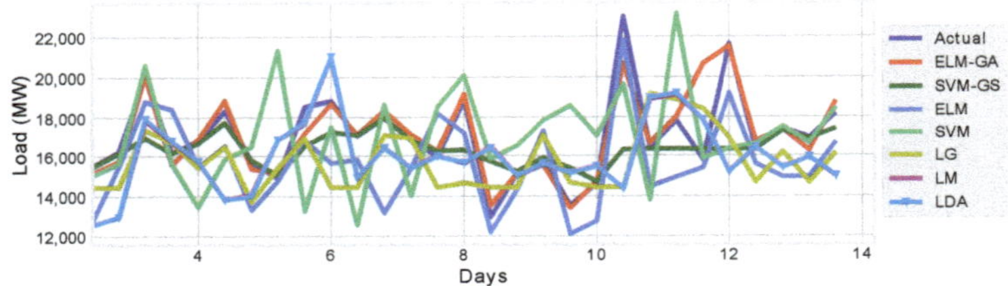

Figure 13. Forecasting of proposed technique with respect to the State of the Art (SOTA).

Figure 14 shows load prediction using optimized ELM based GA; whereas, GA is used for parameter tuning. Also, the proposed technique is compared with SVM in which GS is used for parameter tuning. Simulation results show that the proposed model is more accurate than others, as shown in Table 2.

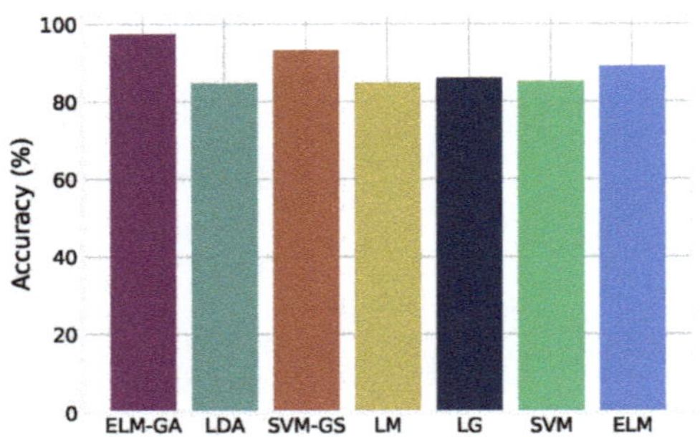

Figure 14. Accuracy of proposed technique with respect to SOTA.

Table 2. Proposed and SOTA techniques results.

Techniques	Accuracy	Performance Metrics			
		MAPE	RMSE	MSE	MAE
ELM-GA	96.42%	2.58	737.35	5.44	4.39
SVM-GS	93.25%	6.75	1811.95	32.83	10.95
LG	86.14%	13.86	2918.49	85.18	22.1
LM	84.88%	15.12	3030.06	91.81	24.46
LDA	84.9%	15.1	3028.09	91.69	24.44
ELM	89.23%	10.77	2014.66	40.59	16.12
SVM	85.31%	14.69	3034.03	92.05	22.75

In addition, performance evaluation metrics are used to demonstrate the productivity of the proposed techniques. At this end, results validate the accuracy of ELM-GA and SVM-GS as 93.25% and 96.42%, respectively. The ELM-GA and SVM-GS have 7% and 8% better results than SOTA techniques.

Performance Evaluation

The performance error of the proposed and SOTA methods are evaluated using performance metrics, i.e., Root Mean Square Error (RMSE) and Mean Absolute Percentage Error (MAPE). The Error rate is calculated by RMSE and MAPE using Equations (8) and (9), respectively [51].

$$\mathrm{RMSE} = \sqrt{\frac{\sum_{\mathrm{i}=1}^{\mathrm{n}} (F_i - A_i)^2}{\mathrm{n}}}. \quad (8)$$

$$\mathrm{MAPE} = \frac{1}{n}\sum_{t=1}^{n}\left|\frac{A_t - F_t}{A_t}\right|. \quad (9)$$

where A, A_t represents the actual value and F, F_t shows forecast value. The values of MAPE and RMSE of proposed techniques are much lower than SOTA as shown in Figure 15.

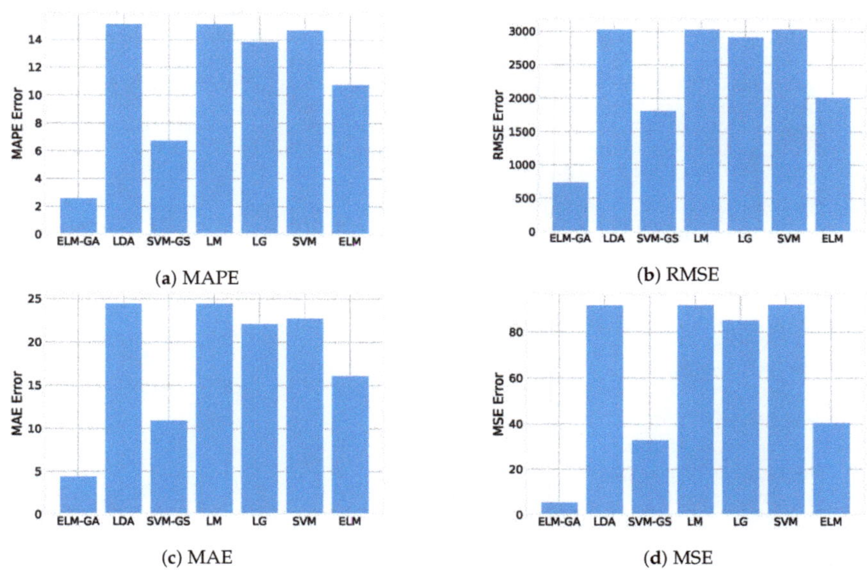

(a) MAPE (b) RMSE

(c) MAE (d) MSE

Figure 15. Performance error.

The detailed results with their performance error are shown in Table 2. It is clearly shown in the table that the proposed techniques outperform SOTA techniques. Here, the performance of ELM-GA is higher than SVM-GS in terms of execution time and forecasting accuracy. ELM-GA and SVM-GS have 7% and 8% better results than benchmark techniques.

Figure 16 describes the asymmetric loss function and accuracy of the model, which shows that the performance of the model is increasing continuously. Moreover, the loss is decreasing by training the model with respect to Epoches.

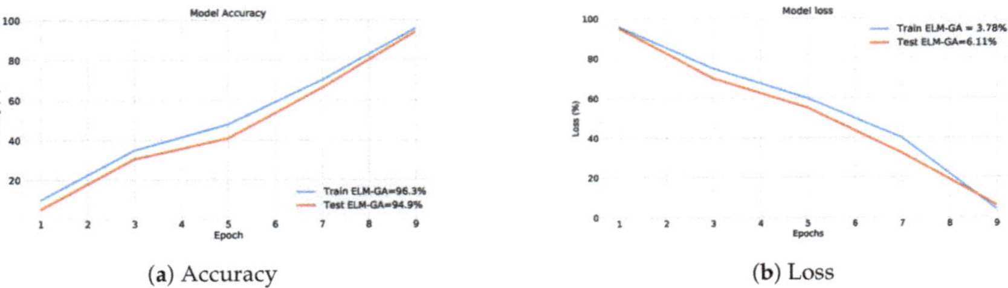

(**a**) Accuracy (**b**) Loss

Figure 16. Performance accuracy and loss function.

6. Conclusions

In this paper, a machine learning and meta-heuristic algorithm based model is proposed for short term electrical load forecasting. The main advantage associated with this model is to achieve better performance and maximum accuracy. Therefore, a real-time electricity load dataset is taken into account, ranging from January 2017 to December 2019 from ISO-NE. Afterwards, data is converted into supervised learning sets, i.e., features (training x and test label x) and labels (training y and test label y). Furthermore, the proposed model is divided into three fundamental steps including feature selection using a hybrid feature selector (XGboost and DT), Redundancy removal using Feature Extraction (RFE) technique and classification/forecasting using ELM-GA and SVM-GS. At this end, parameters of ELM are tuned with the GA and hyperparameters of SVM are tuned with GS. In the end, a comparative analysis is performed with baseline SOTA techniques. It is evident from the results that proposed techniques outperform SOTA techniques in counterparts. In the future, the different low-performance classifiers can further be enhanced with different meta-heuristic techniques to improve forecasting accuracy.

Author Contributions: The research conceptualization and methodology were done by W.A., N.A. and M.A. The technical and theoretical framework was prepared by T.A. and M.I. The technical review and improvement were performed by M.A. and N.A. The overall technical support, guidance, and project administration was done by M.S. and A.G. Finally, responses to the queries of the reviewers were done by N.A. and M.A. All authors have read and agreed to the published version of the manuscript.

Funding: This research received no external funding.

Conflicts of Interest: The authors declare no conflict of interest.

Abbreviations

The following abbreviations are used in this manuscript:

TGs	Traditional Grids
SG	Smart Grid
SMs	Smart Meters
SVM	Support Vector Machine
RBM	Restricted Boltzmann Machine
ReLU	Rectified Linear Unit
DLSTM	Deep Long Short-Term Memory
DAEs	Deep Auto Encoders
GRU	Gated Recurrent Units
CNN	Convolutional Neural Network

ISO NE	Independent System Operator New England
SOTA	State Of The Art
ELM	Extreme Learning Machine
GS	Grid Search
NN	Neural Network
MAPE	Mean Average Percentage Error
RFE	Redundancy removal using Feature Extraction
RMSE	Root Mean Square Error

References

1. Fang, X.; Misra, S.; Xue, G.; Yang, D. Smart grid—The new and improved power grid: A survey. *IEEE Commun. Surv. Tutorials* **2011**, *14*, 944–980. [CrossRef]
2. Samadi, P., Wong, V. W.; Schober, R. Load scheduling and power trading in systems with high penetration of renewable energy resources. *IEEE Trans. Smart Grid* **2015**, *7*, 1802–1812. [CrossRef]
3. Zhao, Z.; Lee, W.C.; Shin, Y.; Song, K.B. An optimal power scheduling method for demand response in home energy management system. *IEEE Trans. Smart Grid* **2013**, *4*, 1391–1400. [CrossRef]
4. Davito, B.; Tai, H.; Uhlaner, R. The smart grid and the promise of demand-side management. *McKinsey Smart Grid* **2010**, *3*, 8–44.
5. Liu, Y. Wireless sensor network applications in smart grid: Recent trends and challenges. *Int. J. Distrib. Sens. Networks* **2012**, *8*, 492819. [CrossRef]
6. Siano, P.; Sarno, D. Assessing the benefits of residential demand response in a real time distribution energy market. *Applied Energy* **2016**, *161*, 533–551. [CrossRef]
7. Aghaei, J.; Alizadeh, M.I. Demand response in smart electricity grids equipped with renewable energy sources: A review. *Renew. Sustain. Energy Rev.* **2013**, *18*, 64–72. [CrossRef]
8. Paterakis, N.G.; Erdinc, O.; Catalao, J. P. An overview of Demand Response: Key-elements and international experience. Renewable and Sustainable Energy Reviews. *Renew. Sustain. Energy Rev.* **2017**, *69*, 871–891. [CrossRef]
9. Pinson, P.; Madsen, H. Benefits and challenges of electrical demand response: A critical review. *Renew. Sustain. Energy Rev.* **2014**, *39*, 686–699.
10. Tabar, V.S.; Jirdehi, M.A.; Hemmati, R. Energy management in microgrid based on the multi objective stochastic programming incorporating portable renewable energy resource as demand response option. *Energy* **2017**, *118*, 827–839. [CrossRef]
11. Zheng, J.; Gao, D.W.; Lin, L. Smart meters in smart grid: An overview. In Proceedings of the 2013 IEEE Green Technologies Conference (GreenTech), Denver, CO, USA, 4–5 April 2013; pp. 57–64.
12. Bessa, R.J. Solar power forecasting for smart grids considering ICT constraints. In Proceedings of the 4th Solar Integration Workshop, Berlin, Germany, 10–11 November 2014.
13. Huang, S.J.; Shih, K.R. Short-term load forecasting via ARMA model identification including non-Gaussian process considerations. *IEEE Trans. Power Syst.* **2003**, *18*, 673–679. [CrossRef]
14. Kandil, N.; Wamkeue, R.; Saad, M.; Georges, S. An efficient approach for shorterm load forecasting using artificial neural networks. In Proceedings of the 2006 IEEE International Symposium on Industrial Electronics, Montreal, QC, Canada, 9–13 July 2006; Volume 3, pp. 1928–1932.
15. Mandal, P.; Senjyu, T.; Urasaki, N.; Funabashi, T. A neural network based several-hour-ahead electric load forecasting using similar days approach. *Int. J. Electr. Power Energy Syst.* **2006**, *28*, 367–373. [CrossRef]
16. Topalli, A.K.; Erkmen, I.; Topalli, I. Intelligent short-term load forecasting in Turkey. *Int. J. Electr. Power Energy Syst.* **2006**, *28*, 437–447. [CrossRef]
17. Mu, Q.; Wu, Y.; Pan, X.; Huang, L.; Li, X. Short-term load forecasting using improved similar days method. In Proceedings of the 2010 Asia-Pacific Power and Energy Engineering Conference, Chengdu, China, 28–31 March 2010; pp. 1–4.
18. Wang, J.M.; Wang, L.P. A new method for short-term electricity load forecasting. *Trans. Inst. Meas. Control* **2008**, *30*, 331–344. [CrossRef]
19. Ruzic, S.; Vuckovic, A.; Nikolic, N. Weather sensitive method for short term load forecasting in electric power utility of Serbia. *IEEE Trans. Power Syst.* **2003**, *18*, 1581–1586. [CrossRef]

20. Haida, T.; Muto, S. Regression based peak load forecasting using a transformation technique. *IEEE Trans. Power Syst.* **1994**, *9*, 1788–1794. [CrossRef]
21. Charytoniuk, W.; Chen, M.S.; Van Olinda, P. Nonparametric regression based short-term load forecasting. *IEEE Trans. Power Syst.* **1998**, *13*, 725–730. [CrossRef]
22. Amjady, N. Short-term hourly load forecasting using time-series modeling with peak load estimation capability. *IEEE Trans. Power Syst.* **2001**, *16*, 498–505. [CrossRef]
23. Park, D.C.; El-Sharkawi, M.A.; Marks, R.J.; Atlas, L.E.; Damborg, M.J. Electric load forecasting using an artificial neural network. *IEEE Trans. Power Syst.* **1991**, *6*, 442–449. [CrossRef]
24. Kandil, M.S.; El-Debeiky, S.M.; Hasanien, N.E. Long-term load forecasting for fast developing utility using a knowledge-based expert system. *IEEE Trans. Power Syst.* **2002**, *17*, 491–496. [CrossRef]
25. Mohandes, M. Support vector machines for short-term electrical load forecasting. *Int. J. Energy Res.* **2002**, *26*, 335–345. [CrossRef]
26. Ayub, N.; Javaid, N.; Mujeeb, S.; Zahid, M.; Khan, W.Z.; Khattak, M.U. Electricity Load Forecasting in Smart Grids Using Support Vector Machine. In Proceedings of the 33rd International Conference on Advanced Information Networking and Applications, Matsue, Japan, 27–29 March 2019; Volume 926, pp. 1–13.
27. Chu, W.; Keerthi, S.S.; Ong, C.J. A general formulation for support vector machines. In Proceedings of the 9th International Conference on Neural Information Processing, Singapore, I8–22 November 2002; Volume 5, pp. 2522–2526.
28. Kumar, V.; Pal, S. A Literature Survey of Load Forecasting Methods and Impact of Different Factors on Load Forecasting. *Int. J. Res. Appl. Sci. Eng. Technol.* **2017**, *5*, 469–472. [CrossRef]
29. Salkuti, S.R. Short-term electrical load forecasting using radial basis function neural networks considering weather factors. *Electr. Eng.* **2018**, *100*, 1985–1995. [CrossRef]
30. Aggarwal, S.K.; Saini, L.M.; Kumar, A. Electricity price forecasting in deregulated markets: A review and evaluation. *Int. J. Electr. Power Energy Syst.* **2009**, *31*, 13–22. [CrossRef]
31. Ahmad, A.; Javaid, N.; Mateen, A.; Awais, M.; Khan, Z.A. Short-term load forecasting in smart grids: An intelligent modular approach. *Energies* **2019**, *12*, 164. [CrossRef]
32. Wang, K.; Xu, C.; Zhang, Y.; Guo, S.; Zomaya, A.Y. Robust big data analytics for electricity price forecasting in the smart grid. *IEEE Trans. Big Data* **2017**, *5*, 34–45. [CrossRef]
33. Mujeeb, S.; Javaid, N.; Ilahi, M.; Wadud, Z.; Ishmanov, F.; Afzal, M.K. Deep long short-term memory: A new price and load forecasting scheme for big data in smart cities. *Sustainability* **2019**, *11*, 987. [CrossRef]
34. Zahid, M.; Ahmed, F.; Javaid, N.; Abbasi, R.A.; Kazmi, Z.; Syeda, H.; Javaid, A.; Bilal, M.; Akbar, M.; Ilahi, M. Electricity price and load forecasting using enhanced convolutional neural network and enhanced support vector regression in smart grids. *Electronics* **2019**, *8*, 122. [CrossRef]
35. Fan, C.; Xiao, F.; Zhao, Y. A short-term building cooling load prediction method using deep learning algorithms. *Appl. Energy* **2017**, *195*, 222–233. [CrossRef]
36. Khan, Z.A.; Zafar, A.; Javaid, S.; Aslam, S.; Rahim, M.H.; Javaid, N. Hybrid meta-heuristic optimization based home energy management system in smart grid. *J. Ambient Intell. Humaniz. Comput.* **2019**, *10*, 4837–4853. [CrossRef]
37. Moghaddass, R.; Wang, J. A hierarchical framework for smart grid anomaly detection using large-scale smart meter data. *IEEE Trans. Smart Grid* **2017**, *9*, 5820–5830. [CrossRef]
38. Samuel, O.; Javaid, S.; Javaid, N.; Ahmed, S.H.; Afzal, M.K.; Ishmanov, F. An efficient power scheduling in smart homes using Jaya based optimization with time-of-use and critical peak pricing schemes. *Energies* **2018**, *11*, 3155. [CrossRef]
39. Ryu, S.; Noh, J.; Kim, H. Deep neural network based demand side short term load forecasting. *Energies* **2017**, *10*, 3. [CrossRef]
40. Zhao, J.; Dong, Z.; Li, X. Electricity price forecasting with effective feature preprocessing. In Proceedings of the 2006 IEEE Power Engineering Society General Meeting, Montreal, QC, Canada, 18–22 June 2006.
41. Javaid, N.; Ahmed, A.; Iqbal, S.; Ashraf, M. Day ahead real time pricing and critical peak pricing based power scheduling for smart homes with different duty cycles. *Energies* **2018**, *11*, 1464. [CrossRef]
42. Luo, F.; Ranzi, G.; Wan, C.; Xu, Z.; Dong, Z.Y. A multistage home energy management system with residential photovoltaic penetration. *IEEE Trans. Ind. Inform.* **2018**, *15*, 116–126. [CrossRef]
43. Khalid, R.; Javaid, N.; Rahim, M.H.; Aslam, S.; Sher, A. Fuzzy energy management controller and scheduler for smart homes. *Sustain. Comput. Inform. Syst.* **2019**, *21*, 103–118. [CrossRef]

44. Ertugrul, Ö.F. Forecasting electricity load by a novel recurrent extreme learning machines approach. *Int. J. Electr. Power Energy Syst.* **2016**, *78*, 429–435. [CrossRef]
45. Khan, M.A.; Javaid, N.; Mahmood, A.; Khan, Z.A.; Alrajeh, N. A generic demand-side management model for smart grid. *Int. J. Energy Res.* **2015**, *39*, 954–964. [CrossRef]
46. Bilalli, B.; Abelló, A.; Aluja-Banet, T.; Wrembel, R. Intelligent assistance for data pre-processing. *Comput. Stand. Interfaces* **2018**, *57*, 101–109. [CrossRef]
47. Fallah, S.N.; Deo, R.C.; Shojafar, M.; Conti, M.; Shamshirb, S. Computational intelligence approaches for energy load forecasting in smart energy management grids: State of the art, future challenges, and research directions. *Energies* **2018**, *11*, 596. [CrossRef]
48. Huang, G.B.; Zhu, Q.Y.; Siew, C.K. Extreme learning machine: Theory and applications. *Neurocomputing* **2006**, *70*, 489–501. [CrossRef]
49. Rojas-Domínguez, A.; Padierna, L.C.; Valadez, J.M.C.; Puga-Soberanes, H.J.; Fraire, H.J. Optimal hyper-parameter tuning of SVM classifiers with application to medical diagnosis. *IEEE Access* **2017**, *6*, 7164–7176. [CrossRef]
50. Li, Z.L.; Xia, J.; Liu, A.; Li, P. States prediction for solar power and wind speed using BBA-SVM. *IET Renew. Power Gener.* **2019**, *13*, 1115–1122. [CrossRef]
51. Morley, S.K.; Brito, T.V.; Welling, D.T. Measures of model performance based on the log accuracy ratio. *Space Weather* **2018**, *16*, 69–88. [CrossRef]

Article

Big Data Analytics for Short and Medium-Term Electricity Load Forecasting Using an AI Techniques Ensembler

Nasir Ayub [1], Muhammad Irfan [2,*], Muhammad Awais [3], Usman Ali [4], Tariq Ali [2], Mohammed Hamdi [5], Abdullah Alghamdi [5] and Fazal Muhammad [6]

1. Department of Computer Science, Federal Urdu University of Arts, Science and Technology, Islamabad 44000, Pakistan; nasir.ayubse@gmail.com
2. Electrical Engineering Department, College of Engineering, Najran University, Najran 61441, Saudi Arabia; tariqhsp@gmail.com
3. School of Computing and Communications, Lancaster University, Bailrigg, Lancaster LA1 4YW, UK; m.awais11@lancaster.ac.uk
4. Department of Computing, RIPHAH University Faisalabad, Faisalabad 38000, Pakistan; r.usmaanali@gmail.com
5. College of Computer Science and Information Systems, Najran University, Najran 61441, Saudi Arabia; mahamdi@nu.edu.sa (M.H.); aaalghamdi@nu.edu.sa (A.A.)
6. Department of Electrical Engineering, City University of Science & Information Technology Peshawar, Peshawar 25000, Pakistan; fazal.muhammad@cusit.edu.pk

* Correspondence: miditta@nu.edu.sa

Received: 14 July 2020; Accepted: 24 September 2020; Published: 5 October 2020

Abstract: Electrical load forecasting provides knowledge about future consumption and generation of electricity. There is a high level of fluctuation behavior between energy generation and consumption. Sometimes, the energy demand of the consumer becomes higher than the energy already generated, and vice versa. Electricity load forecasting provides a monitoring framework for future energy generation, consumption, and making a balance between them. In this paper, we propose a framework, in which deep learning and supervised machine learning techniques are implemented for electricity-load forecasting. A three-step model is proposed, which includes: feature selection, extraction, and classification. The hybrid of Random Forest (RF) and Extreme Gradient Boosting (XGB) is used to calculate features' importance. The average feature importance of hybrid techniques selects the most relevant and high importance features in the feature selection method. The Recursive Feature Elimination (RFE) method is used to eliminate the irrelevant features in the feature extraction method. The load forecasting is performed with Support Vector Machines (SVM) and a hybrid of Gated Recurrent Units (GRU) and Convolutional Neural Networks (CNN). The meta-heuristic algorithms, i.e., Grey Wolf Optimization (GWO) and Earth Worm Optimization (EWO) are applied to tune the hyper-parameters of SVM and CNN-GRU, respectively. The accuracy of our enhanced techniques CNN-GRU-EWO and SVM-GWO is 96.33% and 90.67%, respectively. Our proposed techniques CNN-GRU-EWO and SVM-GWO perform 7% and 3% better than the State-Of-The-Art (SOTA). In the end, a comparison with SOTA techniques is performed to show the improvement of the proposed techniques. This comparison showed that the proposed technique performs well and results in the lowest performance error rates and highest accuracy rates as compared to other techniques.

Keywords: load forecasting; optimization techniques; deep learning; big data analytics

1. Introduction

The electrical industry plays a very vital role in human life from various angles. The electricity demand is increasing day-by-day with the rapid increase in population [1]. The traditional power grid became an old version; which is not efficient enough now, so, the intelligent and smart version of the power grid known as the Smart Grid (SG) is introduced. Through the SG system, it became very easy to manage the distribution of electric load for utility companies and remain in touch with the consumers. The SG also helps to reduce the variations between power demand and supply. The most important task of the SG is to effectively control the consumption, generation, and distribution of electricity. The utility supplies electricity to consumers, according to their demand. Sometimes, the rate of electricity consumption of the user increases and the utility does not have enough energy be supplied. To overcome the issue of balancing between consumption and utility supply, the utility uses the electricity load forecasting model, which is one aspect of SG. The conceptual diagram of the SG is shown in Figure 1.

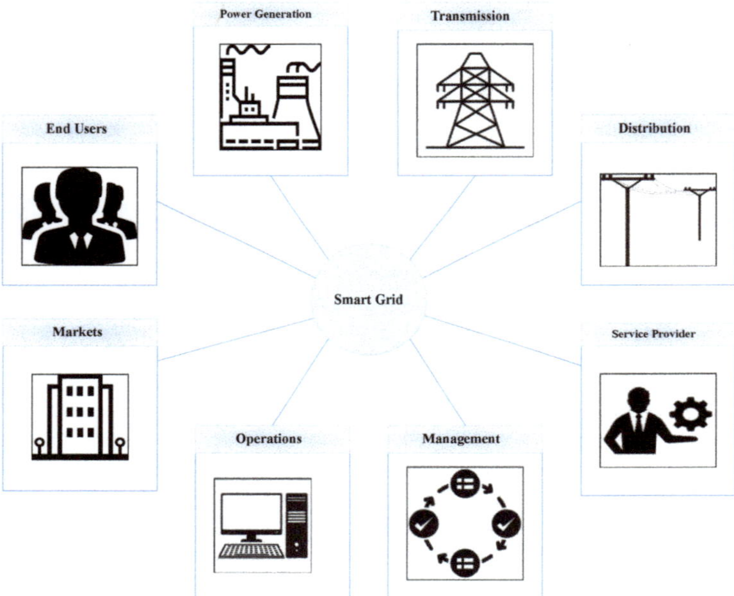

Figure 1. Smart grid infrastructure.

The approximate energy consumption pattern of the user is predicted through load forecasting by their historical data. Generally, forecasting is of four types; Short-Term Forecasting (STF), Very Small Term Forecasting (VSTF), Medium-Term Forecasting (MTF), and Long-Term Forecasting (LTF). In STF, electric load of one day-ahead to some weeks-ahead is predicted, VSTF consists of predictions for some hours to one day, MTF predicts data of one week to one year, and through LTF, one year to several years ahead load can be forecasted [2–4]. In this paper, STF and MTF are performed with an excessive record of electricity dataset. Data analysis is a process of getting useful information from hidden patterns of data. Data analysts measures the price and load consumption by taking historical data in the form of datasets to perform some tasks which allow us to obtain useful information [5]. In [6], the detailed review of data is available. The volume of real-world data is intensively increasing day-by-day and the large volume of data is referred to as big data. Through data analytics, effective information is collected from massive quantities of historical power data to implement analysis over it, which helps to make more enhancements in the market operations management and planning.

Big data is multifaceted and very excessive in volume. The main issue in the big data sets is redundant features; so traditional methods are not very supportive for handling such a large amount of data.

Many techniques are tested and applied to handle big data and extract useful information. Although, big data is still an issue of the current era. Many authors around the globe are working on handling big data using artificial intelligence techniques. The authors in [7,8] improved the price forecasting and load forecasting accuracy; however, the computational time is not considered. Similarly, the issue of load forecasting is addressed in [9]; however, the issue of overfitting is not addressed. Moreover in [10], the author proposed the BPNN model to forecast the day-ahead electricity load; however, the complexity of the proposed model is increased. Additionally, the LSTM-RNN model in [11] is used to forecast an hourly and monthly electricity load. Furthermore in [12], the hybrid of SVM and non-linear regression is used to forecast load; unfortunately, the problem of over-fitting is increased. Hence, conventional simple techniques and methods are not very suitable for a varying electricity load. A better framework and enhanced techniques are required to solve the load forecasting problem. Therefore, the main objective of this paper is to enhance the accuracy rate of electricity load forecasting by optimizing the parameters of machine learning and deep learning techniques on a large amount of electricity load data. As a large amount of data contain redundant and irrelevant features, which increase the time complexity of training. RF and XGB are used as feature selection methods. The RFE technique is used as a feature extraction method to eliminate the redundancy, while SVM with the GWO algorithm and CNN-GRU with the EWO algorithm is used for the classification of electricity load forecasting.

The main contributions of this paper are given below:

- Hybrid of feature selection techniques; Extreme Gradient Boosting (XGB), Random Forest RF and Recursive Feature Elimination (RFE) techniques are applied to clean the huge amount of data.
- Two enhanced classifier techniques, Support Vector Machine with Grey Wolf Optimization (SVM-GWO) and Convolutional Neural Network Gated Recurrent Unit with Earth Worm Optimization (CNN-GRU-EWO) are proposed to forecast the electricity load.
- Grey Wolf Optimization (GWO) and Earth Worm Optimization (EWO) algorithms are used to tune the parameters of SVM and CNN-GRU, respectively.
- The parameters of classifiers are tuned to reduce the computational time efficiently.
- To overcome the overfitting problem, enhanced classifiers are used.
- Our proposed techniques are compared with some State Of The Art (SOTA) to prove the better performance of our enhanced techniques.

The rest of the paper is organized as follows. Section 2 contains the literature review. Section 3 contains the proposed methodology (method). Section 4 contains the results. Section 5 contains the conclusion and policy implications.

2. Related Work

Many techniques and ideas are used and tested to predict the power load and other areas with successive results. In [13], the authors performed load forecasting with various smart home data by applying an analytical approach to the data, however, they were unable to manage a large amount of data properly. Deep Long Short Term Memory (DLSTM) and machine learning-based model are proposed to forecast the price and electricity [14]. The proposed DLSTM outperforms in achieving the accuracy of load forecasting. However, LSTM is not good in terms of training because it needsa memory bandwidth bound calculation and it limits the applicability of neural network solutions.

In [15], the authors performed load forecasting with feature selection and classification models, taking the dataset as input. They used MI to select the best features and discard insignificant features. The authors proposed three-step strategies for load forecasting in [16], in the first step they used Conditional Mutual Information (CMI) for best feature selection. The second step consists of NLSSVM and ARIMA machine learning techniques, which create nonlinear and linear correlations for load forecasting. In the third step, the parameters of NLSSVM are tuned with the ABC algorithm. However, reducing the features with the help of a feature selection method also reduces the forecasting accuracy rate.

For feature selection, IG and MI techniques have been used, which helped measure the redundancy and most relevant features [17]. They also proposed a hybrid wrapper filter-based approach, where the filter part of the method is selected as a little part of the dataset for features by redundancy and iteration of inputs. They introduced a new feature selection method, but failed to maintain high accuracy rates.

The authors in [18] performed hourly forecasting and also defined the uncertainty of the predictive method. They proposed Generalized ELM and Improve WNN techniques to implement on OAE electrical dataset. However, these techniques are outperformed for this development. The parameters of these techniques are manually tuned, however, with dynamically-tuned hyper-parameters, forecasting accuracy rates can be further increased.

In [19], forecasting has been done using a combination of two deep learning techniques; CNN and LSTM. The proposed model is evaluated using the Mean Square Error (MSE). Further, the accuracy rate of the proposed technique also compared with some benchmark techniques, and results show that the proposed technique outclasses all other techniques in terms of the accuracy rate. The proposed technique performed better, but the authors did not consider the feature redundancy. Redundant features can make a negative effect on model accuracy rates.

The authors in [20], performed day-ahead forecasting by increasing the layers of Artificial Neural Networks (ANN) and tuned the hyper-parameters with an optimizing algorithm. To improve the accuracy rate authors in [21] increased the layers of Neural-Network (NN). The enhanced NN is also compared with conventional techniques to demonstrate improved high accuracy rates. The enhanced NN is compared with ARIMA and SVR, showing that it performed better, but failed to avoid overfitting problems.

In [22], each day of the week forecasting is performed by applying the deep CNN for classification. The applied dataset is taken from the Victorian electrical company, Australia. The authors forecast the one day load and analyzed it by comparing the one day load with the same day's load of the previous three months. However, the author used fewer record datasets and was unable to train the CNN model properly.

The hybrid of CNN and LSTM achieved accuracy in terms of electricity load forecasting in [23]. The objective of their work is short-term forecasting and they used MAPE and MAE error metrics for evaluation of results. The hyperparameters of SVM are tuned with a random search algorithm to achieve improved accuracy and a lower error rate [18]. They performed load forecasting and compared the results with manually tuned SVM and CNN. Results show the improvement of the enhanced technique. The authors used eight years of data for load forecasting purposes, however, SVM is not good for classifying large datasets.

In [24], the authors proposed two techniques named enhanced SVM and enhanced ELM to perform short term load forecasting. A grid search optimization algorithm is used to optimize the hyper-parameters of SVM and the hyper-parameters of the Extreme Learning Machine (ELM) tuned with the Genetic Algorithm (GA). The proposed techniques performed better, but the authors failed to avoid overfitting problems for SVM.

The short-term load forecasting is performed using NN and Levenberg Marquardt learning in [25]. The authors used the Tanzanian dataset duration 2000 to 2008. The author used MAE, MAPE, MSE, and MAPE error matrices to calculate the result. However, the calculated results through MPE and MAPE gave good results, but the error rates calculated through MAE and MSE are very high.

The authors performed forecasting based on a feature selection technique and least square SVM technique in [26]. They used ASF to select the most informative input values and least square SVM used to predict the model. Results were evaluated with MAPE and MAE error metrics. The proposed model gave low error rates concerning MAPE values, but through MAE it showed the worst results. The authors in [27] performed forecasting with feature selection and classification model.

Feature selection was based on MI and CA. The classification part was based on the iterative approach of two neural networks. They used the output of the first neural network as an input of the second neural network.

The consumption data of homes are obtained through smart meters and used to perform forecasting with the help of the GRU deep learning technique in [28]. The authors did not consider preprocessing the dataset and did not remove irrelevant features through their applied model. In [29] authors used hourly and historical temperature data for forecasting. The SVM and ANN machine learning techniques are proposed for this purpose. However, the parameters of thr proposed techniques have been tuned manually.

Improved kernel ELM and Cholesky decomposition techniques are used to forecast the electrical load in [30]. The proposed technique is further compared with conventional ELM and GNN. RMSE error evaluator has used to evaluating the results. In their research work, only one evaluator is used to prove the superiority of the proposed model, other error metrics such as MAPE, RMSE, MSE, accuracy, f1-score, and precision, etc. are not included.

In [31,32], authors used the enhanced CNN method to forecast the electricity load. Furthermore, the superiority of the proposed method is shown with different statistical tests. A composite method based on the optimal learning MLP technique is applied to forecast the mid-term electricity load [33,34]. An acceptable accuracy of 85% is achieved to forecast the mid-term electricity load. However, the author has not considered the overfitting problem of MLP and was unable to highlight the issue of disregarding spatial information.

The deep learning techniques CNN and ANN are used for forecasting in [35,36]. The authors tuned the parameters manually and did not eliminate the irrelevant features. The hybrid of CNN and GRU is applied to predict the electricity load in [37–39]. Results are evaluated with MAPE and RMSE values. A comparison of the hybrid technique and conventional techniques also performed. Comparison results show that the hybrid technique outclasses all other techniques. The hybrid model performed well, but parameters have been manually tuned. In our work, we have used the latest heuristic algorithm to automatically allocate the optimum values to the parameters of our proposed techniques. In [40,41], the authors used a framework named feature selection, extraction and classification for load forecasting. They used a hybrid of XGB and DTC techniques to select the most relevant features and eliminated the irrelevant features in the feature extraction step using RFE technique. In the end, classification is performed using SVM. The proposed framework performed well, but the computational complexity of SVM is high and SVM is also not good for processing uncertain data [42–44]. The literature review shows that most authors performed forecasting with machine learning and deep learning Table 1. By finding the optimal value for the hyperparameters of techniques is tough work. Furthermore, the irrelevant features in electricity datasets also have a negative impact on model training. To solve some of the above-mentioned issues, we used heuristic algorithms to find optimal values of hyperparameters automatically and for feature selection, the extraction model was proposed to remove irrelevant features from a dataset.

Table 1. Tabular form of related work.

Proposed Techniques	Objective	Dataset	Limitations
DA [13]	Reduce peak load	PJM	Issue in managing big data
DLSTM [14]	Price and Load forecasting	ISO-NE	Cannot fulfill the requirement of real time data.
CMI, NLSSVM [15,16]	Forecasting with important feature selection method	PJM	Less amount of data is taken into consideration
GELM, IWNN [17]	Hourly price forecasting	PJM	Model complexity is considered
CNN, LSTM [18,19]	Price forecasting	PJM	Redundancy in features are not considered
DNN [20]	Load forecasting	Irish	Overfitting problem needed to improve

Table 1. *Cont.*

Proposed Techniques	Objective	Dataset	Limitations
DCNN [21]	Load forecasting of one day	Victoria	Limited use of dataset
ESVM [22,23]	Short term load forecasting	ISO-NE	SVM is not good to deal big dataset because overfitting problem
ANN [24]	Half hourly load forecasting	Tanzanian	Accuracy rates of their work are not satisfactory.
MI, NN [25]	Short term forecasting	PJM	Maximize the penetration of renewable energy
NARX, ARMAX [26]	Residential based short term load forecasting	IESCO	Model complexity increased
GRU [27]	Load forecasting	PJM	Redundancy of features did not considered
SVM, ANN [28]	Short term forecasting	IITK	Very small dataset is used for experiment
ELM-K [29]	Short term forecasting	Southern China	Only one error metrics used for evaluation.
CNN [20]	Short term forecasting	ISO-NE	Manually tuned the hyper parameters of proposed technique
GRU-CNN [31]	Short term forecasting	Wuwei, Gansu province	Manually tuned the hyper parameters of proposed technique
MI, ANN [32]	Day ahead load forecasting	DAYTOWN, AKPC	Feature selection need more improvement

3. Proposed System Model

After evaluating the literature review and the aforementioned techniques for load forecasting, we propose a framework that is based on average feature selection, extraction, and forecasting. The machine learning techniques, RF, and XGB are used as feature selection techniques, while RFE is used for feature extraction activity. For average feature selection, the average score of RF and XGB is considered for the selection of features as described in Equation (1). Moreover, for classification purposes, machine learning-based technique SVM and deep learning-based technique CNN-GRU are used, respectively. Furthermore, the basic parameters of the CNN-GRU and SVM are tuned with a meta-heuristic algorithm, i.e., EWO and GWO, respectively. The forecasting in Figure 2 displayed the working flowchart of the used model.

3.1. Dataset Description

The latest electricity daily load dataset is used in this paper, which is downloaded from the ISONE website [34]. The columns in the dataset are referred to as "features" in our work. The dataset is organized according to a month-wise pattern, i.e., January 2012, January 2013 up to January 2019 and February 2012, February 2013 up to February 2019, and so on. The benefit of the month-wise organization is to improve the performance and learning rate of training activity on the dataset.

The dataset set contains 14 features. A feature named "System Load" is taken as a label, i.e., target feature. We used 70% of data in the dataset for training and 30% of data for testing our proposed model. Afterwards, the dataset was again divided; 90% for training and 10% for testing. The testing includes the one-week, one-month and four-month prediction, which are shown in the simulation section. The autocorrelation of data is shown in Figure 3. The overview of the dataset is shown in Figure 4.

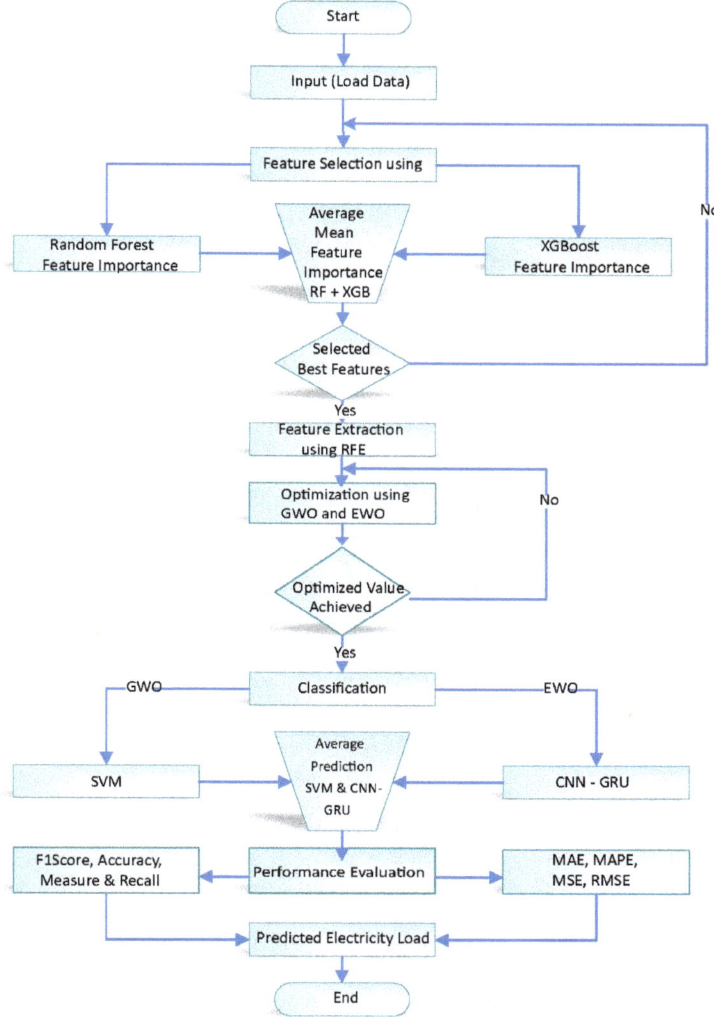

Figure 2. Detailed flowchart of the proposed model.

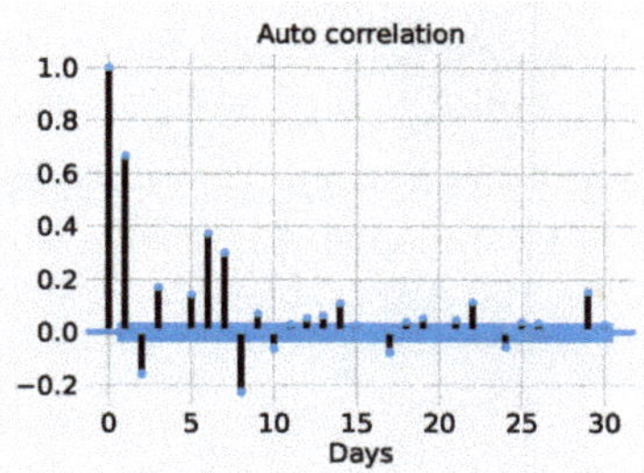

Figure 3. Autocorrelation of input data.

Date	DA_DEMD	DEMAND	DA_LMP	DA_EC	DA_CC	DA_MLC	RT_LMP	RT_EC	RT_CC	RT_MLC	DryBulb	DewPnt	System_Load	RegCP
1/1/2012	12,735.00	12,238.00	39.36	44.27	4.15	0.76	32.28	32.23	0.00	0.05	35.00	32.00	12,435.00	7.03
1/2/2012	12,398.00	12,346.00	39.12	39.23	0.00	0.11	45.62	45.72	0.00	0.10	42.00	41.00	12,539.00	9.18
1/3/2012	15,343.00	16,076.00	48.63	49.17	0.00	0.54	41.35	41.41	0.00	0.06	27.00	12.00	16,352.00	8.00
1/4/2012	16,137.00	16,894.00	48.64	49.43	0.49	0.30	56.11	56.10	0.00	0.01	22.00	14.00	17,200.00	6.85
1/5/2012	15,694.00	16,742.00	54.14	54.57	0.00	0.43	59.59	59.67	0.00	0.08	25.00	18.00	17,059.00	6.53
1/6/2012	16,611.00	17,242.00	51.94	52.21	0.00	0.27	55.99	55.97	0.00	0.02	19.00	11.00	17,565.00	9.00
1/7/2012	15,875.00	17,158.00	58.00	58.04	0.00	0.04	71.38	71.43	0.00	0.05	21.00	13.00	17,483.00	10.67
1/8/2012	14,335.00	14,820.00	52.98	52.54	0.00	0.44	50.05	49.80	0.00	0.25	24.00	21.00	15,083.00	7.40
1/9/2012	13,765.00	14,024.00	50.30	49.87	0.00	0.43	47.88	47.69	0.00	0.19	25.00	16.00	14,206.00	7.60
1/10/2012	16,379.00	17,255.00	62.60	62.70	0.00	0.10	65.72	65.36	0.00	0.36	21.00	10.00	17,515.00	5.59
1/11/2012	16,992.00	17,413.00	68.03	68.00	0.00	0.03	79.31	78.85	0.00	0.46	16.00	10.00	17,673.00	5.43
1/12/2012	17,383.00	15,503.00	78.76	78.23	0.00	0.53	61.69	61.66	0.00	0.03	26.00	24.00	15,814.00	10.00
1/13/2012	16,501.00	16,836.00	85.37	85.92	0.00	0.55	69.24	68.99	0.00	0.25	22.00	13.00	17,103.00	8.70
1/14/2012	16,833.00	17,632.00	84.67	85.69	0.00	1.02	81.67	82.02	0.00	0.35	13.00	7.00	17,940.00	6.50
1/15/2012	14,619.00	16,258.00	83.59	84.38	0.00	0.79	100.79	101.00	0.00	0.21	6.00	1.00	16,567.00	7.51
1/16/2012	13,690.00	14,019.00	53.08	53.65	0.57	0.00	55.45	55.57	0.00	0.12	23.00	13.00	14,275.00	6.80
1/17/2012	16,506.00	16,927.00	75.53	77.93	2.23	0.17	91.67	91.75	0.00	0.08	10.00	2.00	17,233.00	14.00
1/18/2012	16,271.00	17,521.00	77.93	80.30	1.96	0.41	93.36	93.31	0.00	0.05	19.00	15.00	17,817.00	13.34
1/19/2012	15,942.00	16,397.00	61.99	62.41	0.00	0.42	69.77	69.91	0.00	0.14	33.00	31.00	16,693.00	11.13

Figure 4. Dataset Overview (Features and Labels).

3.2. Feature Engineering

Machine learning techniques XGB lnd RF are used for the selection of relevant features. RF and XGB calculate the features' importance, i.e., the impact of all features on the target feature. The values are calculated in decimals between 0 and 1. To make the feature selection better, the average of feature importance is taken as given in Equation (1). The feature engineering step removes the unnecessary features and reduces the complexity of the proposed model by providing exact and relevant features for training.

$$Fs = \frac{Fi(XGB) + Fi(RF)}{2} \tag{1}$$

Whereas, *Fs* defines feature selection and *Fi* describes the feature importance.

After the selection of relevant features, the most redundant features are extracted using the RFE technique. The RFE technique calculates the dimension and priority of features in terms of true/false and positive integer numbers. After calculating the feature importance through feature selection and dimensional conversion with feature extraction, the drop-out rate is set to eliminate unimportant features. According to Equation (2), those features are selected/reserved, whose average feature importance/weight is greater than the defined threshold and the priority of feature is higher than the defined priority threshold. Moreover, those features are rejected/dropped whose feature weight is fewer than defined feature importance selection threshold and priority are greater than the defined feature priority threshold. The selection threshold of features using average feature selection is 0.6. Furthermore, the features with a priority greater than 5 are considered for selection. The overall selection of features is carried out according to the Equation (2).

$$Fos = \sum_{fr=0}^{n} \begin{cases} reserveFeature, & avg_{imp}(f) \geq \alpha \& \, \mathrm{RFE}(f)_{pr} \leq \beta pr \\ dropFeature, & avg_{imp}(f) \geq \alpha \& \, \mathrm{RFE}(fr)_{pr} > \beta pr \end{cases} \tag{2}$$

Whereas, *Fos* denotes the overall feature selection and *f* indicates the feature. *avgimp* denotes the average feature importance while *pr* represents the priority of the feature. The α and βpr describe the feature importance threshold and feature priority threshold. After feature selection and extraction, the most relevant features are passed to the classifier for classification and forecasting.

3.3. Classification and Forecasting

The classification is carried out using machine learning, i.e., SVM and deep learning CNN-GRU techniques set tuned with optimization techniques, i.e., GWO and EWO, respectively. The tuning parameters of SVM are loss function (gamma), cost incentive (C), and kernel function. The tuning parameters of CNN-GRU are numbers of hidden layers, numbers of neurons on each layer, dropout value. The tuning step will provide optimum values to the classifier, which results in the best training of the model and reduces the chance of overfitting the model on a large amount of data.

3.3.1. CNN-GRU-EWO

The hybrid of CNN and GRU has been used, further, the parameters of this hybrid model are tuned with the EWO technique. The output shape of tuned CNN-GRU layers is shown in Figure 5. The detailed description of CNN and GWO is given below.

Convolutional Neural Network: CNN is a type of deep learning algorithm. This technique is widely used in text and image recognition [29]. CNN might have multiple hidden layers between a single input and the output layer. The hidden layers are convolutional, dense, max-pooling, dropout, and flatten.

Input Layer: This layer is used as the beginning of the workflow of proposed CNN. It is used as a 1st layer of the network. It has no previous layer, nor any weight input. The number of neurons and dataset features is equal at this stage.

Hidden Layer: There can be multiple hidden layers on CNN. The output of the first layer is given to these hidden layers. Each hidden layer can have a different number of neurons. The output of these layers is evaluated with the multiplication of matrices and with previous layers output.

Output Layer: The hidden layer's output becomes the input of this layer. Softmax or sigmoid and logistic functions are used to transfer this input into the probability score.

Convolutional Layer: This layer has multiple filters and performs the most computational work. The convolutional operation is performed through this layer and results are given to the next layer.

Dense Layer: It acts as a conventional MLP. It is directed to connect the neuron of one layer with any other layers' neurons.

Pooling Layer: This layer is used for combining the output of neurons. Further, it is divided into three types; average-pooling, max-pooling, sum-pooling. In our model, max-pooling is used to minimize the parameters and reduce calculation. Generally, this layer is used between convolutional and drop layers.

Activation Function: The activation function Rectified-Linear-Unit (ReLU) is used in the convolutional layer.

Layer (type)	Output Shape	Param #
conv1d_1 (Conv1D)	(None, 8, 64)	192
gru_1 (GRU)	(None, 8, 150)	96750
dense_1 (Dense)	(None, 8, 10)	1510
dropout_1 (Dropout)	(None, 8, 10)	0
max_pooling1d_1 (MaxPooling1	(None, 4, 10)	0
dense_2 (Dense)	(None, 4, 50)	550
max_pooling1d_2 (MaxPooling1	(None, 2, 50)	0
flatten_1 (Flatten)	(None, 100)	0
dense_3 (Dense)	(None, 2)	202
dense_4 (Dense)	(None, 1)	3

Figure 5. Output shape of tuned Convolutional Neural Network (CNN) layers.

3.3.2. Gated Recurrent Unit (GRU)

GRU is an updated version of the Recurrent Neural Network (RNN). RNN has a problem with short term memory, in this context, LSTM and GRU have been proposed. Both GRU and LSTM are useful for maintaining long term information with the help of a gating mechanism. LSTM consists of three gates and GRU just has two gates, named update gate and reset gate [30].

Update Gate: This gate helps the model to calculate how much previous information is needed to be passed in the future. The update gate is very useful for eliminating the risk factor of the vanishing gradient problem because it remembers past information and decides which information is useful and which is not.

Reset Gate: This gate is used to decide how much of the previous information to forget.

CNN-GRU: CNN is useful to handle high-dimensional datasets and GRU is useful to process sequence data in minimum time. With the hybridization of these techniques, we can achieve both qualities. In this paper, a hybrid of CNN and GRU is proposed. In this proposed model, the output of the feature selection and extraction is given to CNN. After the input layer of CNN, a GRU is placed. After that, fully connected hidden layers are placed. In the end, the model is compiled and trained to get predicted results.

EWO: To tune the hyperparameters of CNN-GRU, the EWO optimization technique is used as shown in Figure 6. EWO is a nature-inspired heuristic algorithm that is used to solve the optimization problem. In this technique, every earthworm can produce offspring of only two kinds. The child earthworm contains the same length gene as his parent earthworm has. Some earthworms have the best fitness and forward this best fitness to the next generation without any change.

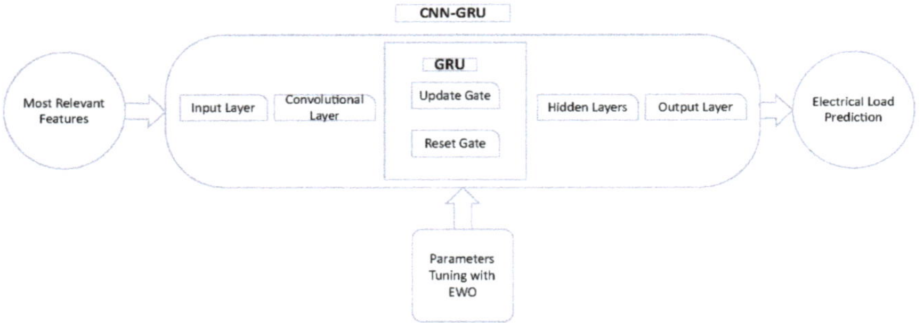

Figure 6. CNN-GRU-GWO.

3.3.3. SVM-GWO

To improve the performance of the machine learning technique SVM, the hyperparameters of SVM are tuned with the GWO optimization algorithm.

SVM: It is a type of supervised machine learning algorithm. It is widely used to solve the classification and regression problems. In SVM, a hyperplane line is drawn as in Figure 7, to divide the features into two classes; Linear and Non-Linear. In our proposed SVM, the parameter "gamma", i.e., loss function used with kernel RBF. After tuning the SVM, the optimization technique GWO calculated an optimum value for the SVM parameters; the value of gamma is 0.1, and the value of C is 1.0.

GWO: It is the part of the metaheuristic and swam optimization family. To tune the parameters of the SVM technique, GWO is used. This technique is developed by [31] in 2014, the authors were inspired by the grey wolf's social behavior and named this technique based on the "Grey Wolf Optimization". The hybrid of SVM and GWO is shown in Figure 8.

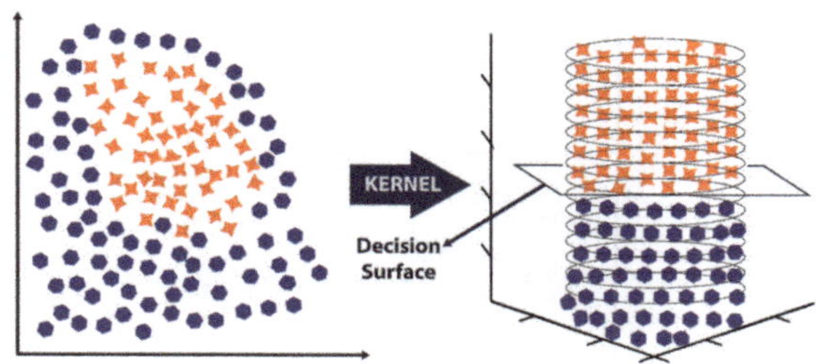

Figure 7. Support Vector Machine (SVM).

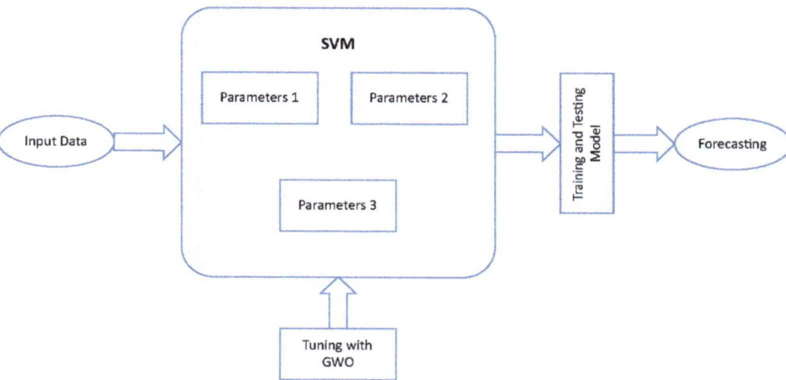

Figure 8. SVM-Grey Wolf Optimization (GWO).

4. Simulation Results

The complete implementation of our proposed framework was carried out on the system with specification; Intel Core i7, 8 GB RAM, dual 2.4 GHz Processor, built-in GPU Intel 5200, and 1TB SSD. The simulation was carried out on software named Anaconda Spyder and the python language environment with packages including Keras v2.3.1, Tensorflow v2.0, pyswarm v1.1.0, mealpy v0.3.0, pandas v1.0.3, and numpy v1.18.4. The simulation results are shown in figures and tables.

4.1. Average Feature Selection Based on RF and XGB

Feature importance calculated by XGB and RF is shown in Figure 9. Features with more importance have high values and less important features have low values. To make selection feasible and effective, the average feature importance was calculated from RF and XGB feature importance as shown in Figure 10. The features which had higher importance than the threshold were selected and low importance features were rejected. The feature importance calculated by RF is shown in Figure 9a, which shows it's Demand and the Dewpoint feature was most important in the dataset and high impact on the target data. Figure 9b shows the feature score calculated using XGB, which gives RT_CC and Demand as the most relevant feature. The average of Figure 9a,b is taken to calculate the average importance, which is shown in Figure 10. According to the average calculation of features, Demand, RT_MLC, DA_MLC and Dewpoint are the most relevant feature with high influence on the target feature.

RFE calculates the dimensions of the features, i.e., true/false, and thus it removes the redundant features from the dataset. The threshold set for feature selection and extraction is the features with average importance greater than 0.6 and dimension true, which were selected as the best features, while importance less than 0.6 and false dimension were rejected features. Table 2 shows the feature dimensions calculated by RFE. Furthermore, the abbreviation of features is also described.

Table 2. Features overview and dimensions calculated by Recursive Feature Elimination (RFE).

Target Feature	Features	Short Name	Dimension
System Load	Day-Ahead Cleared Demand	DA_Demand	TRUE
	Regulation Market Service clearing price	Reg_Capacity_Price	TRUE
	Real-Time Demand	RT_Demand	TRUE
	The dewpoint temperature	Dew_Point	FALSE
	Day-Ahead Locational Marginal Price	DA_LMP	FALSE
	The dry-bulb temperature	Dry_Bulb	FALSE
	Energy Component of Day-Ahead	DA_EC	FALSE
	Marginal Loss Component of Real-Time	RT_MLC	FALSE
	Congestion Component of Day-Ahead	DA_CC	FALSE
	Congestion Component of Real-Time	RT_CC	FALSE
	Marginal Loss Component of Day-Ahead	DA_MLC	FALSE
	Energy Component of Real-Time	RT_EC	TRUE
	Real-Time Locational Marginal Price	RT_LMP	TRUE
	Regulation Market Capacity clearing	Reg_Service_Price	FALSE

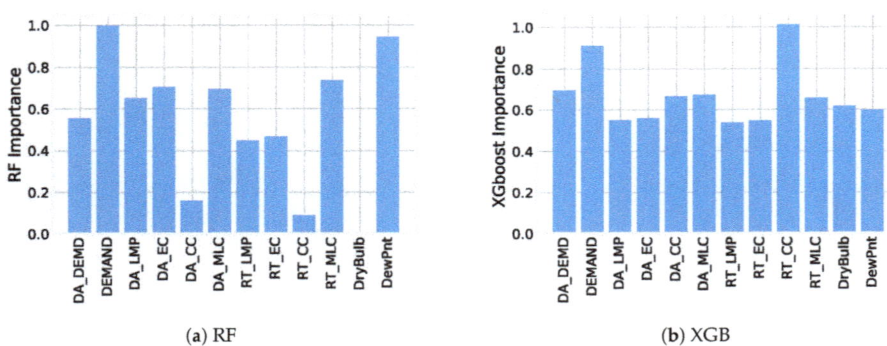

(**a**) RF (**b**) XGB

Figure 9. Feature importance calculated by (**a**) Random forest (**b**) Extra Gradient Boosting.

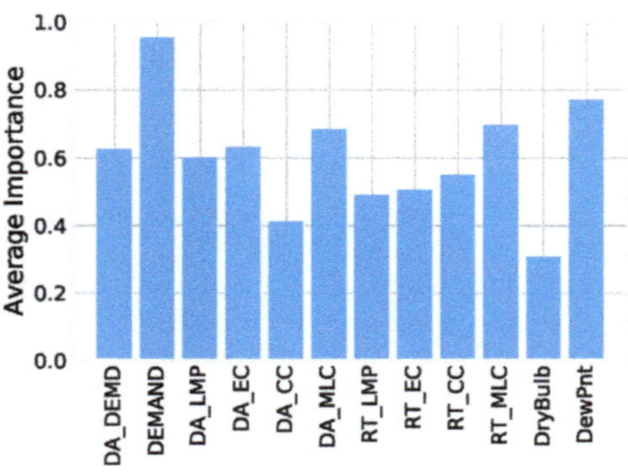

Figure 10. Average feature importance.

4.2. Classification and Forecasting Using SVM-GWO and CNN-GRU-EWO

Figure 11 shows the normal electricity load data from 1 January 2019 to 31 December 2019, which is provided by ISONE.

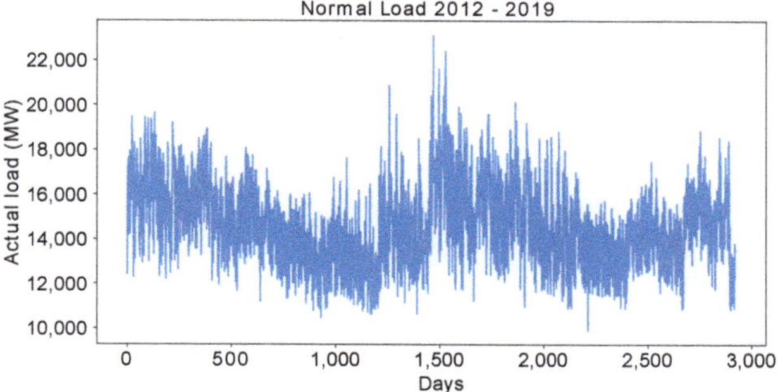

Figure 11. Normal electricity load.

The data is arranged month-wise, as the load pattern of similar months are approximately the same, which is shown in Figure 12. The monthly load of the Jan 2018 and Jan 2019 is approximately the same. Similarly, the load pattern of Dec 2018 and Dec 2019 are almost the same. The same pattern of load helps in the training of our model better.

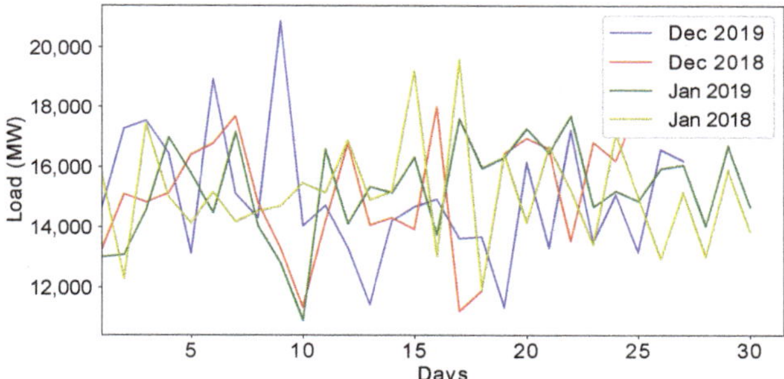

Figure 12. Similar months load pattern.

Figures 13 and 14 describe the prediction of one week, i.e., 1 Feburary 2019 to 7 Feburary 2019 and one month prediction, i.e., March 2019, respectively. The STF and MTF are covered in this paper. While forecasting the first week. All data, except the first week of Feburary 2019, were considered for training. The same case was applied for March 2019. During forecasting the electricity load of March 2019, all data, except March 2019, were considered for training. Figure 13 shows weekly forecast and Figure 14 shows monthly electricity load forecast. Our proposed algorithm performed better in achieving forecasting accuracy.

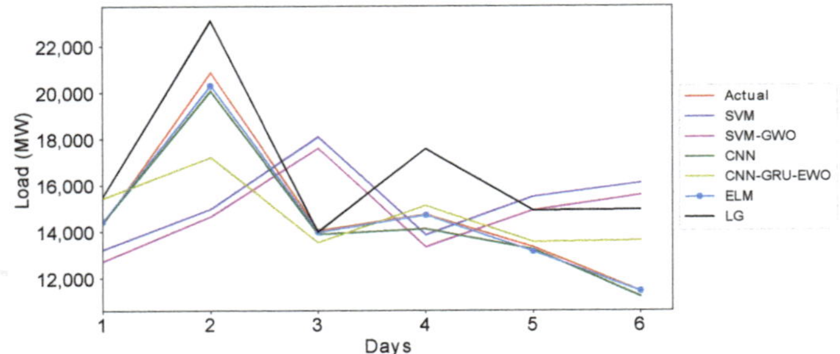

Figure 13. One-week prediction.

In Figures 13 and 14, the prediction values of our proposed techniques were nearly the same as the actual values of electricity load.

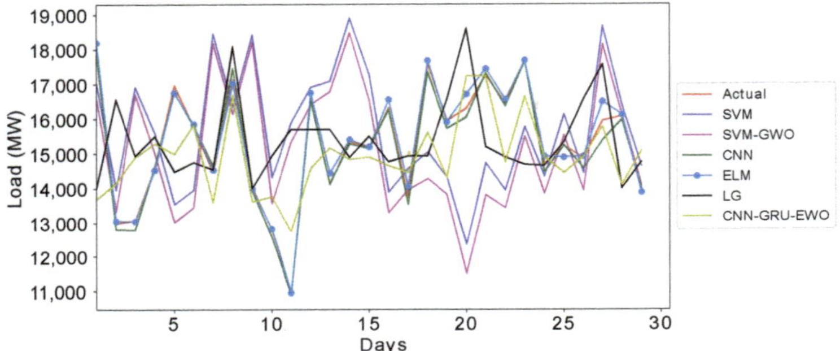

Figure 14. One-month prediction.

Due to efficient training of our proposed techniques on a large amount of data, our proposed model can forecast the load of the upcoming four months, i.e., September 2019, October 2019, November 2019, and December 2019 as shown in Figure 15. While forecasting the electricity load of the last four months of the year 2019, all data except the last four months of the year 2019 are considered for training. The trained model is then tested in the last four months of the year 2019. Our proposed techniques outperform SOTA as shown in Figure 15.

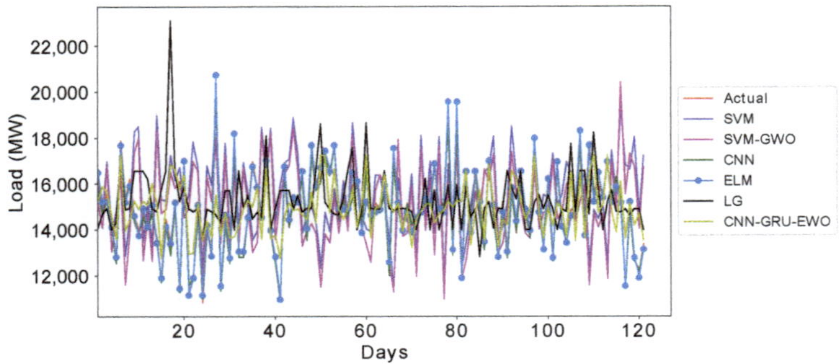

Figure 15. Four-month prediction.

The accuracy of our proposed techniques is higher than the SOTA as described in Figure 16. The enhanced version of the technique outperformed the actual technique. The optimization techniques, i.e., GWO and EWO found the best optimum solutions for the hyperparameters of the techniques, which enhance the accuracy and reduce the time complexity of training the model. The accuracy of our proposed techniques CNN-GRU-EWO and SVM-GWO is 93% and 90%, respectively. The accuracy of SOTA techniques SVM, CNN, LR and ELM is 87.98%, 89%, 78.34%, and 78.98%, respectively, as shown in Figure 16.

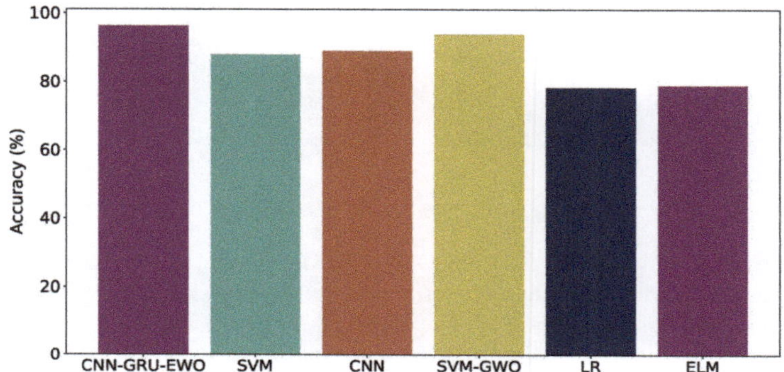

Figure 16. Accuracy of proposed techniques vs the State-Of-The-Art (SOTA).

5. Performance Metrics

The performance of our proposed model and SOTA techniques were evaluated using MAPE, MAE, RMSE, MSE, precision, re-call, f-measure, and accuracy. The performance errors of our proposed techniques were much lower than the SOTA as shown in Figure 17. The performance evaluation metrics accuracy was higher than benchmark techniques as shown in Figure 18.

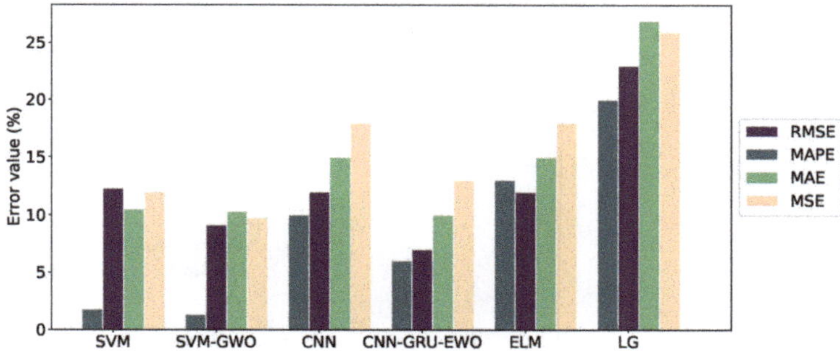

Figure 17. Performance error.

The performance evaluation metrics like precision and recall are calculated using Equations (3) and (4).

$$\text{Precision} = \frac{\text{True Positive}}{\text{True Positive} + \text{False Positive}} \tag{3}$$

$$\text{Recall} = \frac{\text{True Positive}}{\text{True Positive} + \text{False Negative}} \tag{4}$$

The performance values i.e., F1-score, accuracy, precision, and recall of CNN-GRU-EWA and CNN-GWO is greater than LG, CNN, SVM, and ELM.

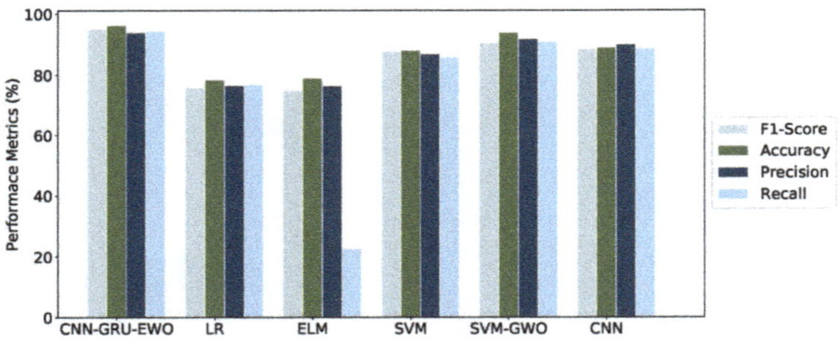

Figure 18. Performance evaluation.

The MAPE error of SVM-GWO and CNN-GRU-EWA is 1.33% and 6%. The LG technique has the highest performance error of 20%. The high-performance error reduces the forecasting accuracy

The training and testing accuracy of our proposed model is shown in Figures 19 and 20. The graph of accuracy is gradually increases with the increase in training on an excessive amount of data. The loss graph is gradually decreased, which shows that our model is well trained and test.

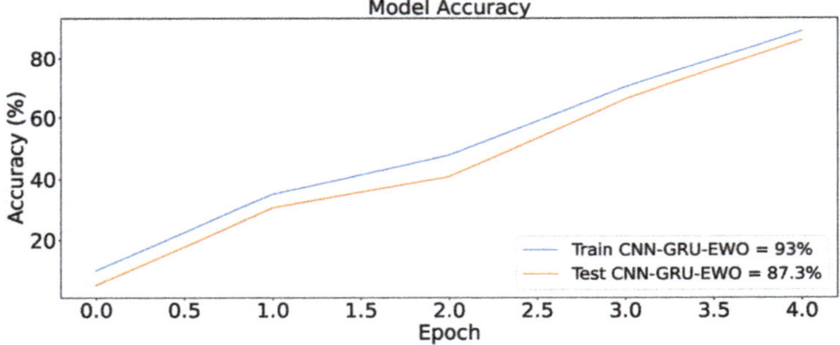

Figure 19. Train–test accuracy.

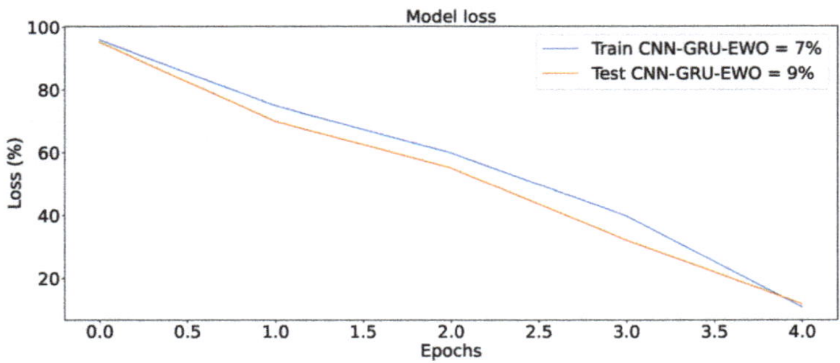

Figure 20. Train–test loss.

The performance evaluation with the performance metric values of our model is described in Table 3.

Table 3. Performance evaluation results of proposed and benchmark methods.

Techniques	Performance Metrics							
	F1-Score	**Accuracy**	**Precision**	**Recall**	**MAPE**	**RMSE**	**MAE**	**MSE**
CNN_GRU_EWA	95.23	96.33	94.00	94.62	6.00	7.00	10.00	13.00
LR	75.88	78.35	76.56	76.98	20.00	23.00	27.00	26.00
ELM	75.00	78.98	76.45	22.78	13.00	12.00	15.00	18.00
SVM	87.88	87.99	86.91	85.99	1.79	12.30	10.50	12.00
SVM_GWO	90.67	93.99	91.87	90.99	1.33	9.12	10.31	9.75
CNN	88.66	89.00	90.00	88.76	10.00	12.00	15.00	18.00

Table 4 shows different correlation-based tests, a parametric statistical hypothesis-based tests and non-parametric statistical hypothesis based statistical tests of proposed techniques and state-of-the-art techniques.

Table 4. Statistical analysis of proposed techniques and benchmark algorithms.

Techniques and Tests		Correlation Tests				Parametric Statistical Hypothesis Tests			Non-Parametric Statistical Hypothesis Tests		
		Pearson's Test	Spearman's Test	Kendalla's Test	Chi-Squared Test	Student's Test	Paired Student's Test	ANOVA Test	Mann-Whitney Test	Wilcoxon Test	Kruskal Test
SVM	F-stastistic	−0.0404	−0.0549	−0.0362	157,449.28	−5.5019	−5.3941	30	225,955	104,549	26.0883
	p-value	0.2753	0.1379	0.1429	0.0000	0.0000	0.0000	0.0000	0.0000	0.0000	0.0000
SVM-GWO	F-stastistic	−0.0376	−0.0553	−0.0362	164,404.40	0.2530	0.2484	0.0640	262,798	132,003	0.2949
	p-value	0.3106	0.1349	0.1436	0.0000	0.8003	0.8039	0.8003	0.2936	0.8054	0.5871
CNN	F-stastistic	0.9964	0.9963	0.9499	575.09	1.1820	19.2812	1.3971	257,449	37,953	1.4537
	p-value	0.0000	0.0000	0.0000	1.0000	0.2374	0.0000	0.2374	0.1140	0.0000	0.2279
CNN-GRU-EWA	F-stastistic	0.7367	0.7208	0.5321	37,815.93	−0.8087	−1.4750	0.6539	267,085	131,225	0.0001
	p-value	0.0000	0.0000	0.0000	0.0000	0.4188	0.1406	0.4188	0.4953	0.6555	0.9906
ELM	F-stastistic	0.9887	0.9856	0.9143	1865.32	−0.1100	−1.0303	0.0121	26,4803	124,235	0.0868
	p-value	0.0000	0.0000	0.0000	0.0000	0.9124	0.3032	0.9124	0.3842	0.1538	0.7683
LG	F-stastistic	0.2411	0.2033	0.1415	89,538.00	−6.0994	−6.9077	37	218,561	94,749	36.3238
	p-value	0.0000	0.0000	0.0000	0.0000	0.0000	0.0000	0.0000	0.0000	0.0000	0.0000

6. Conclusions

In this paper, a deep learning, machine learning, and optimization techniques-based model is used for short and medium-term electricity load forecasting. The eight-year electricity load data set was downloaded from the ISONE website. The ISONE provides electricity to different cities in England. To deal with such a huge amount of data, normal forecasting models are unable to perform well. A framework consists of feature selection, extraction process, and classification, which is proposed to forecast electricity load. The feature engineering process removes the redundancy and selects the most relevant features which have a high impact on the target feature. Furthermore, it also reduces the complexity of the model by providing the most important features to the classifiers. The RF and XGB techniques are used as a feature selection and RFE as a feature extraction method. The feature engineering activity refined the data and passed it to the classifiers. The techniques CNN-GRU and SVM were used as classifiers. To enhance the performance of classifiers, the parameters of CNN-GRU and SVM were tuned with an optimization algorithm EWO and GWO, respectively. The optimization algorithm finds the best optimum values for the techniques of hyperparameters. Moreover, the tuning of parameters provide optimum values to the classifiers, which reduces the chances of model overfitting and helps to increase the accuracy of the model. Our proposed techniques—CNN-GRU-EWO and SVM-GWO—outperform SOTA. The accuracies of CNN-GRU-EWO and SVM-GWO are 96.33% and 93.99%, respectively. Our proposed techniques perform 7% and 3% better than CNN and SVM classifiers. In the future, other optimization techniques will be applied to the machine learning classifiers to enhance the accuracy of electricity load forecasting.

Author Contributions: The research conceptualization and methodology were done by N.A., M.I. and M.A. The technical and theoretical framework was prepared by U.A. and T.A. The technical review and improvement were performed by M.H. and A.A. The overall technical support, guidance, and project administration was done by A.A. and F.M. Finally, responses to the queries of the reviewers were done by N.A., M.H. and M.A. All authors have read and agreed to the published version of the manuscript.

Funding: This research received no external funding.

Acknowledgments: The authors acknowledge the Deanship of Scientific Research, Najran University. Kingdom of Saudi Arabia, for their technical support.

Conflicts of Interest: The authors declare no conflict of interest.

Abbreviations

The following abbreviations are used in this manuscript:

CMI	Conditional Mutual Information
NLSSVM	Nonlinear Least Square Support Vector Machine
ABC	Artificial Bee Colony
ARIMA	Autoregressive Integrated Moving Average
IWNN	Wavelet Neural Network
ELM	Extreme Learning Machine
CNN	Convolutional Neural Network
LSTM	Long Short Term Memory
ASF	Auto Correlation Function
IITK	India Institution of Technology Kanpoor
ELM	Extreme Learning Machine
XGB	Extreme Gradient Boosting
DTC	Decision Tree Classifier
MAE	Mean Absolute Error
RMSE	Root Mean Square Error
MSE	Mean Square Error
MAPE	Mean Absolute Percentage Error
PJM	Pennsylvania New Jersey Maryland
CA	Correlation Analysis

References

1. Zhu, Z.; Tang, J.; Lambotharan, S.; Chin, W.H.; Fan, Z. An integer linear programming based optimization for home demand-side management in smart grid. In Proceedings of the Innovative Smart Grid Technologies (ISGT), Washington, DC, USA, 16–20 January 2012; pp. 1–5.
2. Samadi, P.; Wong, V.W.S.; Schober, R. Load Scheduling and Power Trading in Systems with High Penetration of Renewable Energy Resources. *IEEE Trans. Smart Grid* **2016**, *7*, 1802–1812. [CrossRef]
3. Chen, X.; Zhou, Y.; Duan, W.; Tang, J.; Guo, Y. Design of intelligent De-mand Side Management system respond to varieties of factors. In Proceedings of the China International Conference on Electricity Distribution (CICED), Nanjing, China, 13–16 September 2010; pp. 1–5.
4. Hahn, H.; Meyer-Nieberg, S.; Pickl, S. Electric load forecasting methods: Tools for decision making. *Eur. J. Oper. Res.* **2009**, *199*, 902–907. [CrossRef]
5. Wang, K.; Yu, J.; Yu, Y.; Qian, Y.; Zeng, D.; Guo, S.; Xiang, Y.; Wu, J. A survey on energy internet: Architecture, approach and emerging technologies. *IEEE Syst. J.* **2017**, *12*, 2403–2416. [CrossRef]
6. Jiang, H.; Wang, K.; Wang, Y.; Gao, M.; Zhang, Y. Energy big data: A survey. *IEEE Access* **2016**, *4*, 3844–3861. [CrossRef]
7. Fatima, A.; Shabbir, S. Data Analytics for Load and Price Forecasting via Enhanced Support Vector Regression. In *Advances in Internet, Data and Web Technologies: The 7th International Conference on Emerging Internet, Data and Web Technologies (EIDWT-2019)*; Springer: Cham, Switzerland, 2019; Volume 29, p. 259.
8. Naz, A.; Javed, M.U.; Javaid, N.; Saba, T.; Alhussein, M.; Aurangzeb, K. Short-term electric load and price forecasting using enhanced extreme learning machine optimization in smart grids. *Energies* **2019**, *12*, 866. [CrossRef]
9. Gao, X.; Li, X.; Zhao, B.; Ji, W.; Jing, X.; He, Y. Short-term electricity load forecasting model based on EMD-GRU with feature selection. *Energies* **2019**, *12*, 1140. [CrossRef]
10. Liu, Z.; Sun, X.; Wang, S.; Pan, M.; Zhang, Y.; Ji, Z. Midterm power load forecasting model based on kernel principal component analysis and back propagation neural network with particle swarm optimization. *Big Data* **2019**, *7*, 130–138. [CrossRef]
11. Bouktif, S.; Fiaz, A.; Ouni, A.; Serhani, M.A. Multi-Sequence LSTM-RNN Deep Learning and Metaheuristics for Electric Load Forecasting. *Energies* **2020**, *13*, 391. [CrossRef]
12. Cai, M.; Pipattanasomporn, M.; Rahman, S. Day-ahead building-level load forecasts using deep learning vs. traditional time-series techniques. *Appl. Energy* **2019**, *236*, 1078–1088. [CrossRef]
13. Jindal, A.; Singh, M.; Kumar, N. Consumption-aware data analytical demand response scheme for peak load reduction in smart grid. *IEEE Trans. Ind. Electron.* **2018**, *65*, 8993–9004. [CrossRef]
14. Mujeeb, S.; Javaid, N.; Ilahi, M.; Wadud, Z.; Ishmanov, F.; Afzal, M.K. Deep long short-term memory: A new price and load forecasting scheme for big data in smart cities. *Sustainability* **2019**, *11*, 987. [CrossRef]
15. Chitsaz, H.; Zamani-Dehkordi, P.; Zareipour, H.; Parikh, P.P. Electricity price forecasting for operational scheduling of behind-the-meter storage systems. *IEEE Trans. Smart Grid* **2018**, *9*, 6612–6622. [CrossRef]
16. Ghasemi, A.; Shayeghi, H.; Moradzadeh, M.; Nooshyar, M. A novel hybrid algorithm for electricity price and load forecasting in smart grids with demand-side management. *Appl. Energy* **2016**, *177*, 40–59. [CrossRef]
17. Abedinia, O.; Amjady, N.; Zareipour, H. A New Feature Selection Technique for Load and Price Forecast of Electrical Power Systems. *IEEE Trans. Power Syst.* **2017**, *32*, 62–74. [CrossRef]
18. Rafiei, M.; Niknam, T.; Khooban, M.H. Probabilistic Forecasting of Hourly Electricity Price by Generalization of ELM for Usage in Improved Wavelet Neural Network. *IEEE Trans. Ind. Inform.* **2017**, *13*, 71–79. [CrossRef]
19. Kuo, P.H.; Huang, C.J. An electricity price forecasting model by hybrid structured deep neural networks. *Sustainability* **2018**, *10*, 1280. [CrossRef]
20. Wang, J.; Liu, F.; Song, Y.; Zhao, J. A novel model: Dynamic choice artificial neural network (DCANN) for an electricity price forecasting system. *Appl. Soft Comput. J.* **2016**, *48*, 281–297. [CrossRef]
21. Shi, H.; Xu, M.; Li, R. Deep Learning for Household Load Forecasting-A Novel Pooling Deep RNN. *IEEE Trans. Smart Grid* **2018**, *9*, 5271–5280. [CrossRef]
22. Khan, S.; Javaid, N.; Chand, A.; Rashid, F. Electricity load forecasting for each day of the week using deep CNN. In Proceedings of the Workshops of the International Conference on Advanced Information Networking and Applications, Taipei, Taiwan, 27–29 March 2019.
23. Tian, C.; Ma, J.; Zhang, C.; Zhan, P. A Deep Neural Network Model for Short-Term Load Forecast Based on Long Short-Term Memory Network and Convolutional Neural Network. *Energies* **2018**, *11*, 3493. [CrossRef]

24. Ali, U.; Rauf, A.; Iqbal, U.; Shoukat, I.A.; Hassan, A. Big data analytics for a novel electrical load forecasting technique. *Int. J. Inf. Technol. Secur.* **2019**, *11*, 33–40.
25. Ahmad, W.; Ayub, N.; Ali, T.; Irfan, M.; Awais, M.; Shiraz, M.; Glowacz, A. Towards short term electricity load forecasting using improved support vector machine and extreme learning machine. *Energies* **2020**, *13*, 2907. [CrossRef]
26. Houimli, R.; Zmami, M.; Ben-Salha, O. Short-term electric load forecasting in Tunisia using artificial neural networks. *Energy Syst.* **2020**, *11*, 357–375. [CrossRef]
27. Yang, A.; Li, W.; Yang, X. Short-term electricity load forecasting based on feature selection and Least Squares Support Vector Machines. *Knowl. Based Syst.* **2019**, *163*, 159–173. [CrossRef]
28. Zheng, S.; Zhong, Q.; Peng, L.; Chai, X. A simple method of residential electricity load forecasting by improved Bayesian neural networks. *Math. Probl. Eng.* **2018**, *2018*, 4276176. [CrossRef]
29. Abbas, F.; Feng, D.; Habib, S.; Rahman, U.; Rasool, A.; Yan, Z. Short term residential load forecasting: An improved optimal nonlinear auto regressive (NARX) method with exponential weight decay function. *Electronics* **2018**, *7*, 432. [CrossRef]
30. Heghedus, C.; Chakravorty, A.; Rong, C. Energy Load Forecasting Using Deep Learning. In Proceedings of the 2018 IEEE International Conference on Energy Internet (ICEI), Beijing, China, 21–25 May 2018; pp. 146–151. [CrossRef]
31. Zhang, Z.; Ding, S.; Sun, Y. A support vector regression model hybridized with chaotic krill herd algorithm and empirical mode decomposition for regression task. *Neurocomputing* **2020**, *410*, 185–201. [CrossRef]
32. Zhang, Z.; Ding, S.; Jia, W. A hybrid optimization algorithm based on cuckoo search and differential evolution for solving constrained engineering problems. *Eng. Appl. Artif. Intell.* **2019**, *85*, 254–268. [CrossRef]
33. Rafati, A.; Joorabian, M.; Mashhour, E. An efficient hour-ahead electrical load forecasting method based on innovative features. *Energy* **2020**, *201*, 117511. [CrossRef]
34. RAskari, M.; Keynia, F. Mid-term electricity load forecasting by a new composite method based on optimal learning MLP algorithm. *IET Gener. Transm. Distrib.* **2019**, *14*, 845–852. [CrossRef]
35. Khan, A.R.; Dewangan, C.L.; Srivastava, S.C.; Chakrabarti, S. Short Term Load Forecasting using SVM Models. In Proceedings of the 8th IEEE Power India International Conference PIICON, Kurukshetra, India, 10–12 December 2018; pp. 1–5. [CrossRef]
36. Cheng, Y.; Jin, L.; Hou, K. Short-Term Power Load Forecasting based on Improved Online ELM-K. In Proceedings of the 2018 International Conference on Control, Automation and Information Sciences (ICCAIS), Hangzhou, China, 24–27 October 2018; pp. 128–132. [CrossRef]
37. Mujeeb, S.; Javaid, N.; Akbar, M.; Khalid, R.; Nazeer, O.; Khan, M. Big data analytics for price and load forecasting in smart grids. In *International Conference on Broadband and Wireless Computing, Communication and Applications*; Springer: Cham, Switzerland, 2018; pp. 77–87. [CrossRef]
38. ISO NE Electricity Market Data. Available online: https://www.iso-ne.com/isoexpress/web/reports/load-and-demand (accessed on 28 April 2020). [CrossRef]
39. Xin, M.; Wang, Y. Research on image classification model based on deep convolution neural network. *Eurasip J. Image Video Process.* **2019**, *2019*, 40. [CrossRef]
40. Wu, L.; Kong, C.; Hao, X.; Chen, W. A Short-Term Load Forecasting Method Based on GRU-CNN Hybrid Neural Network Model. *Math. Probl. Eng.* **2020**, *2020*, 1428104. [CrossRef]
41. Mirjalili, S.; Mirjalili, S.M.; Lewis, A. Grey Wolf Optimizer. *Adv. Eng. Softw.* **2014**, *69*, 46–61. [CrossRef]
42. Ahmad, A.; Javaid, N.; Mateen, A.; Awais, M.; Khan, Z.A. Short-term load forecasting in smart grids: An intelligent modular approach. *Energies* **2019**, *12*, 164. [CrossRef]
43. Ayub, N.; Javaid, N.; Mujeeb, S.; Zahid, M.; Khan, W.Z.; Khattak, M.U. Electricity Load Forecasting in Smart Grids Using Support Vector Machine. In Proceedings of the International Conference on Advanced Information Networking and Applications, Matsue, Japan, 27–29 March 2019; Springer: Cham, Switzerland, 2019; pp. 1–13. [CrossRef]
44. Zhu, Q.; Han, Z.; Başar, T. A differential game approach to distributed demand side management in smart grid. In Proceedings of the 2012 IEEE International Conference on Communications (ICC), Ottawa, ON, Canada, 10–15 June 2012; pp. 3345–3350. [CrossRef]

Article

An Optimal Energy Optimization Strategy for Smart Grid Integrated with Renewable Energy Sources and Demand Response Programs

Kalim Ullah [1], Sajjad Ali [2], Taimoor Ahmad Khan [1], Imran Khan [1], Sadaqat Jan [3], Ibrar Ali Shah [3] and Ghulam Hafeez [1,4,*]

1 Department of Electrical Engineering, University of Engineering and Technology, Mardan 23200, Pakistan; kalimullahbtk1@gmail.com (K.U.); taymourkahn@gmail.com (T.A.K.); imran@uetmardan.edu.pk (I.K.)

2 Department of Telecommunication Engineering, University of Engineering and Technology, Mardan 23200, Pakistan; sajjad@uetmardan.edu.pk

3 Department of Computer Software Engineering, University of Engineering and Technology, Mardan 23200, Pakistan; sadaqat@uetmardan.edu.pk (S.J.); ibrar@uetmardan.edu.pk (I.A.S.)

4 Department of Electrical and Computer Engineering, COMSATS University Islamabad, Islamabad Campus, Islamabad 44000, Pakistan

* Correspondence: ghulamhafeez393@gmail.com or ghulamhafeez@uetmardan.edu.pk

Received: 28 July 2020; Accepted: 25 September 2020; Published: 2 November 2020

Abstract: An energy optimization strategy is proposed to minimize operation cost and carbon emission with and without demand response programs (DRPs) in the smart grid (SG) integrated with renewable energy sources (RESs). To achieve optimized results, probability density function (PDF) is proposed to predict the behavior of wind and solar energy sources. To overcome uncertainty in power produced by wind and solar RESs, DRPs are proposed with the involvement of residential, commercial, and industrial consumers. In this model, to execute DRPs, we introduced incentive-based payment as price offered packages. Simulations are divided into three steps for optimization of operation cost and carbon emission: (i) solving optimization problem using multi-objective genetic algorithm (MOGA), (ii) optimization of operating cost and carbon emission without DRPs, and (iii) optimization of operating cost and carbon emission with DRPs. To endorse the applicability of the proposed optimization model based on MOGA, a smart sample grid is employed serving residential, commercial, and industrial consumers. In addition, the proposed optimization model based on MOGA is compared to the existing model based on multi-objective particle swarm optimization (MOPSO) algorithm in terms of operation cost and carbon emission. The proposed optimization model based on MOGA outperforms the existing model based on the MOPSO algorithm in terms of operation cost and carbon emission. Experimental results show that the operation cost and carbon emission are reduced by 24% and 28% through MOGA with and without the participation of DRPs, respectively.

Keywords: multi-objective energy optimization; smart grid; renewable energy sources; wind; photovoltaic; demand response programs

---

1. Introduction

Energy optimization is an indispensable task in energy management of the smart grid (SG) [1,2]. Optimal energy optimization is possible only by actively engaging consumers in demand response programs (DRPs) offered by electric utility companies (ECUs) [3]. DRPs enable the ECUs to shift a load of consumers from on-peak to off-peak hours by giving economic incentive to the consumers [4]. A market overview model consisting of 10 chapters is presented [5] for economic cost reduction, which consists of market forecasting, market management, and market monitoring to schedule energy, ancillary services, and transmission. Moreover, DRPs provide ease and flexibility to consumers to actively participate in the electricity market for energy optimization.

Energy demand is rising, and conventual energy sources are limited and depleting gradually. Therefore, renewable energy is a hot topic for researchers due to high energy potential and continually replenished nature. According to [6], the increase in population results in an increase in energy demand, and it is predicted that demand will increase to 50% or more at the end of 2030. Renewable energy is free of economic cost and emission. It is most suitable to use. Thus, in [7], energy optimization of a model considering RESs is discussed due to the low price and environment-friendly advantages of RESs. The integration of RESs was studied recently in [8], which provides a massive comfort to SG technology in terms of cost. There is always a variation issue in RESs, to overcome these issues, and an energy optimization model was proposed, which consists of a mathematical tool PDF discussed in [9], to model solar and wind sources. Intermittent behaviour of wind and solar energy is modelled in [10] by using PDF and Rayleigh distribution; in this case, the proposed method was a tree optimization. Intermittency in RESs is one of the significant issues, RESs integration is discussed [11] by taking a survey of models all around the world. The author concluded that communication system, specifically two-way communication plays a significant role in the energy optimization of the SG.

An article [12] was discussed to reduce economic cost and carbon emission simultaneously as a multi-objective function. The proposed model was accurate in all perspectives as implemented on SCADA software and also the hardware of the proposed model was tested; in addition, the economic cost and carbon emission are successfully reduced through a dynamic programming-based algorithm. In this research, the author avoids the use of solar energy. A central controller is designed for a micro-grid to improve the efficiency of the micro-grid and predict the performance of a dynamic system [13]. The implementation of this model shows that, when a micro-generator is less than seven, the proposed model is not working properly. In this paper, the author used an economic model predictive control (EMPC) method to reduce economic cost. The traditional grid is not applicable for energy optimization due to lake of communication infrastructure. The SG has advanced metering and bi-directional communication infrastructures, which enables RESs accommodation and active participation of consumers in DRPs to ensure low operation cost and gives carbon emission [14]. In addition, SG enables us to perform optimization from all perspectives like energy, cost, carbon emission, and maintaining a balance between demand and supply. The DRPs in SG reduce cost and provide relief to the end-users; similarly, DRPs are used to overcome uncertainty in RESs. Implementation of DRPs for operating cost minimization and efficiency improvement is discussed in [15]. The proposed energy optimization model based on teaching and learning-based optimization (TLBO) and shuffle frog leaping (SFL) techniques is tested on four types of residential consumers in the centre of Tehran in Iran. The focus of the authors is only on residential load, and no value is assigned to commercial and industrial consumers. A model predictive control (MPC) based work is presented for sharing distributed energy resources (DERs) in micro-grids in order to optimize the available energy [16]. The implementation of this model reduces the economic cost. A fuzzy logic controller based model is studied in [17] to reduce economic costs. In this model, the aim is to manage the charging and discharging rate of an energy storage system to minimize consumers' operational cost. For RES forecasting, the authors used a new method to take the difference between the current REEs and load rather than forecasting approach to predict RESs.

Similarly, an energy optimization model for the residential load is presented in [18]. The authors perform optimization of economic cost function by managing the operation of appliances in the low generation and on-peak period using a robust optimization algorithm. The authors compared the proposed model based on a math-heuristic optimization algorithm with the existing one, which was based on a mixed-integer nonlinear (MILP) method and achieved better results than the current model. However, the carbon emission is not discussed, which is a very critical future challenge. In [19], the authors addressed both operating cost and carbon emission using a mixed-integer nonlinear programming (MINLP) technique of a microgrid including microturbine, fuel cell, battery, and utility as a back-up source. The proposed model based on MINLP results is compared to the genetic algorithm (GA) and particle swarm optimization (PSO) algorithm based models in terms of operation cost and emission. However, the uncertainty accompanied by RESs is catered either by DRPs or PDF/CDF. Economic cost and peak to average ratio (PAR) are reduced by using two heuristic based demand side management techniques [20], a bacterial Foraging Optimization Algorithm (BFOA) and a Flower Pollination Algorithm (FPA). In order to optimize the results of both proposed techniques, a novel heuristic algorithm is introduced. Demand side management (DSM) is a very important aspect in the energy management system (EMS). Economic cost is reduced by shifting load from peak to off-peak hours; moreover, peak to average ratio (PAR) is also minimized in the proposed model by using the concept of DSM to control the load at user end [21]. In order to implement DSM, BFOA and FPA are proposed. Home appliances are scheduled in such a way to minimize economic cost, PAR, and provide user comfort to consumers while using the home energy management system (HEMS) concept [22]. In this research, the authors used GA, FPA, and the combination (hybrid) of these two techniques, the genetic-flower pollination algorithm (GFPA). Similarly, HEMS focuses on scheduling home appliances in such a way to reduce the peak load, as a result, it reduces electricity cost, PAR, optimizes user comfort as well as the time of execution [23]. The time of use (TOU) concept was used to obtain the required results; moreover, the proposed techniques are GA, biogeography-based optimization (BBO). For EMS, it is important to create a balance between energy consumption and user comfort. This paper proposes three techniques GA, Pigeon Inspired Optimization (PIO), and hybrid of GA and PIO techniques to distribute appliances in off-peak time and reduce load at peak time to optimize electricity cost and PAR [24]. In this research [25], few methods are used for energy optimization, such as TOU to reduce economic cost by shifting load from peak to off peak hours, real time pricing (RTP), and DRPs. The proposed model is beneficial for both electricity market and consumers. The multi-agents network can provide ease and reduce economic cost. In this study [26], a nano-biased system engages multi-agents that consist of residential, commercial, and industrial consumers. The operational cost is reduced by using the concept of real time tariff while purchasing/sell electricity. Moreover, incentive-based packages are provided to the different consumers. The proposed model is designed with machine learning and reinforcement intelligence.

In this work, an optimal energy optimization strategy is developed to optimize the operation of SG integrated with RESs in terms of operation cost and carbon emission. In addition, the concept of incentive-based DRPs is introduced as price offer packages to overcome the uncertainty factor in power generation by RESs like solar and wind. In this method, end-users can select an offered price package to participate in energy optimization. In this model, the Rayleigh PDF is proposed to model variation in energy generation caused by RESs like solar and wind. Units, distributed generations (DGs), fuel cell (FC), wind turbine (WT) and battery are intended to provide relief to SG. Residential, commercial and industrial consumers are considered the end-users in the proposed model. The multi-objective problem is solved through a programming technique, multi-objective genetic algorithm (MOGA), taking Pareto fronts into account for achieving the desired optimization results. In brief, the main contributions of this paper are as follows:

- The uncertainty in renewable energy generation like solar and wind is covered by DRP implementation by considering operation cost and carbon emission as multi-objective functions.
- Incentive-based DRPs introduced to encourage end-users, commercial, residential and industrial sectors to participate in energy optimization actively.
- Probability density function is proposed to predict wind and solar RES behaviour.
- The multi-objective optimization problem of operating cost and carbon emission is solved through a multi-objective genetic algorithm (MOGA).

The remaining sections of this work are organized as: the problem statement is discussed in Section 2, objective function is introduced in Section 3, the proposed system model is discussed in Section 4, the proposed technique is presented in Section 5, simulation results and discussion are illustrated in Section 6, followed by the conclusions in Section 7.

2. Problem Statement

In this paper, an energy management system model is proposed to optimize operational cost and pollution emission with and without the implementation DRPs in SG. Due to the stochastic behaviour of RESs, the prediction of wind and solar are not possible; in order to solve this issue, PDF is proposed. Moreover, there is always uncertainty in RESs; in order to resolve this problem in wind and solar, an incentive based DRPs are proposed. However, the balance between generation and consumption is a very important factor; the proposed model discussed these frameworks while applying demand response programs in the smart grid.

3. Objective Function

The two main objectives of the proposed energy optimization model are operation cost and carbon emission reduction.

3.1. Operation Cost

The operational cost is divided into two categories, certain operational costs which include start up and running cost of DGs, reserve costs of DGs and the power costs provide or taken to/from utility, and uncertain operational cost by taking the probability of the proposed scenarios (probs) in time slot t = 1 to T and kth scenarios, which is affected by uncertainty in the wind and solar parameters in each case. The uncertain operational cost function normally includes the reduction in load and expected energy not served (EENS) for consumers at the user end. The operation cost objective function is defined as follows:

$$\begin{aligned} MinF_1(X) &= \sum_{t=1}^{T} f^{\cos t}(t) \\ &= \sum_{t=1}^{T} C_{op}(t) + \sum_{t=1}^{T}\sum_{k=1}^{K} probs \times UC_{op} \end{aligned} \tag{1}$$

where C_{op} and UC_{op} are certain and uncertain cost of operation, respectively, and arranged according to the proposed scenario, *probs* is the probability of the proposed scenarios k, and t is the time period starts from t = 1 to T. A certain operation cost function is given below:

$$\begin{aligned} C_{op} = &\sum_{j=1}^{N_{DG}} \left[W_j(t)\sigma_j(t)Y_j(t) + R_j(t)\left|Y_j(t) - Y_j(t-1)\right| + \mathrm{R}_e C_j{}^{DGs}(t) \right] \\ &+ \sum_{k=1}^{K} \mathrm{R}_e C_k{}^{DRPs}(t) Y_{Buy}(t) W_{Grid-Buy}(t)\sigma_{Grid-buy}(t) - Y_{Sell}(t) W_{Grid-Sell}(t)\sigma_{Grid-Sell}(t) \end{aligned} \tag{2}$$

where $W_j(t)$ and $\sigma_j(t)$ show output power and offered price for different units, $Y_i(t)$ indicates on, off mode of the jth DGs in time slot t. $R_j(t)$ indicates running and shutdown costs for different units in time period t. $\mathrm{R}_e C_j{}^{DGs}(t)$ and $\mathrm{R}_e C_k{}^{DRPs}(t)$ are the reserve cost of DGs and DRPs in time slot t, $W_{Grid-Buy}(t)$, and $W_{Grid-sell}(t)$ shows energy exchange with a utility in time period t. Similarly, the uncertain operating cost function is defined as:

$$UC_{op} = \sum_{i=1}^{N_{DG}} RC_{i,s}{}^{DG}(t) + \sum_{k=1}^{K} RC_{k,s}{}^{DR}(t) + ENSs(t) \tag{3}$$

where $RC_{i,s}{}^{DG}(t)$, $RC_{k,s}{}^{DR}(t)$ shows the running cost of DGs, DRPs in time slot t and ENSs(t) is the expected energy not served in time period t, which is also categorized in uncertain operating costs, respectively.

3.2. Carbon Emission

The carbon emission produced by DGs and grid during the energy generation is minimized through the following equation:

$$\begin{aligned} MinF_2(X) &= \sum_{t=1} f^{CE}(t) \\ &= \sum_{t=1} CE^{DGs}(t) + \sum_{t=1} CE^{Grid}(t) \end{aligned} \tag{4}$$

where $CE^{DGs}(t)$ and $CE^{Grid}(t)$ are carbon emission produced by DGs and Grid in time period t, respectively. The carbon emission produced by DGs is defined by the following equation:

$$CE^{DGs}(t) = \sum_{k=1}^{N_{DGs}} Emission_{CO_2}{}^{DGs}(j) \times Po^{DGs}(t) \tag{5}$$

where $Emission_{CO_2}{}^{DGs}(t)$ is the carbon dioxide emission produced by jth DGs in time slot t; these carbon emissions produced during power generation and $Po^{DGs}(t)$ is the output power produced by DGs in time slot t. The carbon emission generated by grid is calculated as follows:

$$CE^{Grid}(t) = Emission_{CO_2}{}^{Grid} \times Po^{Grid}(t) \tag{6}$$

where $Emission_{CO_2}{}^{Grid}$ is carbon emissions produced by grid during power generation and $Po^{Grid}(t)$ is the output power produced by grid in time period t, respectively.

4. System Model

A programming-based energy optimization model is proposed to minimize operating cost and carbon emission with and without DRPs implementation in SG integrated with RESs. The proposed model consists of wind energy system, solar energy system, hybrid energy system, and demand response programs, which is shown in Figure 1. The detailed description is as follows:

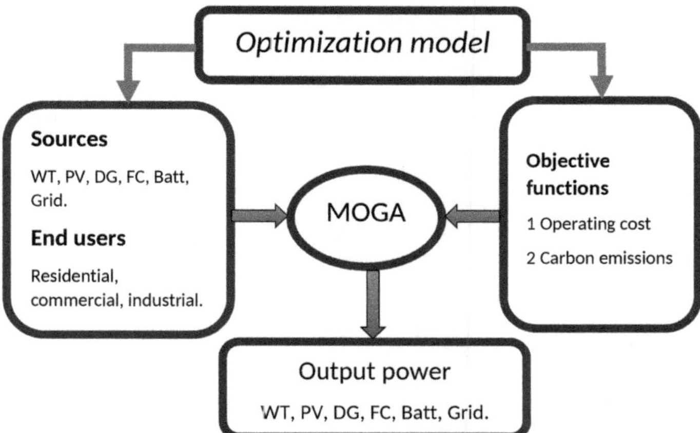

Figure 1. Proposed multi-objective energy optimization model.

4.1. Wind Based Renewable Energy Generating System

The wind turbine produces electrical energy from the potential of wind. With the demand of energy increasing day by day, RE is suitable to use because of low cost and emission. The online web site data for wind is predicted through Rayleigh distribution [27] is proposed for modeling wind energy. The PDF function is as follows:

$$F_E(S_{wind}) = 1 - \exp\left[-\left(\frac{Swind}{\beta\omega}\right)^2\right] \tag{7}$$

where S_{wind} is speed of wind and and $\beta\omega$ is the scale parameter. The function for output power of wind energy system is as follows:

$$P(S_{wind}) = \begin{cases} 0 S_{wind} < S_{c\_i} \\ P_{rated}\frac{(S_{wind}-S_{c\_i})}{(2S_r-S_{c\_i})} S_{c\_i} \leq S_{wind} < S_r \\ P_{rated} S_r \leq S_{wind} < S_{c\_o} \\ 0 S_{wind} \geq S_{c\_o} \end{cases} \tag{8}$$

where P_{rated}, S_{wind}, S_{c_i}, S_r *and* S_{c_o} are turbine rated power, wind speed, cut-in speed, rated speed, and cut-off speed, respectively. The wind turbine in this study is used as the vertical axis wind turbine [28], where P_{rated} = 18kw, S_{c_i} = 4 m/s, S_{c_o} = 20 m/s, S_r = 19 m/s.

4.2. Solar Energy System

The solar generator converts solar energy into electrical energy. To model the behaviour of solar irradiance, a mathematical tool PDF is proposed [29]. The solar irradiance modeling through PDF is shown in Equation (9):

$$F_{beta} = \begin{cases} \frac{\Gamma(\delta+\gamma)}{\Gamma(\delta)\Gamma(\gamma)}(s\_i)^{\delta-1}(1-s\_i)^{\gamma-1} \\ 0\ otherwise \end{cases}$$
$$where\ 0 \leq s\_i \geq 1, \delta \geq 0, \gamma \geq 0 \tag{9}$$

where s_i shows solar radiation coming from the sun, δ and γ are the parameters of beta PDF, which is calculated from solar radiation. Beta PDF parameters can be calculated from solar radiation data as follows:

$$\delta = \psi\left(\frac{\psi(1+\psi)}{\sigma 2} - 1\right), \tag{10}$$

$$\gamma = 1 - \psi\left(\frac{\psi(1+\psi)}{\sigma 2} - 1\right). \tag{11}$$

The output power depends on the amount of solar irradiance. The following equation is used to convert solar radiation into solar energy [30]:

$$W_{p\_v}(s\_i) = A \times \eta \times s\_i \tag{12}$$

where A shows the effective area of the PV array, $W_{p\_v}(s\_i)$ shows total output energy of the PV system in kW, η shows efficiency of PV energy system and s_i indicates solar radiation which comes from the sun in kW/m^2.

4.3. Hybrid Energy System

The hybrid energy system is the combination of output power of the solar and wind energy system:

$$P_{hyb} = P_{wind} + P_{photovoltaics} \tag{13}$$

where P_{hyb}, P_{wind}, and $P_{photovoltaics}$ is total output power, wind power and solar power, respectively. It seems to be difficult to use PDF in a mathematical way because the process is so lengthy; therefore, authors used Monte Carlo simulation for solving this problem [31]. This technique is used to forecast different models and is beneficial while making decisions. In general, this method is used to predict such models which can not be predicted easily.

4.4. Incentive-Based DRPs

In this study, the consumers are residential, commercial and industrial which are taking part in DRPs. In Equations (14)–(16), constraints show that energy reduction by each consumer at each hour should be less than or equal to the total amount of energy offered to each consumer. The following equations are used to model the behaviour of different types of consumers:

$$\text{Res}(res,t) = R_C(res,t) \times \sigma_{r,t}, R_C(res,t) \leq R_C{}^{max} \tag{14}$$

$$Com(c,t) = C_C(com,t) \times \sigma_{com,t}, C_C(com,t) \leq C_C{}^{max} \tag{15}$$

$$Ind(i,t) = I_C(ind,t) \times \sigma_{ind,t}, I_C(ind,t) \leq I_C{}^{max} \tag{16}$$

where *Res(r,t)*, *Com(c,t)* and *Ind(i,t)* show cost due to load minimization by each consumers in time *t*, *res*, *com* and *ind* are industrial, residential and commercial consumers, respectively. $R_C(res,t)$, $C_C(com,t)$, $I_C(ind,t)$ are load minimization according to the plans by each consumers. $R_C{}^{max}$, $C_C{}^{max}$ and $I_C{}^{max}$ are max minimization of load in time t. $\sigma_{r,t}$, $\sigma_{c,t}$ and $\sigma_{i,t}$ are incentive based payments, respectively.

5. Proposed Multi-Objective Genetic Algorithm

In this study, the MOGA technique is proposed for operating cost and Carbon emission reduction. The MOGA algorithm used the position and velocity of the particle and used Pareto-fronts for positioning best possible results [32,33]. MOGA uses the non-dominated classification of the GA population, also maintaining diversity in non-dominated solutions. The nearest solutions to Pareto-front are ranked equal to 1, and the ranked equal to 1 solutions are pick-up solutions. Similarly, the other solutions than the ranked one solution are ranked accordingly, based on its location. The equation for finding rank as follows:

$$R_j = 1 + N_j \tag{17}$$

where R_j and N_j show the rank of solutions and the number of solutions which dominate j, respectively. If a large number of solutions dominates, its mean rank is higher. To combine more than one objective, the equation is as follows:

$$F(Y) = m_1 \cdot F_1(Y) + \cdots + m_i \cdot F_i(Y) + \cdots + m_N f_N(Y) \tag{18}$$

where Y, $F(Y)$, and $F_i(Y)$ are string of the rank, fitness function and j-th objective function, respectively, m_i represents the variable weight for $F_i(Y)$ and N indicates objectives function. MOGA steps are:

step 1: Assigning rank according to R_j.
step 2: Using a linear mapping function (LMF) to assign row fitness to each solution. Linear mapping functions will assign the row fitness and also assign the row fitness function for the worst solutions.
step 3: Calculating the average of row fitness values for each rank solutions. If the rank is one, then check the number of solutions having rank 1, and take the average of row fitness value of these solutions.
step 4: Applying crossover to each of the assigned values to produce a new string.
step 5: Applying to mutate.
step 6: The algorithm returns to step number 1, if satisfying conditions are not valid.

Now, here we discussed how to assign fitness values to MOGA.

MOGA fitness assignment: Assign fitness values are calculated as follows:

step 1: Choose σ share, which is a constant variable and denotes how much distance is considered between two solutions. If σ share has a lower value, then we say that the solutions are near.
step 2: Compute the number of solutions N_j and rank of solution R_j as shown in Equation (19).
step 3: If $j \propto N$, $j = j + 1$ back to step 1. Otherwise, go to step 4.
step 4: Identify max rank R_j.

The assigned fitness value is called average fitness value, and given as follows:

$$F_j = N_j - \sum_{j=1}^{R_j+1} \mu(j) - 0.5[\mu(R_j) - 1] \tag{19}$$

Equation (20) will give average fitness to each solution j. Where N_j is total number of solutions, μ_j is number of solutions of the rank R_j and $\mu(R_j)$ are the number of solutions in the current rank. For every solution in the rank, we have to calculate the niche count [34], which is calculated as follows:

$$Nc_j = \sum_{i=1}^{\mu(R_j)} sh(dji) \tag{20}$$

where j and i are two different solutions which must be in the same rank, and $sh(dji)$ is a share fitness value. The Pareto-fronts determine the best possible solutions [35]. The proposed technique implementation is shown in Algorithm 1.

Algorithm 1 MOGA code.

```
procedure
    Inputs:Population size, max iteration, Boundary conditions, Crossover, mutation;
    Output: Minimization of operational cost and Carbon emissions using RE;
    Initialization:
    Nj= No of solutions assigned;
    Rj= Rank of solutions;
    Sigma share= A constant which determines distance between two solutions;
    Step1: Assigning Rank Rk=s, where s=1,2,3,.........................n;
    Step2: Using LMF to assign row fitness to each solution to number of best and worst solutions;
    Implementation:
    Step3: Choose solutions having rank 1;
    Step4: Calculate the average of row fitness value for each rank solutions;
    Step5: Assigning fitness to each rank;
    step6: Applying mutate;
    Fitness assignment to MOGA;
    Choose σ share;
    Compute Rj and Nj using Equation (19);
    while iter <Maximum iterations do
        for Rj=1 do
            using equation Rj=1+Nj, then how many solutions for rank 1?;
            Take average of row fitness value of these solutions, These are assigned fitness values;
            if j∝N, j=j+1 then
                back to step 1;
                otherwise, Go to step 4;
            End If
            if Rj= other then 1 then
                move to step 1;
            End If
            Apply crossover to each assigned values to produce new string;
            if conditions satisfied then
                pareto ranks are checked;
                then Apply mutate;
            End If
        End For
        for Rj=1 do
            for i = 1 do
                Getting desired pareto-fronts ranks;
                Algorithm will back to step 1 if the following conditions are not satisfied;
            End For
        End For
    End While
End Procedure
```

6. Simulation Results and Discussion

A programming-based energy optimization model is proposed to reduce operational cost and carbon emission with and without DRPs in SG using RESs. To predict the behavior of RESs, wind, and solar energy, a mathematical tool PDF is proposed. Monte Carlo simulation is proposed for PDF implementation because PDF is difficult to use in a mathematical way. DRPs are proposed to overcome the uncertainty factor in renewable energy generation like solar and wind. To implement DRPs in SG, we recommend an incentive-based payment method as price offered packages. In this method, industrial, residential, and commercial consumers can participate in DRPs by taking an offered package at a time. The proposed model consists of different participants, such as sources and end-users and objective functions, operating cost, and carbon emission, which are shown in Figure 1. The data for wind energy generation are taken from [36,37]. The behavior of wind speed is shown in Figure 2. The solar energy system used in the proposed model has the following specifications: 25 kW SOLAREX MSX type, compose d of solar array 10×2.5 kW with h = 18.6% and s = 10 m^2 [38]. The average solar irradiance profile utilized by the solar energy system is shown in Figure 3.

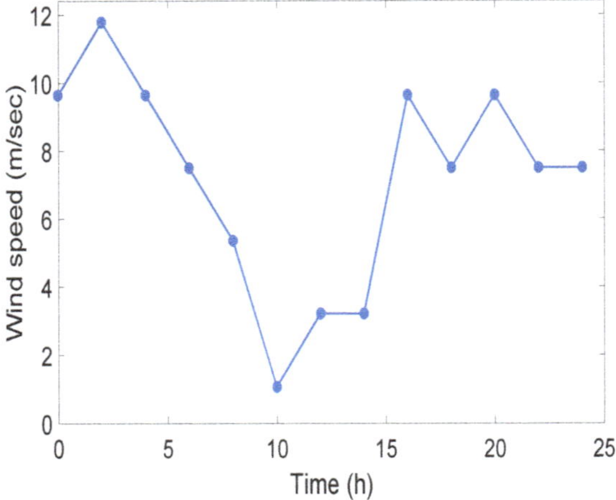

Figure 2. Hourly wind speed curve.

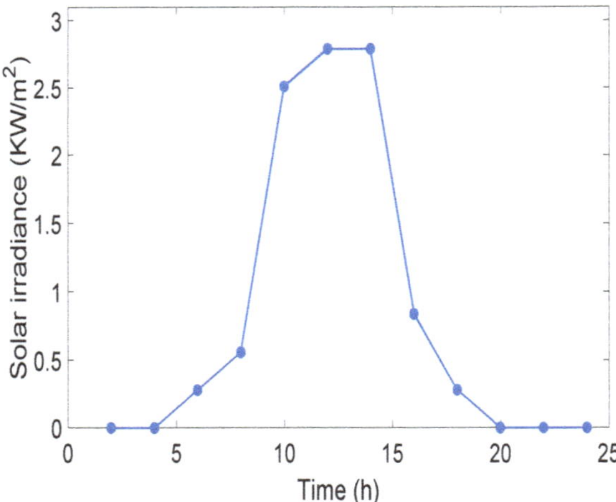

Figure 3. Hourly solar irradiance curve.

A battery used in this study, having high and low charges 10% and 100% with efficiency 94% [39,40]. Residential, commercial and industrial consumers' daily load demand profile is illustrated in Figure 4 [41].

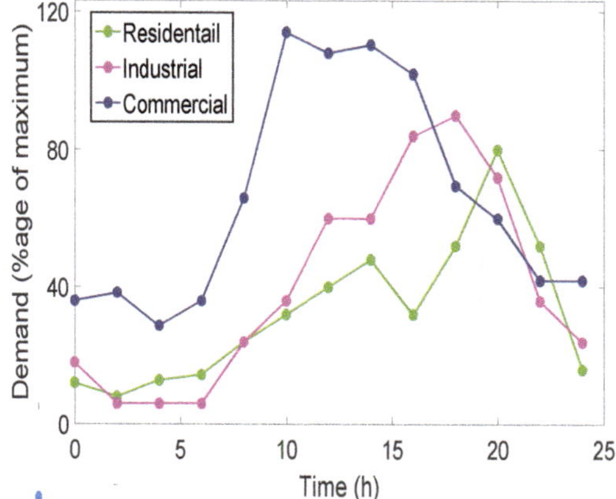

Figure 4. Daily load curves for different consumers like industrial, residential and commercial.

Figure 5 shows that DRPs optimize operation cost and carbon emission by actively engaging consumers in the electricity market to efficiently utilize RESs (Wind/Solar) and shift their load from on-peak hours to off-peak hours. The consumers' participation in DRPs reduce the burden on ECUs in terms of not turning peak power plants. The system operator reduces unscheduled power generation and is capable of managing demand with scheduled power generating units. The results before and after DRPs implementation are shown in Figure 5. To implement the proposed scenarios, we divided the proposed study into three steps.

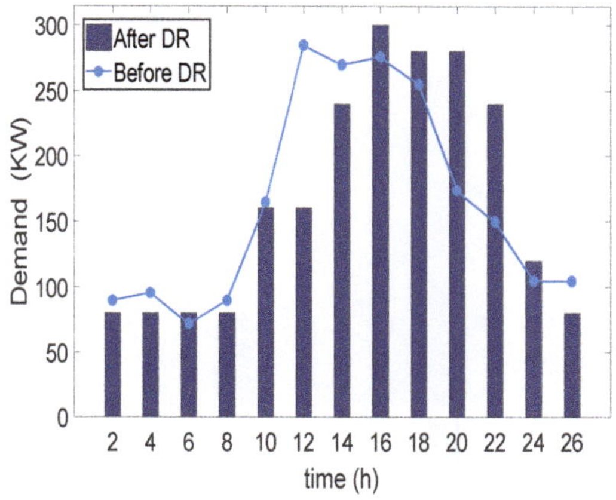

Figure 5. Load demand with and without involvement of demand response programs.

Step 1: Operation cost and carbon emission minimization with DRPs.

Step 2: Operation cost and carbon emission minimization without DRPs

Step 3: The MOGA technique is proposed to solve the problem by taking operation cost and carbon emission as multi-objective functions.

All the units are taking part in the SG operation to provide relief to SG. In this study, the multi-objective problem is solved by taking operational cost and carbon emission as objective functions and are implemented in MATLAB 2017b.

Step 1: Operation cost and carbon emission minimization with DRPs.

In this step, the cost of operation and carbon emissions are separately reduced without the involvement of DRPs. Table 1 results indicate that, in working hours, the emissions produced by sources are high. In this case, the utility takes energy from SG. While taking carbon emission function into account, both wind and solar energy produce low emissions during power production. In this case, the power reaches the maximum level. Wind and solar energy have high operating costs, and this scenario is only suitable for carbon emission function.

The results of Figure 6a,b show the simultaneous reduction of cost of operation and Carbon emission with and without the implementation of DRPs. The cost of operation and Carbon emission can be reduced by 24.5% and 28%, respectively. Figure 7a,b indicates wind and solar power generation while taking cost of operation and Carbon emission reduction and simultaneous reduction of wind and solar function with and without DRPs.

Step 2: Operation cost and carbon emission without DRPs.

In this step, the cost of operation and carbon emission are separately reduced with the involvement of DRPs. The cost of operation and carbon emission are separately reduced successfully, and the results are shown in Tables 2 and 3, respectively. Comparing Tables 1–4 shows that, when DRPs are implemented, the wind and power generation are reduced from 9.72 kW to 8.12 kW and from 5.54 kW to 4 kW, respectively. The optimization model proposed that the DRPs are taken with incentive-based payments. In the case of using DRPs with incentive-based payments, it converts loads from on-peak to off-peak hours and helps in operating cost reduction. When load shifts from peak to off-peak hours the cost of operation starts to reduce in this period.

Step 3: Solving multi-objective energy optimization problem using a MOGA technique by taking Pareto optimal fronts into account.

In this step, the simultaneous minimization of multi-objective functions, cost of operation, and carbon emission with and without DRPs takes place and is based on the MOGA technique. Both functions start to reduce simultaneously and result in the best possible solution by using a Pareto optimal path. Figure 6a,b show that operation cost and carbon emission are plotted on the x-axis and y-axis, respectively. Moving from an initial point towards the finishing point along the Pareto path represents a change in operation behavior from low operation cost and high emission to high operation cost and low emission, respectively. The optimize operation point is obtained via a fuzzy mechanism of MOGA.

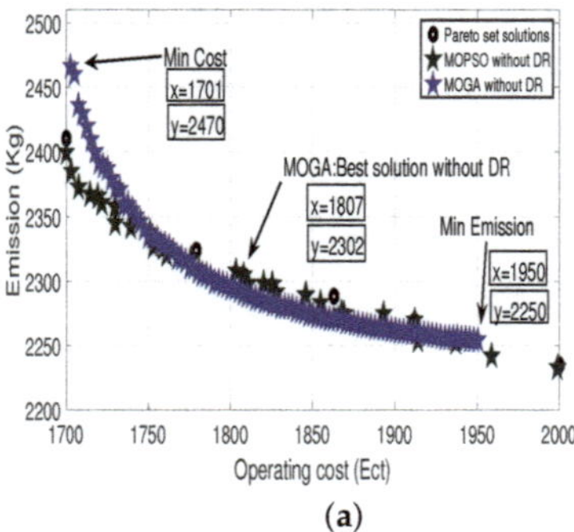

(a)

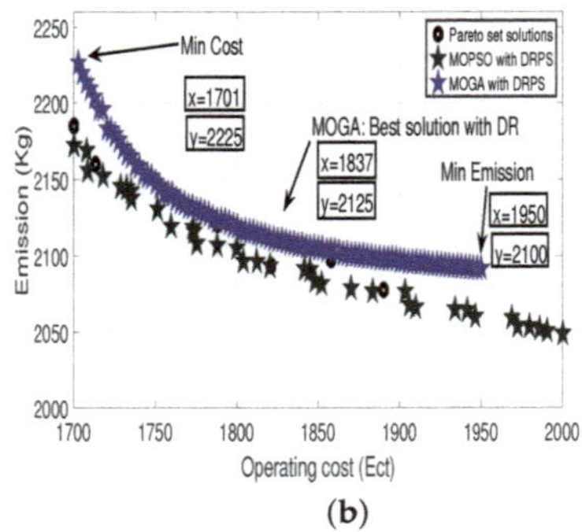

(b)

Figure 6. Multi-objective energy optimization of the proposed MOGA and existing MOPSO algorithm using Pareto-fronts criterion for operation cost and carbon emission optimization (**a**) without DRPs; (**b**) with DRPs

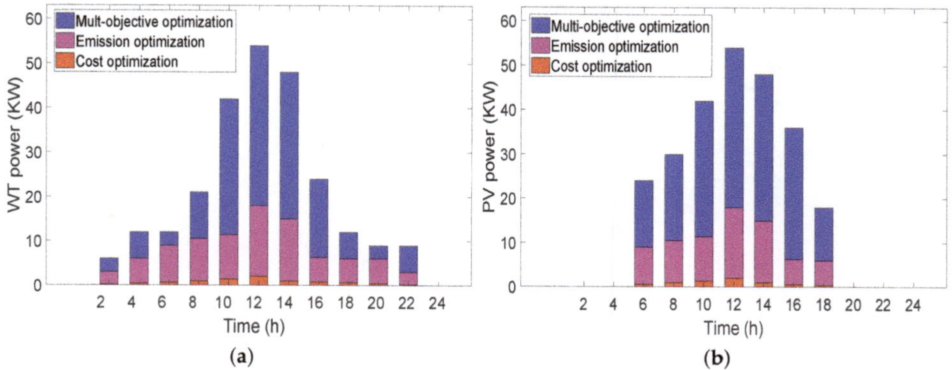

Figure 7. Renewable energy sources generation. (**a**) wind energy system; (**b**) solar energy system

Figure 7a,b results show that maximum solar and wind energy production are utilized considering carbon emission and operation cost to establish a balance between them by simultaneous optimization.

Table 1. Carbon emission optimization different energy generating units without DRPs.

Hours	Distributed Generations (kW)	Fuel Cell (kW)	Wind Turbine (kW)	Solar Array (kW)	Battery (kW)	Utility (kW)
1	39.64	30.91	1.92	0.00	31.08	−30.00
2	32.00	25.00	0.78	0.00	29.58	−26.56
3	32.00	27.17	1.98	0.00	30.95	−25.54
4	32.05	30.55	0.98	0.00	30.56	−23.33
5	32.00	28.00	1.95	0.00	31.98	−26.91
6	32.00	32.00	0.99	0.00	31.01	−26.28
7	35.27	30.68	1.90	0.00	30.35	−24.26
8	52.71	30.00	1.95	0.27	29.25	0.00
9	49.52	31.56	1.93	4.12	30.00	−9.00
10	115.04	30.00	1.89	12.00	31.95	−8.00
11	127.77	30.99	10.75	13.81	30.44	−30.00
12	152.40	30.99	11.41	25.00	30.00	−30.00
13	143.10	30.98	4.95	21.65	29.25	−26.96
14	165.92	32.00	2.94	8.42	29.98	−20.84
15	176.90	32.00	1.98	4.19	30.05	−10.91
16	171.37	30.09	1.99	0.93	31.80	−15.00
17	160.97	28.93	1.99	0.50	31.00	−21.22
18	154.91	31.00	1.98	0.00	31.06	−26.00
19	117.21	29.81	1.90	0.00	31.40	28.87
20	110.76	29.98	1.98	0.00	25.86	27.35
21	99.16	31.00	1.80	0.00	31.80	21.92
22	74.69	31.00	1.88	0.00	31.10	28.23
23	46.10	30.98	1.11	21.65	29.25	−25.96
24	40.99	29.00	0.75	8.32	29.98	−20.83

Table 2. Operating cost optimization of different energy generating units with DRPs.

Hours	Distributed Generations (kW)	Fuel Cell (kW)	Wind Turbine (kW)	Solar Array (kW)	Battery (kW)	Utility (kW)
1	33.64	7.18	0.38	0.00	0.68	30.00
2	34.00	12.00	0.28	0.00	−16.51	26.57
3	30.00	12.31	0.38	0.00	−19.97	29.47
4	30.15	20.5	0.71	0.00	−29.57	29.13
5	30.00	4.000	0.58	0.00	−29.99	3.81
6	30.00	7.000	0.14	0.00	−21.30	16.28
7	31.77	25.67	0.50	0.00	−29.73	28.22
8	35.71	28.000	0.05	0.25	−3.525	2.00
9	80.56	21.456	0.60	0.12	−20.00	−32.00
10	161.60	16.000	0.80	0.00	26.75	−32.00
11	233.37	5.99	0.75	0.51	28.44	−32.00
12	203.64	14.98	0.10	0.00	18.00	−30.00
13	260.21	26.98	0.91	0.65	−1.25	−30.66
14	223.99	4.00	0.34	0.42	−4.88	−34.83
15	214.90	22.00	0.90	0.19	−7.05	−32.71
16	171.36	23.09	0.38	0.38	−3.90	−32.00
17	125.09	26.93	1.79	0.55	11.00	7.52
18	44.01	28.00	0.68	0.00	5.07	29.00
19	52.25	28.12	1.11	0.00	30.34	31.79
20	90.26	24.98	1.01	0.00	27.88	31.35
21	129.16	22.00	0.22	0.00	25.80	−29.72
22	100.69	18.00	0.35	0.00	30.81	29.33
23	29.21	26.98	0.65	0.51	−1.24	−30.66
24	28.99	29.00	0.35	0.21	−4.98	−34.43

Table 3. Carbon emission optimization of different energy generating units with DRPs.

Hours	Distributed Generations (kW)	Fuel Cell (kW)	Wind Turbine (kW)	Solar Array (kW)	Battery (kW)	Utility (kW)
1	29.00	10.98	0.08	0.00	26.08	−30.000
2	29.00	23.00	0.28	0.00	28.18	−24.567
3	30.00	26.17	0.00	0.00	29.75	−27.547
4	30.15	2.65	0.00	0.00	19.76	−29.133
5	30.00	3.00	2.78	0.00	30.88	−31.981
6	29.00	26.00	0.98	0.00	5.31	−29.288
7	31.77	26.68	1.01	0.00	21.75	−28.262
8	30.77	29.00	0.05	0.00	29.25	−29.000
9	34.56	24.46	2.30	3.11	29.00	−28.00
10	37.64	28.00	7.09	7.00	23.75	−20.00
11	78.75	26.99	9.77	9.51	26.44	20.00
12	105.60	30.99	3.40	11.10	21.60	15.00
13	40.21	30.88	2.15	21.65	29.45	29.96
14	128.99	29.00	1.35	23.32	27.98	1.83
15	124.98	28.00	2.80	7.89	23.25	15.71
16	70.37	30.19	1.74	5.98	29.80	29.00
17	57.07	29.53	1.93	0.50	28.60	3.22
18	65.01	30.00	1.85	0.00	30.07	22.00
19	100.15	30.30	0.00	30.30	23.79	−19.66
20	95.26	28.96	1.30	0.00	26.86	18.75
21	56.16	28.00	1.20	0.00	27.80	27.72
22	35.99	30.00	1.55	0.00	26.80	15.33
23	50.10	29.98	2.91	21.65	29.45	16.96
24	30.92	29.00	1.34	23.32	27.98	−11.43

Table 4. Operation cost optimization of different energy generating units without DRPs.

Hours	Distributed Generations (kW)	Fuel Cell (kW)	Wind Turbine (kW)	Solar Array (kW)	Battery (kW)	Utility (kW)
1	30.43	8.98	0.48	0.00	12.78	30.00
2	33.00	29.00	0.48	0.00	−16.51	27.57
3	37.00	20.31	0.28	0.00	−16.97	23.57
4	39.66	25.65	0.81	0.00	−29.57	30.13
5	33.00	14.70	0.78	0.00	−26.99	27.91
6	31.00	24.91	0.00	−21.30	−10.28	23.00
7	32.12	26.67	0.50	0.00	−19.73	23.22
8	34.61	25.00	0.305	0.27	−0.55	23.00
9	106.66	3.46	0.30	0.12	30.00	-32.00
10	243.60	19.00	0.09	0.00	−13.95	−32.00
11	215.77	5.99	0.75	0.81	18.44	−32.00
12	291.64	6.98	0.10	0.00	−26.00	−30.00
13	277.21	9.98	0.95	0.95	−25.25	−30.96
14	223.99	25.00	5.00	0.345	22.32	−30.88
15	218.90	29.00	0.98	0.89	32.20	−27.91
16	228.36	7.09	0.78	0.38	30.00	−27.00
17	117.09	27.93	1.99	0.50	31.00	27.22
18	95.01	30.00	0.98	0.00	31.07	29.00
19	107.25	32.82	1.31	0.00	30.34	31.79
20	125.76	31.96	1.31	0.00	27.88	31.35
21	157.11	29.00	0.32	0.00	30.80	−29.72
22	89.69	25.00	0.55	0.00	30.81	29.33
23	56.21	26.98	0.95	0.65	30.24	−30.66
24	33.99	25.00	0.35	0.32	29.98	26.83

7. Conclusions

A programming-based energy optimization model is proposed to optimize operation cost and carbon emission with and without the involvement of DRPs in SG with integrated RESs like solar and wind. The incentive-based DRPs in SG as price offered packages is introduced to overcome the uncertainty factor in renewable energy generation like solar and wind. Moreover, to achieve optimized results, a PDF is intended to predict the behaviour of solar and wind RESs. Simulations are conducted in three steps: (i) multi-objective energy optimization problem of operation cost and carbon emission is solved using MOGA, (ii) operation cost and carbon emission optimization without DRPs, and (ii) operation cost and carbon emission optimization with DRPs. The proposed model is tested on a smart sample grid serving consumers of residential, commercial, and industrial sectors. The proposed energy optimization model based on MOGA outperforms the existing models in terms of operation cost and carbon emission. Experimental results show that the operation cost and carbon emission with and without DRPs are reduced by 24% and 28% using the proposed technique MOGA, respectively.

Author Contributions: Conceptualization, K.U. and G.H.; methodology, G.H.; software, I.K.; validation, S.A., S.J.; formal analysis, T.A.K.; investigation, I.A.S.; resources, G.H.; data curation, K.U.; writing—original draft preparation, K.U.; writing—review and editing, G.H., T.A.K; visualization, S.A.; supervision, I.K., G.H. All authors have read and agreed to the published version of the manuscript.

Funding: This research received no external funding.

Conflicts of Interest: The authors declare no conflict of interest.

References

1. Rahman, S. An efficient load model for analyzing demand side management impacts. *IEEE Trans. Power Syst.* **1993**, 8, 1219–1226. [CrossRef]

2. Cohen, A.I.; Wang, C.C. An optimization method for load management scheduling. *IEEE Trans. Power Syst.* **1988**, *3*, 612–618. [CrossRef]
3. Palensky, P.; Dietrich, D. Demand side management: Demand response, intelligent energy systems, and smart loads. *IEEE Trans. Ind. Inform.* **2011**, *7*, 381–388. [CrossRef]
4. Renzhi, L.; Hong, S.H. Incentive-based demand response for smart grid with reinforcement learning and deep neural network. *Appl. Energy* **2019**, *236*, 937–949.
5. Shahidehpour, M.; Yamin, H.; Li, Z. *Market Operations in Electric Power Systems: Forecasting, Scheduling, and Risk Management*; John Wiley and Sons: Hoboken, NJ, USA, 2003.
6. Shuba, E.S.; Kifle, D. Microalgae to biofuels: 'Promising'alternative and renewable energy, review. *Renew. Sustain. Energy Rev.* **2018**, *81*, 743–755. [CrossRef]
7. Reinhard, P.; Rusdianasari, R.; Eviliana, E. Renewable Energy: Advantages and Disadvantages. In Proceedings of the Forum in Research, Science, and Technology (FIRST) 2016, Politeknik Negeri Sriwijaya, Palembang, Indonesia, 18–19 October 2016.
8. Rehmani, M.H.; Reisslein, M.; Rachedi, A.; Erol-Kantarci, M.; Radenkovic, M. Integrating renewable energy resources into the smart grid: Recent developments in information and communication technologies. *IEEE Trans. Ind. Inform.* **2018**, *14*, 2814–2825. [CrossRef]
9. Domenico, M.; Oliveti, G.; Labonia, E. Estimation of wind speed probability density function using a mixture of two truncated normal distributions. *Renew. Energy* **2018**, *115*, 1260–1280.
10. Ran, X.; Shihong, M. Three-phase probabilistic load flow for power system with correlated wind, photovoltaic and load. *Iet Gener. Transm. Distrib.* **2016**, *10*, 3093–3101. [CrossRef]
11. Hafeez, G.; Alimgeer, K.S.; Wadud, Z.; Khan, I.; Usman, M.; Qazi, A.B.; Khan, F.A. An Innovative Optimization Strategy for Efficient Energy Management with Day-ahead Demand Response Signal and Energy Consumption Forecasting in Smart Grid using Artificial Neural Network. *IEEE Access* **2020**, *8*, 84415–84433. [CrossRef]
12. Kanchev, H.; Colas, F.; Lazarov, V.; Francois, B. Emission reduction and economical optimization of an urban microgrid operation including dispatched PV-based active generators. *IEEE Trans. Sustain. Energy* **2014**, *5*, 1397–1405. [CrossRef]
13. Tedesco, F.; Mariam, L.; Basu, M.; Casavola, A.; Conlon, M.F. Economic model predictive control-based strategies for cost-effective supervision of community microgrids considering battery lifetime. *IEEE J. Emerg. Sel. Top. Power Electron.* **2015**, *3*, 1067–1077. [CrossRef]
14. Kakran, S.; Chanana, S. Smart operations of smart grids integrated with distributed generation: A review. *Renew. Sustain. Energy Rev.* **2018**, *81*, 524–535. [CrossRef]
15. Derakhshan, G.; Shayanfar, H.A.; Kazemi, A. The optimization of demand response programs in smart grids. *Energy Policy* **2016**, *94*, 295–306. [CrossRef]
16. Parisio, A.; Wiezorek, C.; Kyntäjä, T.; Elo, J.; Strunz, K.; Johansson, K.H. Cooperative MPC-based energy management for networked microgrids. *IEEE Trans. Smart Grid* **2017**, *8*, 3066–3074. [CrossRef]
17. Teo, T.T.; Logenthiran, T.; Woo, W.L.; Abidi, K. Intelligent Controller for Energy Storage System in Grid-Connected Microgrid. *IEEE Trans. Syst. Man Cybern. Syst.* **2018**. [CrossRef]
18. Melhem, F.Y.; Grunder, O.; Hammoudan, Z.; Moubayed, N. Energy management in electrical smart grid environment using robust optimization algorithm. *IEEE Trans. Ind. Appl.* **2018**, *54*, 2714–2726. [CrossRef]
19. Kamboj, A.; Chanana, S. Optimization of cost and emission in a Renewable Energy micro-grid. In Proceedings of the 2016 IEEE 1st International Conference on Power Electronics, Intelligent Control and Energy Systems (ICPEICES), Delhi, India, 4–6 July 2016; IEEE: Piscataway, NJ, USA, 2016; pp. 1–6.
20. Hafeez, G. Energy Efficient Integration of Renewable Energy Sources in the Smart Grid for Demand Side Management. Master's Thesis, COMSATS University Islamabad, Islamabad, Pakistan, 2017.
21. Khan, T.A.; Ullah, K.; Hafeez, G.; Khan, I.; Khalid, A.; Shafiq, Z.; Usman, M.; Qazi, A.B. Closed-Loop Elastic Demand Control under Dynamic Pricing Program in Smart Microgrid Using Super Twisting Sliding Mode Controller. *Sensors* **2020**, *20*, 4376. [CrossRef]
22. Pedram, S.; Mohsenian-Rad, H.; Schober, R.; Wong, V.W.S. Advanced demand side management for the future smart grid using mechanism design. *IEEE Trans. Smart Grid* **2012**, *3*, 1170–1180.
23. Imran, A.; Hafeez, G.; Khan, I.; Usman, M.; Shafiq, Z.; Qazi, A.B.; Khalid, A.; Thoben, K.D. Heuristic-Based Programable Controller for Efficient Energy Management Under Renewable Energy Sources and Energy Storage System in Smart Grid. *IEEE Access* **2020**, *8*, 139587–139608. [CrossRef]

24. Abdul-Rahman, A.; Zualkernan, I.A.; Rashid, M.; Gupta, R.; Alikarar, M. A smart home energy management system using IoT and big data analytics approach. *IEEE Trans. Consum. Electron.* **2017**, *63*, 426–434.
25. Linas, G.; Gamage, K.A.A. Demand side management in smart grid: A review and proposals for future direction. *Sustain. Cities Soc.* **2014**, *11*, 22–30.
26. Saifuddin, M.R.B.M.; Logenthiran, T.; Naayagi, R.T.; Woo, W.L. A nano-biased energy management using reinforced learning multi-agent on layered coalition model: Consumer sovereignty. *IEEE Access* **2019**, *7*, 52542–52564. [CrossRef]
27. Rahimi, E.; Rabiee, A.; Aghaei, J.; Muttaqi, K.M.; Nezhad, A.E. On the management of wind power intermittency. *Renew. Sustain. Energy Rev.* **2013**, *28*, 643–653. [CrossRef]
28. Seki, K.; Shimizu, Y.; Kato, Y. Vertical Axis Wind Turbine. U.S. Patent 4,247,253, 27 January 1981.
29. Ettoumi, F.Y.; Mefti, A.; Adane, A.; Bouroubi, M.Y. Statistical analysis of solar measurements in Algeria using beta distributions. *Renew. Energy* **2002**, *26*, 47–67. [CrossRef]
30. Deshmukh, M.K.; Deshmukh, S.S. Modeling of hybrid renewable energy systems. *Renew. Sustain. Energy Rev.* **2008**, *12*, 235–249. [CrossRef]
31. Tina, G.; Gagliano, S.; Raiti, S. Hybrid solar/wind power system probabilistic modelling for long-term performance assessment. *Sol. Energy* **2006**, *80*, 578–588. [CrossRef]
32. Murata, T.; Ishibuchi, H.; Sanchis, J.; Blasco, X. MOGA: Multi-objective genetic algorithms. In *IEEE International Conference on Evolutionary Computation*; IEEE: Piscataway, NJ, USA, 1995; Volume 1, pp. 289–294.
33. Herrero, J.M.; Manuel Martínez, J.S.; Blasco, X. Well-Distributed Pareto Front by Using the $\epsilon\nearrow$—*MOGA* Evolutionary Algorithm. In Proceedings of the International Work-Conference on Artificial Neural Networks, San Sebastián, Spain, 20–22 June 2007; Springer: Berlin/Heidelberg, Germany, 2007; pp. 292–299.
34. Miller, B.L.; Shaw, M.J. Genetic algorithms with dynamic niche sharing for multimodal function optimization. In *Proceedings of IEEE International Conference on Evolutionary Computation*; IEEE: Piscataway, NJ, USA, 1996; pp. 786–791.
35. Yang, J.; Ji, Z.; Liu, S.; Jia, Q. Multi-objective optimization based on pareto optimum in secondary cooling and EMS of continuous casting. In *2016 International Conference on Advanced Robotics and Mechatronics (ICARM)*; IEEE: Piscataway, NJ, USA, 2016; pp. 283–287.
36. Available online: https://wind.willyweather.com.au/ (accessed on 15 September 2019).
37. Reconstruction and Short-term Forecast of the Solar Irradiance. Available online: https://www.lpc2e.cnrs.fr/en/scientific-activities/plasmas-spatiaux/projects/other-projects-anr-europe-etc/fp7-soteria/ (accessed on 20 October 2019).
38. The Solar Power Group Company. Available online: https://www.enfsolar.com/directory/installer/67081/the-solar-power-group (accessed on 1 November 2019).
39. Chen, C.; Duan, S.; Cai, T.; Liu, B.; Hu, G. Smart energy management system for optimal microgrid economic operation. *IET Renew. Power Gener.* **2011**, *5*, 258–267. [CrossRef]
40. Mirzaei, M.J.; Kazemi, A.; Homaee, O. RETRACTED: Real-world based approach for optimal management of electric vehicles in an intelligent parking lot considering simultaneous satisfaction of vehicle owners and parking operator. *Energy* **2014**, *76*, 345–356. [CrossRef]
41. Luo, C.; Ukil, A. Modeling and validation of electrical load profiling in residential buildings in Singapore. *IEEE Trans. Power Syst.* **2014**, *30*, 2800–2809.

Publisher's Note: MDPI stays neutral with regard to jurisdictional claims in published maps and institutional affiliations.

Article

An Optimization Based Power Usage Scheduling Strategy Using Photovoltaic-Battery System for Demand-Side Management in Smart Grid

Sajjad Ali [1,2], Imran Khan [3], Sadaqat Jan [4] and Ghulam Hafeez [3,5,*]

1. Department of Telecommunication Engineering, University of Engineering and Technology, Peshawar 25000, Pakistan; sajjad@uetmardan.edu.pk
2. Department of Telecommunication Engineering, University of Engineering and Technology, Mardan 23200, Pakistan
3. Department of Electrical Engineering, University of Engineering and Technology, Mardan 23200, Pakistan; imran@uetmardan.edu.pk
4. Department of Computer Software Engineering, University of Engineering and Technology, Mardan 23200, Pakistan; sadaqat@uetmardan.edu.pk
5. Department of Electrical and Computer Engineering, COMSATS University Islamabad, Islamabad 44000, Pakistan

* Correspondence: ghulamhafeez393@gmail.com or ghulamhafeez@uetmardan.edu.pk; Tel.: +92-300-5003574 or +92-348-8818497

Academic Editor: Herodotos Herodotou

Received: 18 March 2021; Accepted: 6 April 2021; Published: 15 April 2021

Abstract: Due to rapid population growth, technology, and economic development, electricity demand is rising, causing a gap between energy production and demand. With the emergence of the smart grid, residents can schedule their energy usage in response to the Demand Response (DR) program offered by a utility company to cope with the gap between demand and supply. This work first proposes a novel optimization-based energy management framework that adapts consumer power usage patterns using real-time pricing signals and generation from utility and photovoltaic-battery systems to minimize electricity cost, to reduce carbon emission, and to mitigate peak power consumption subjected to alleviating rebound peak generation. Secondly, a Hybrid Genetic Ant Colony Optimization (HGACO) algorithm is proposed to solve the complete scheduling model for three scenarios: without photovoltaic-battery systems, with photovoltaic systems, and with photovoltaic-battery systems. Thirdly, rebound peak generation is restricted by using Multiple Knapsack Problem (MKP) in the proposed algorithm. The presented model reduces the cost of using electricity, alleviates the peak load and peak-valley, mitigates carbon emission, and avoids rebound peaks without posing high discomfort to the consumers. To evaluate the applicability of the proposed framework comparatively with existing frameworks, simulations are conducted. The results show that the proposed HGACO algorithm reduced electricity cost, carbon emission, and peak load by 49.51%, 48.01%, and 25.72% in scenario I; by 55.85%, 54.22%, and 21.69% in scenario II, and by 59.06%, 57.42%, and 17.40% in scenario III, respectively, compared to without scheduling. Thus, the proposed HGACO algorithm-based energy management framework outperforms existing frameworks based on Ant Colony Optimization (ACO) algorithm, Particle Swarm Optimization (PSO) algorithm, Genetic Algorithm (GA), Hybrid Genetic Particle swarm Optimization (HGPO) algorithm.

Keywords: energy management; battery energy storage systems; photovoltaic; demand response; scheduling; smart grid

1. Introduction

Electrical energy is one of the most indispensable needs of human life. Developing countries cannot optimally meet this basic need for residents due to limited financial budgets and scarce generating stations. The electric utility companies involuntary move towards load shedding to partially satisfy their consumers. However, load shedding is not a solution because it causes frustration to consumers. Considering the above limitations, one viable solution is Demand Side Management (DSM) via consumer load scheduling. Optimal DSM is only possible by actively engaging consumers in the DR programs via the two-way communication infrastructure of a Smart Grid (SG). Moreover, current power-generating plants are operated on conventional sources, which are limited, scarce, and expensive and cause pollutant emissions that lead to climate change. Hence, the Renewable Energy Sources (RESs) are an alternative solution that is abundant, cheap, environment friendly, and continually replenished [1].

In the literature, three approaches are adopted to perform optimal DSM in SG via DR programs: mathematical methods, game theory-based methods, and heuristic algorithms. The mathematical techniques are classified into two classes, namely, deterministic and stochastic methods. The difference between stochastic and deterministic methods is the initialization of the initial solution, deterministic methods generate the same initial solution when addressing the same problem, and stochastic methods randomly initialize solutions that permit various solutions for the given problem in each run [2]. A novel modular modelling Energy Management System (EMS) is developed for urban multi-energy systems [3]. To evaluate the proposed EMS's saving potential, an extensive case study is conducted compared to the traditional control strategies. It can be evident from the results that an annual cost-saving potential between 3 and 6 percent can be attained when the proposed EMS model is used in combination with additional components such as battery and thermal energy storage. However, there are still no open source solvers available to handle large-scale Mixed Integer Linear Programming (MILP). Furthermore, in large-scale EMS applications (e.g., city districts) such as IBM ILOG CPLEX or GUROBI, the monetary benefit is usually high enough to justify commercial solvers' costs. A Mixed Integer Linear Programming (MILP)-based scheme is introduced for energy-efficient management in SG [4]. The developed scheme optimally plans smart devices and the charging and discharging of electric vehicles to reduce energy costs. In the developed model, users can produce their energy from microgrids containing wind turbines, solar panels, and Energy Storage System (ESS). The results confirm that the proposed MILP-based energy management is more effective and productive than legacy models. However, how to handle the intermittent nature of renewable energy sources is not discussed. A combination of Sequential Quadratic Programming algorithm (SQP) and Binary Particle Swarm Optimization (BPSO)-based optimization model is proposed in [5]. The proposed combination model is applied for energy management of the residential sector. The results validated that the proposed model performed energy management efficiently. However, for reliability in DR programming, the irregular and indeterminate renewable energy characteristics will raise substantial challenges. Thus, to improve the reliability of the renewable energy uncertainty, an appropriate uncertainty processing technology such as chance-constrained technique and Model Predictive Control (MPC) can be incorporated into the MINLP model presented in the proposed work. The authors in [6] proposed a smart home energy management system model including power generation, solar panels, small wind turbine, battery, and appliances. The proposed model efficiently schedules household thermal and electrical appliances using time-varying pricing to minimize financial expenses and to ensure peak demand clipping. A DR program is employed at different levels [7]: single home, combined home, network level, and market level for consumers to actively participate in DSM. Electric Vehicles (EVs), Energy Storage Systems (ESSs), and RESs are actively engaged in the DR program in each scenario. The mathematical formulation of each level is implemented with uncertainty consideration. However, the technology barriers such as sensing, controlling, monitoring, and communication infrastructure and markets such as policies, regulation, and structure are not considered. The authors considered the DSM of a home including smart appliances, EVs, ESS, and PV microgeneration in [8]. The resources and loads are coordinated using indexed pricing models for DSM to maximize self-generation and to minimize cost by reducing utility purchase. However, an investigation on minimum software and hardware requirements for the unit hosting the Smart Home Controller (SHC) module (an issue that is strictly related to the task of developing efficient and tailored solution algorithms) and a comprehensive and detailed cost–benefit analysis of the proposed residential EMS were performed using approaches such as the ones presented to generate highly realistic sequences of events. EMS's modular design for a grid-connected battery-based microgrid was presented in [9]. The developed model was for power generation-side using MILP to handle charging and discharging of batteries, to encourage self-consumption, and to reduce operating costs. The proposed model has been tested and experimentally validated at Alborg University at the Microgrid Research Laboratory. However, this optimization problem can be enhanced if power losses and DR programs are considered. In [10], a complete scheduling scheme including distributed generation and residential load was modeled using RESs. However, the carbon emission was not considered, which is interdependent in distributed generation. To resolve the problems accompanied by mathematical methods, game-theoretic-based approaches are adopted. For example, in [11], an adaptive game-theoretic-based model is proposed to solve the DSM problem by reducing the Peak to Average Ratio (PAR) and saves the nearby cost location. A Nash game theory-based optimization model is developed for scheduling building energy consumption under utility and RESs. The proposed model minimizes energy cost, maintains consumers' comfort levels, and reduces peak demands within the imposed constraints. The model is verified with the case studies of the nearby location in Sydney. It also makes sure that the user does not make a profit if the user diverges from the consumption pattern assigned to the user. The numerous algorithms' performance parameters are assessed, and their effects have been discussed on the PAR and energy costs. However, the mechanism of Common Storage Facility (CSF) discharge on a bidding process has not been considered. In [12], a game theory-based, dynamic pricing strategy for the commercial and residential sector electricity market of Singapore was proposed. The model was evaluated with five loads and datasets prices to cover all events, i.e., weekdays, weekends, public holidays, and the highest/lowest demand in the year. Three strategies of the prices were compared and evaluated, i.e., the half-hourly Real-Time Pricing (RTP), Time-of-Use Price (TOUP), and Day-Night (DN) pricing. The outcomes show that the RTP increases the peak load reduction for the residential and commercial sectors by 10% and 5% percent. Furthermore, an increase in the profit by 15.5 percent and 18.7 percent has been observed, and a reduction in the total load is minimized for a realistic scenario. Moreover, tests on the satisfaction functions that are different from one another were performed, which includes many user types who knows the impact of RESs on DSM by taking the dynamic pricing technique and know that the impact of dynamic pricing on short-term Priced Elasticity Demands (PEDs) for the long term is not catered. The Stackelberg game approach and the DR model for trading electricity between utility companies were developed in [13]. The purpose was to balance the supply and demand by smoothing the load demand curve. The 1-leader was expressed by the interaction between the leader utility company and follower N-follower. The Stackelberg game was employed for solving the optimization problem. The RTP as a pricing function was adopted with the primary aim to encourage the user to join the game to determine the optimal power generation and power demands. However, the distribution network evaluation with the nodal pricing approach was not discussed. A hierarchical system model was introduced in [14], where multiple providers and prosumers interacted to set the best price and demands for the electricity market. The purpose of the model for overall energy management was to increase a prosumer's capacity to produce more energy and to reduce the dependency on the utility electricity service providers. Thus, in the proposed work, the price and demands have been improved considering the limitations and various providers and prosumers by creating a Stackelberg game to model two types of interactions: (i) provider–consumer and (ii) prosumer–prosumer. Hence, it is verified that there is a unique equilibrium solution, and it is evident from the results that the proposed method improves energy consumption and cost. A voltage and frequency relaying scheme-based DSM model is proposed in [15]. The purpose is to minimize the load on 11 kV distribution feeders in peak hours without blanket load shedding. This model's development is a crucial step towards cleaner production that minimizes load during peak hours and Green House Gas (GHG) emission. However, classification modeling to learn anomalous events and vitals are not considered. Game theory methods have their inherent limitations, which deprived them of optimal solutions. Thus, heuristic algorithms are adopted to cater to DSM problems. For instance, DSM framework is presented in [16], PSO algorithm is used to minimize the PAR and cost of the electricity, which results in achievement of the optimal sharing schedule and sharing the PEV battery with those nearby. However, the resultant system is complex. In [17], an optimal energy scheduling strategy for Home Energy Management (HEM) was developed. The purpose is to acquire the desired tradeoff between cost and benefit. However, the user comfort in terms of waiting time is increased while reducing cost and increasing benefits. The authors presented an efficient DSM system that minimizes the cost of electricity and PAR in [18]. However, users face frustration in terms of waiting time while reducing their costs. In [19], the authors presented a generic DSM model for HEM to minimize electricity cost, appliance waiting time, and PAR. However, the power grid's reliability and sustainability, and balance between supply and demand were not catered to, which are indispensable in DSM. The residential loads were controlled and monitored for a smart home in [20], in which the author has proposed the Grey Wolf Accretive Satisfaction Algorithm (GWASA) and have compared the achieved results with other algorithms that resultantly reduced the cost of electricity. However, the PAR was not addressed. In [21], an efficient heuristic algorithm-based EMC utilizing RESs, and TOUP and Incline Block Rate (IBR) tariffs was developed for DSM. The DSM problem formulated was mapped to reduce the electricity cost, user discomfort, and PAR. The simulation results confirmed that the required objectives were obtained and that the sustainability of the SG increased. However, the authors do not address the following issues: (i) appliances waiting time, (ii) frequency of interruption, and (iii) demand curve smoothing. An innovative home appliance scheduling framework was proposed in [22]. The Grey Wolf and Crow Search Optimization (GWCSO) algorithm was used and reduced the PAR and electricity cost. However RESs were not addressed. To reduce the PAR, appliance waiting time, and electricity cost for residential consumers, the generic DSM model was studied in [23]. A GA-based Energy Management Controller (EMC) was developed for DSM under DR program. The GA-based EMC scheduled the residential load, keeping in view the operation limitations of users. To avoid rebound peaks, a combined RTP and IBR tariff was used. The developed model is useful for both single and multiple users. However, the balance between supply and demand, user comfort, reliability, and sustainability of the grid was not catered. In [24], the authors proposed an incentive-based optimal energy consumption scheduling algorithm in HEM. The energy demand during peak hours were minimized using the TOUP and DR program. To enhance the cost-saving, a heuristic BPSO technique was used for appliance scheduling based on HEM and RES. However, Consumers' comfort were ignored, which is essential in energy consumption scheduling. An EMC was designed based on heuristic algorithms such as GA, Bacterial Foraging Optimization Algorithm (BFOA), Wind Driven Optimization (WDO), PSO, and Genetic Binary Particle Swarm Optimization (GBPSO) algorithms in [25] to reduce electricity price and PAR. However, the system model was more complicated. In [26], an optimal control strategy was proposed using GA to obtain the minimum cost of electricity based on the user's response, dynamic pricing, and equipment operating power. The proposed technique determines the optimal operating parameters of each piece of equipment. However, a limited number of appliances were taken into account. The residential consumer power scheduling for DR was discussed in [27]. For scheduling, two types of electric appliances, power-flexible and time-flexible, were considered. However, the PAR was ignored, which is strongly related to electricity cost. In SG, an advanced RTP algorithm for DSM was presented in [28] to reduce electricity cost. The main aim was to communicate among smart meters and utility by exchanging control messages, including real-time cost information and consumer energy consumption. However, the DSM was performed on increased system complexity. A device and ESS scheduling schemes were presented in [29]. The consumers scheduled their devices by observing the low peak hours and by extending the electricity cost using smart electricity storage. The latest and updated comprehensive review of the literature in terms of proposed algorithms, objectives, and limitations are conducted and listed in Table 1.

Table 1. A comprehensive review of the latest literature in terms of proposed algorithms, objectives, and limitations.

Reference(s)	Algorithms(s)	Objective(s)	Limitation(s)
A novel modular modelling EMS is developed in [3]	MILP	An annual cost savings potential between 3 and 6 percent can be attained when the proposed EMS model is used in combination with additional components such as battery and thermal energy storage	No open source solvers available to handle large-scale MILP
Energy-efficient management in SG [4]	MILP	Proposed MILP-based energy management is more effective and productive than legacy models	How to handle the intermittent nature of renewable energy sources is not discussed
The proposed combination model is applied for energy management of the residential sector [5]	SQP and BPSO	The proposed model performed energy management efficiently	The irregular and indeterminate renewable energy characteristics will raise substantial challenges
Smart home energy management system model [6]	MILP	Efficiently schedules the household thermal and electrical appliance using time-varying pricing to minimize financial expenses and ensure peak demand clipping	The system is more complicated
A DR program is employed in different levels [7]	MILP	Each level is implemented with uncertainty consideration	The technology barriers such as sensing, controlling, monitoring, and communication infrastructure and markets such as policies, regulation, and structure are not considered.
DSM of a home including smart appliances, EVs, ESS, and PV in [8]	Indexed pricing models	Maximize self-generation and minimize cost by reducing utility purchase	Investigation on minimum software and hardware requirements for the unit hosting the SHC module is not addressed
EMS modular design for a grid-connected battery-based microgrid in [9]	MILP	Promote self-consumption and reduce operating cost	Optimization problem can be enhanced if power losses and DR programs are considered
Optimal scheduling scheme for smart residential community [10]	MILP	To minimize electricity cost and peak load	carbon emission is not considered
An adaptive game-theoretic-based model is proposed to solve the DSM problem in [11]	Game-theoretic model	Minimizes energy cost, maintains consumers' comfort level, and reduces peak demands within the imposed constraints	The mechanism of CSF discharge on a bidding process has not been considered
Dynamic pricing strategy for the commercial and residential sector is proposed in [12]	Game-theoretic model	Peak load is reduced for the residential and commercial sectors	Impact of dynamic pricing on short-term PEDs for the long term is not catered
DR model for trading electricity between utility companies are developed in [13]	Stackelberg game	Balance the supply and demand by smoothing the load demand curve	Distribution network evaluation with the nodal pricing approach is not discussed
A hierarchical system model is introduced in [14]	Stackelberg game	Increased a prosumer's capacity to produce more energy and reduce the dependency on the utility electricity service providers	The system is more complex
A voltage and frequency relaying scheme based DSM model is proposed in [15]	Game-theoretic model	Minimizes load during peak hours and GHG emission	Classification modeling to learn anomalous event and vitals are not considered.
DSM framework is presented[16]	PSO	To reduce PAR and electricity cost	The system is more complex

Table 1. *Cont.*

Reference(s)	Algorithms(s)	Objective(s)	Limitation(s)
For HEM, an optimal energy scheduling strategy is developed in [17]	Polyblock approximation algorithm	To obtain the desired tradeoff between cost and benefit	DTR is increased
An efficient DSM system is presented in [18]	Scheduling algorithm	Reduce the cost of electricity and PAR	Increased waiting time
A generic DSM model for HEM is presented [19]	GA	To reduce PAR, appliances waiting time and electricity cost	Power grid reliability and sustainability are not cosidered
Residential load for smart home is presented [20]	GWASA	To reduce electricity cost	PAR is not addressed
An efficient heuristic algorithm is developed in [21]	Heuristic algorithms	To reduce user discomfort, PAR, and electricity cost	Demand curve smoothing is not considered
Home appliance scheduling framework is proposed [22]	GWCSO	To reduce electricity cost and PAR	RESs are not considered
A DSM model is studied in [23]	GA	To reduce PAR, electricity cost, appliances waiting time and avoid rebound peaks	Sustainability of power grid is considered
An incentive-based optimal energy scheduling for HEM [24]	BPSO	To Minimize electricity cost and shift the load from high peak to low peak	RESs integration is not addressed
EMC is designed based on heuristic algorithms [25]	Four heuristic algorithms	To reduce electricity cost and PAR	The system is more complex
Ref. [26]	GA and loop search optimization algorithm	To minimize electricity cost	Limited number of appliances are considered
DR program is presented in [27]	Power scheduling strategy	To reduce electricity cost	PAR is not considered
DSM is presented in [28]	Novel real-time pricing algorithm	To minimize the electricity cost	The system is more complex
An intelligent energy management framework is presented in [29]	Aggregator	To reduce electricity cost by storing power during off-peak hours	The system is more complex

The above-discussed methods are valuable assets of literature, and all are capable of solving DSM problems. However, the mathematical techniques suffer from several drawbacks such as the incapability to cater to stochastic and nonlinear effects, the necessity of considering all the periods at once, shifting the load to unfeasible hours, and the risk of high-dimensionality of the problem. Furthermore, the mathematical models are much more complex and exhaust too much time, and returned solutions are not robust in a real-world context. The game theory-based methods suffer due to the techniques of solving games involving mixed strategies, particularly in a large pay-off matrix, being very complicated, and the competitive problems cannot be analyzed with the help of game theory. Furthermore, the assumption that players know their pay-offs and pay-offs of others is not practical. Similarly, heuristic algorithms suffer from the vast majority of cases being unable to deliver an optimal solution to the scheduling problem. For example, heuristic algorithms suffer from premature convergence that leads to losing population diversity and does not have standard parameters adjustment and termination criteria. Moreover, the above-discussed literature did not cater to electricity cost, carbon emission, user comfort, and PAR simultaneously. In this work, a novel HGACO algorithm-based energy management framework is proposed to simultaneously solve the DSM problems and to cater to all objectives. The novelty and major contribution of this work is as follows.

- An optimal energy management framework is developed utilizing the SG's two-way communication infrastructure under utility and RESs for solving the DSM problems.
- The developed energy management framework has an EMC based on our proposed HGACO algorithm that schedules smart home loads.
- The DR program's RTP scheme is mathematically modeled and implemented to actively engage consumers in DSM to facilitate both consumers and utility.
- A hybrid generation system comprises PV, ESS, and electric utility companies to resolve the energy management problems.

- In addition to electricity cost and PAR, carbon emission and user discomfort in terms of waiting time are mathematically modeled and catered simultaneously.
- To handle the power consumption of smart home appliances in a hybrid generation system, an objective function and constraints are developed that reduce the electricity cost, PAR, and carbon emissions and the maximize user comfort.
- To demonstrate the efficiency of the proposed HGACO-based optimal energy management framework through simulations by comparing it with existing energy management frameworks based on GA, BPSO, HGPSO, and Ant Colony Optimization (ACO).

The organization of this work is as follows: The system model is described in Section 2, and the simulation results, performance, evaluation, and discussions are illustrated in Section 3. At the end, this work is concluded in Section 4.

2. System Model

In this work, a system model is proposed for optimal energy management in SG to simultaneously cater electricity cost, carbon emission, user discomfort, and PAR. An abstract view diagram of the proposed model is shown in Figure 1. The proposed model comprises Advanced Metering Infrastructure (AMI), smart meters, ESS, PV, smart home appliances, power grid, and In-Home Display (IHD). The DSM framework utilizing two-way communication infrastructure for optimal energy management is shown in Figure 1.

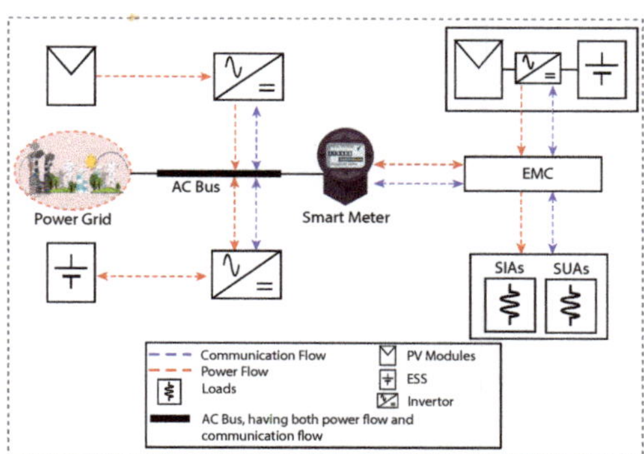

Figure 1. Proposed system model for optimal energy management in smart grid.

2.1. Advanced Metering Infrastructure/Two-Way Communication Infrastructure

AMI is the central processing unit of the proposed model, including Communication Module (CM) and Meter Data Management System (MDMS), which is shown in Figure 2. The AMI is used for automated bidirectional communication between the power supplier and smart meter. Moreover, the AMI is responsible for collecting and transmitting consumption data transmitted from distributed smart meters to power supplier company and for transmitting the DR signal in real-time to smart meters from the utility company. The energy consumption data is received by a concentrator and fed to the MDMS. The MDMS analyzes the received energy consumption data and extracts useful information from the data. The extracted favorable information is delivered to the power supply company. The detailed and useful information provided by AMI in real-time empowers the power supplier company to detect power outages, to measure electricity bill, to schedule maintenance, and to manage assets. The power supplier company provides price-based DR programs to AMI to encourage consumer participation in DSM via scheduling their energy usage pattern to reduce electricity cost, PAR, and carbon emission. The power supply company can also turn on and off household appliances to optimize energy consumption. Between the AMI and the EMC, a smart meter is mounted and used to read and to process the consumption data transmitted to the power grid. For further utilization, the

DR signal is sent to the EMC. The EMC in the DSM framework receives the DR signal, power from RESs and utility company, and consumer's preferences to schedule consumer energy usage as shown in Figure 1. The EMC transmits the generated energy usage schedule to consumers via a two-way communication infrastructure.

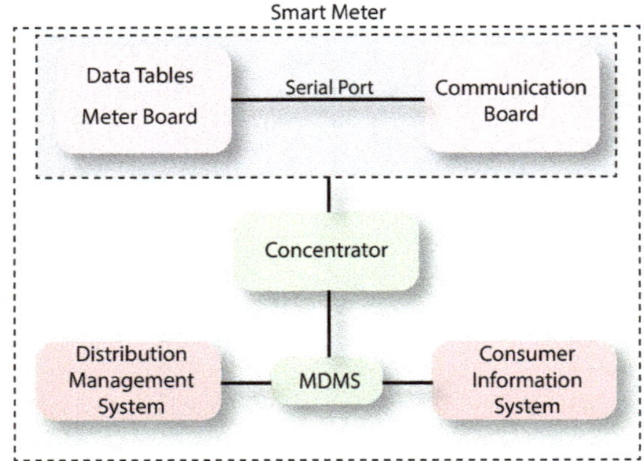

Figure 2. Advanced metering infrastructure.

2.2. Price-Based Demand Response Programs

Various types of price-based DR programs such as RTP, Critical Peak Pricing (CPP), TOUP, and Day-Ahead Pricing (DAP) exist in the literature [30]. Among these pricing signals, RTP provides better flexibility and encourages consumers to participate actively in DSM. Thus, we used the RTP signal in our work, where the day time horizon is divided into three price hours: on-peak price hours, mid-peak price hours, and off-peak price. The RTP signal is mathematically modeled as in Equation (1).

$$\phi(t) = \begin{cases} \phi_1, & if \quad t_7 \leq t \leq t_{10} \\ \phi_2, & if \quad t_{10} < t < t_{15}, \\ & \quad t_4 \leq t < t_7 \\ \phi_3, & if \quad t_1 \leq t < t_4, \\ & \quad t_{15} \leq t \leq t_4 \end{cases} \tag{1}$$

where ϕ_1 is on-peak hours, ϕ_2 is mid-peak hours, ϕ_3 is off-peak hours, and t_1 to t_{24} are 24 day time slots. The EMC receives this RTP signal and power signals of RESs and utility to create consumer's energy usage schedule.

2.3. Renewable Energy Sources

In nature, RESs include PV, wind, fuel cell, tidal, biogas, etc. However, among RESs, PV is abundant, free (small operation and maintenance cost), and at ease for all consumers. Therefore, in this work, PV is considered as a renewable source. The houses and power grids are equipped with PV panels. The purpose is to effectively utilize energy from the PV system to minimize electricity cost, PAR, and carbon emission. The following mathematical equation determines the output power form PV system (2) [31].

$$E^{pv}(t) = \partial^{pv} \times A^{pv} \times Irr(t) \times (1 - 0.005(Temp(t) - 25)) \tag{2}$$

where E^{pv} is the output energy produced in an hour, ∂^{pv} is the energy efficiency of PV panel, and A^{pv} is the area of PV panel. The outdoor temperature and solar irradiation are represented by $Temp(t)$ and $Irr(t)$, respectively, for the time interval t, and 0.005 is the temperature correction factor. The solar irradiation for an hour are modeled using the Weibull probability density function, which is illustrated in Equation (3).

$$F\,(Irr(t)) = \omega \left(\frac{\psi 1}{\lambda 1}\right) \left(\frac{Irr(t)}{\lambda 1}\right)^{(\psi 1-1)e^{-\left(\frac{Irr}{\lambda 1}\right)^{\psi 1}}} + (1-\omega) \left(\frac{\psi 2}{\lambda 2}\right) \left(\frac{Irr(t)}{\lambda 2}\right)^{(\psi 2-1)e^{-\left(\frac{Irr}{\lambda 2}\right)^{\psi 2}}} \tag{3}$$

where $0 < Irr(t) < \infty$, ω is a weighted factor, $\psi 1$ and $\psi 2$ are the shape factors, and $\lambda 1$ and $\lambda 2$ are the scale factors. The EMC utilized the energy generated from a PV system for scheduling during on-peak hours and for storing it in ESS during off-peak hours or when the energy is surplus.

2.4. Energy Storage System

ESS is touted as a "Holy Grail" in alleviating greenhouse gas emission due to high penetration in electric vehicles, in plug-in electric vehicles, as backup generators during grid outages, and as storage with RESs. The purpose of ESS is to convert the power grid into a carbon-free system. The ESS when used with RESs such as PV panels, wind turbine, etc. takes the generated clean energy during daytime (sky is clear, shinny, and sunny) and stores it. The stored energy is sent back to the power grid on cloudy days, at night, or whenever required. Moreover, the ESS smoothes out the fluctuating nature of RESs up to some extent. Thus, the ESS significantly lowers greenhouse gas emissions and potentially reduces the electricity bills of prosumers exchanging energy with the power grid when demand is at peak and energy prices are highest [32]. In this work, the ESS stores energy during off-peak hours, when power is in surplus, or when the battery is empty (storage level is lower than the lower charging cutoff). The ESS is mainly to efficiently utilize the output of the PV system. During charging and discharging of ESS, some energy is lost; thus, the turn around in efficiency of ESS is mathematically modeled in Equation (4).

$$PS\,(t) = PS\,(t-1) + \eta \cdot \mu^{ESS} \cdot EE^{Ch}\,(t) - \frac{\eta \cdot EE^{Dch}\,(t)}{\mu^{ESS}} \forall t \tag{4}$$

where PS is stored energy (kWh) at time t, μ^{ESS} is ESS efficiency, η is time duration in our hours, EE^{Ch} is power supplied (kW) from PV to ESS, and EE^{Dch} is the power supplied (kW) from ESS to the load. Avoiding deep discharging/overcharging the following limits are set.

$$EE^{Ch}\,(t) \leq EE^{Ch}_{UB} \tag{5}$$

$$EE(t)^{Dch} \leq EE^{Dch}_{LB} \tag{6}$$

$$PS\,(t) \leq PS^{Ch}_{UB} \tag{7}$$

The EMC receives the RTP signal, the power signal of RESs and utility, and consumer's priorities of appliances operation to create an energy usage schedule.

2.5. Users' Smart Appliances

In the DSM framework, users' smart appliances, instead of interacting with each other, directly interact with EMC. The EMC would schedule the smart appliances' operation using the RTP signal, the power signal of RESs and utility, and consumers' priorities. In this work, home appliances are categorized into three types: Shiftable Interruptible Appliances (SIAs), Shiftable Uninterruptible Appliances (SUAs), and Regular Appliances (RAs). The RAs are not considered for the proposed model because RAs operated for 24 h consistently. Thus, the delayed or advanced operation for such appliances is not possible. That is why RAs do not participate in scheduling. Only the SIAs and SUAs interact with the proposed model's EMC and participate in scheduling for DSM to reduce electricity cost, PAR, and carbon emission. SIAs include humidifiers, water heaters, and dishwashers, and SUAs include clothes dryers, EVs, and washing machines. The parameter details of these appliances are listed in Table 2. This classification and parameters are adopted from [33]. The total energy consumption per day for SIAs can be determined using Equation (8) adopted from [34].

$$E^{shi} = \sum \delta_{shi} \in D_{shi} \left(\sum_{t=1}^{24} \Omega_{shi} \times \delta_{shi}(t) \right) \tag{8}$$

where E^{shi} is total energy consumption, A_s combination of SIAs, $D_{shi} \in A_s$ represents all the appliances from SIAs, Ω_{shi} represents the power consumption of each appliance, δ_{shi} shows the on/off status of appliances, and β represents the unit price.

The total cost per day of all the SIAs in time interval T can be calculated using Equation (9).

$$\alpha_a \xi_{shi} = \sum \delta_{shi} \in D_{shi} \left(\sum_{t=1}^{24} \Omega_{shi} \times \beta(t) \times \delta_{shi}(t) \right) \tag{9}$$

The total energy consumption per day for SUAs can be calculated using Equation (10).

$$E^{shu} = \sum \delta_{shu} \in D_{shu} \left(\sum_{t=1}^{24} \Omega_{shu} \times \delta_{shu}(t) \right) \tag{10}$$

where E^{shu} is total energy consumption, A_s is the combination of SUAs, $D_{shu} \in A_s$ represents all the appliances from SUAs, Ω_{shu} defines the power consumption of each appliance, δ_{shu} shows the on/off status of the appliances, and β represents the unit price. Total cost per day of all the SUAs for time interval T is obtained using Equation (11).

$$\alpha_a \xi_{shu} = \sum \delta_{shu} \in D_{shu} \left(\sum_{t=1}^{24} \Omega_{shu} \times \beta(t) \times \delta_{shu}(t) \right) \tag{11}$$

In this work, the EMC is programmed based on our proposed HGACO algorithm to generate an optimal power usage schedule using the received the RTP signal, the power signal of RESs and utility, consumer's priority, and appliance power ratings. The developed power usage schedule is shared with smart appliances using a two-way communication infrastructure.

Table 2. Description of appliances.

Category	Appliances	Power Rating (kW)	Daily Use (Hours)
SIAs	Humidifier	0.07	11
	Water heater	0.15	9
	Dish washer	0.0372	12
SUAs	Washing Machine	0.2377	3
	Cloth Dryer	0.1	8
	Electric Vehicle	0.206	7

2.6. Proposed HGACO Algorithm

A novel algorithm is proposed, namely HGACO algorithm. The EMC controller is programmed using our proposed HGACO algorithm to perform DSM via optimal power usage scheduling, resulting in reduced electricity cost, carbon emission, PAR, and user discomfort. The HGACO algorithm control parameters are listed in Table 3. The detailed description of HGACO algorithm is as follows. HGACO algorithm is developed by hybridization of the ACO algorithm and GA. The motivation for this hybridization is that the ACO algorithm is capable of electricity cost and carbon emission reduction, while GA is effective in PAR reduction. Thus, we developed HGACO algorithm by applying a crossover and mutation operation of GA on the optimal global results obtained from the ACO algorithm to simultaneously reduce electricity cost, alleviate PAR, and minimize carbon emission. ACO is based on the following three steps: (i) problem definition, (ii) parameters initialization, and (iii) position update. In problem definition, the proposed model minimizes electricity cost, carbon emission, and PAR subjected to total energy consumption being less than the grid capacity. Different parameters such as swarm size represent the number of ants in our proposed model that are appliances, then initialize evaporation rate ∂ = 0.5, and then update the ACO position. Initially, each ant can choose any path, so we have six ants (6 appliances) and 24 paths (24 h). Each ant has a 1/24 probability to choose a path out of 24 paths. According to the ACO algorithm, one path is the minimum distance compared to other paths out of 24 paths. It is assumed that all paths have equal pheromones because we do not have previous knowledge of the solution paths at the start. During operation, knowledge is gained about the fitness of each path. Therefore, the pheromone on each path is updated first. The pheromone update formula confirms that the pheromone associated with the best solution increases while it decreases for other solutions, in other words, evaporates. The pheromone updates the formula as follows:

$$\chi_{ab}^{new} = (1 - \partial)\, \chi_{ab}^{old} + \sum\nolimits_k \Delta\chi_{ab}^k \quad (12)$$

where Equation (12), $\partial \in (0, 1)$ is a user-defined parameter known as evaporation rate, and the value of ∂ is 0.5. χ_{ab}^{old} is the pheromone amount in the beginning when the a_{th} variable attains b_{th} value. $\Delta\chi_{ab}^k$ is the pheromone amount laid by the k_{th} ant and is given by the following formula.

$$\Delta\chi_{ab}^k = Q\frac{f_{best}}{f_{worst}} \quad (13)$$

In Equation (13), Q is a constant. Usually $Q = 2$, and it is clear that, in initial iterations, when the difference between fbest and fworst is large, the ratio of fbest/fworst is small and Q is fbest/fworst. As the iterations progress, the difference between fbest and fworst will be small, the ratio fbest/fworst will tend to 1, and Q fbest/fworst will tend to $Q = 2$. This pheromone deposition property confirms that less pheromone is deposited in early iterations to avoid the suboptimal solution's stagnation. In Equation (12), only the best ants can deposit the pheromone. If there is more than one best ant at any iteration, then the summation extends the best ants. In this iteration, we have only one best ant, and therefore, the summation has only one term and the evaporation rate ∂ = 0 for the best ant. The pheromone update formula for best ant is as follows:

$$\chi_{ab}^{new} = \chi_{ab}^{old} + \sum\nolimits_k \Delta\chi_{ab}^k \quad (14)$$

For other ants, the updated formula is as follows:

$$\chi_{ab}^{new} = (1 - \partial)\, \chi_{ab}^{old} \quad (15)$$

When the best path is selected, then the crossover operator is applied to further improve the global best derived from ACO.

Table 3. Parameters of the Hybrid Genetic Ant Colony Optimization (HGACO) algorithm.

Parameters	Values
antsh	6
evaph	0.5
iteration	200
Qh	2
Nantsh	10
insiteh	2
sumphh	1

In the end, the global best results from the HGACO algorithm is passed through MKP, which further improves our desired results by checking the grid capacity if the scheduled load is greater than the grid capacity. Thus, the proposed algorithm will switch off the appliances that were operated more than the rest of the appliances and when the scheduled load is less than the grid capacity. Then, the proposed algorithm will switch on some appliances to achieve the optimal solution. The optimal global best schedule result is achieved after MKP. Then the appliances will be scheduled as follows: schedule the appliances without photovoltaic-battery systems, with PV, and with photovoltaic-battery systems, where the electricity cost is computed in Equations (9) and (11), where carbon emission is from Equation (17), and where PAR is from Equation (18). Then, check the condition: if the value for hour is less than 24, the program will run again for another hour to schedule all the smart appliances; if not, the program is terminated. The overall implementation flowchart of the proposed model is depicted in Figure 3.

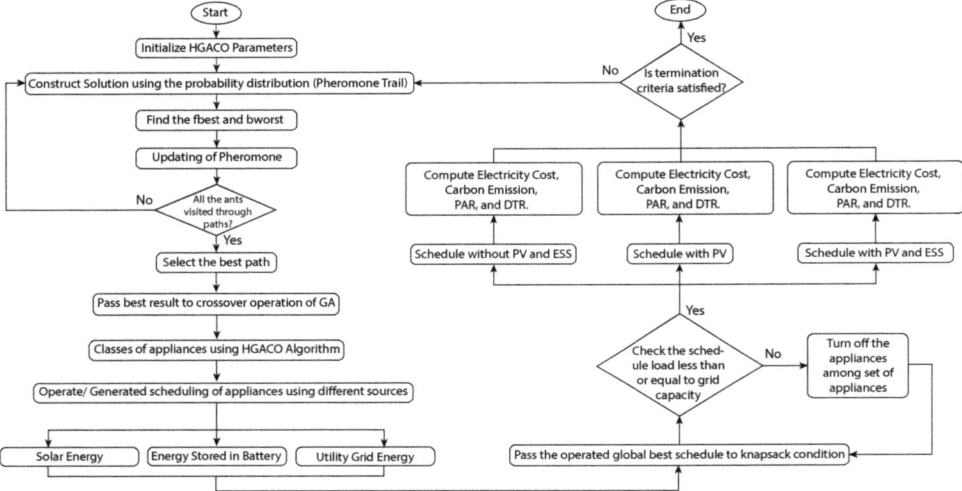

Figure 3. Complete implementation flowchart of the proposed system model.

2.7. Problem Formulation

The proposed model's main objectives are to minimize electricity cost by scheduling power consumption patterns consumers, to alleviate carbon emission, to maximize user comfort, to mitigate PAR, and to cope with the gap between demand and supply. First, every objective is elaborated and formulated individually. Then, the overall DSM problem is formulated. Electricity cost is the bill that a utility company issues to a consumer for the consumed electricity during the specified time interval. The proposed model includes two types of appliances: SIAs and SUAs. The cost paid by consumers for using SIAs is $\alpha_a\xi_{shi}$ and is formulated using Equation (9); the cost paid by consumers for using SUAs is $\alpha_a\xi_{shu}$ and is formulated as in Equation (10); and the total cost is mathematically modeled as in Equation (16).

$$\Pi = \alpha_a\xi_{shi} + \alpha_a\xi_{shu} \tag{16}$$

where Π represents total electricity cost paid by the consumers for operating both SIAs and SUAs.

Carbon emission is the release of carbon into the atmosphere while operating both SIAs and SUAs. The mathematical formula for carbon emission as follows.

$$\Upsilon = \frac{avgEP}{\varepsilon \times \varsigma \times \Im} \tag{17}$$

Equation (17) shows carbon emission in pounds, where $avgEP$ represents the average amount of electricity price, ε shows the price per kWh, ς shows electricity emission factor, and $\Im$ represent hour in a day.

PAR is the ratio of peak power consumption to average power consumption for specific time slots. The reduction of PAR is beneficial for consumers and utility because it helps to minimize the gape between demand and supply. It is represented by PAR and mathematical formulated as follows:

$$PAR = \frac{\max_{t \in T(\Phi_t)}}{\frac{1}{T}\sum_{t=1}^{T} \Phi_t} \tag{18}$$

where the power used by appliance is Φ_t in time t.

User comfort is measured in terms of various aspects such as waiting time, energy consumption, temperature, air quality, illumination, humidity, and demographic profile of the smart home users [35]. This study measures user comfort in aspects of Delay Time Rate (DTR). The DTR is the waiting time that an appliance faces before starting operation. The consumers face delay because the EMC shifted the consumers' load from on-peak hours to off-peak hours by giving an incentive to them. The consumers' power usage pattern is different with and without scheduling because consumers' activities are shifted from high-price hours to low-price hours subjected to rebound peak avoidance. A tradeoff exists between electricity cost and DTR: those consumers who tolerate high DTR will pay less utility bill, and those who cannot accept DTR will pay a high utility bill. The user comfort in terms of DTR is formulated as follows in Equation (19).

$$w_b = \frac{\sum_{t=1}^{T}\sum_{b=1}^{n} \left| \left(T_{b,t}^{o,unsch} - T_{b,t}^{o,sch} \right) \right|}{T_b^{lo}} \tag{19}$$

The term w_b coveys a DTR that each appliance faces due to delay or advances in operation, $T_{b,t}^{o,unsch}$ represents the appliance status without scheduling, $T_{b,t}^{o,sch}$ depicts appliance status with scheduling, and T_b^{lo} represents length of total operation timeslots. The heuristic-based EMC schedules the power usage pattern of consumers in response to the RTP signal and consumers priority. The maximum delay that an appliance can tolerate is determined by Equation (20).

$$w_b^d = T_b^t - T_b^{lo} \tag{20}$$

where w_b^d represents the maximum delay that an appliance may face while shifting operation from on-peak hours to off-peak hours and T_b^t conveys the appliances total time interval. The user comfort is negatively related to the maximum delay, i.e., with the increase in w_b^d, the user comfort is compromised. The percentage discomfort can be computed using the Equation (21).

$$D = \frac{w_b}{w_b^d} \times 100 \tag{21}$$

The overall objective function DSM problem is to alleviate electricity cost, to minimize carbon emission, to reduce user frustration, and to mitigate PAR by scheduling consumers energy consumption. The objective function is formulated as a minimization problem as follows.

$$Min\,(\Pi + \Upsilon) \tag{22}$$

The minimization problem in Equation (22) has the following constraints:

$$PAR = \frac{\max_{t \in T(\Phi_t)}}{\frac{1}{T}\sum_{t=1}^{T}\Phi_t} \le P_c \tag{23}$$

$$T_{\min \le t \le T_{\max}} \tag{24}$$

$$\sum_{t=1}^{T}\Lambda_t^{unsh} = \sum_{t=1}^{T}\Lambda_t^{sh} \tag{25}$$

$$Z_t^{unsch} <= Z_t^{sch} \tag{26}$$

$$I_t^{unsch} = I_t^{sch} \tag{27}$$

Constraint (23) shows that the total PAR must be less than or equal to grid capacity P_c. The grid capacity is the amount of power available from the power grid. This constraint helps to avoid power shortage or the blackout problem. Constraint (24) shows the scheduling interval. Constraint (25) shows the power consumption constraint in which the total power consumption before and after scheduling remains the same. Constraint (26) shows that appliance status before and after scheduling is not the same. Similarly, constraint (27) shows that the operation time of appliances before and after scheduling are the same, where I is the operation time of appliances and there are t_1 to t_{24} time slots in a day.

3. Simulation Results, Performance Evaluation, and Discussions

The proposed model's simulations based on the HGACO algorithm are conducted in MATLAB R2013b compared to the benchmark algorithms GA, PSO, ACO, and HGPO. These algorithms are chosen as benchmark due to architectural similarities to the proposed algorithm. The control parameters for both proposed and benchmark algorithms are chosen subjected to pair assessment. The proposed model based on the HGACO algorithm and knapsack problem formulation to solve DSM problems result in electricity cost, carbon emission, PAR, and user discomfort reduction to enhance the reliability and sustainability of the power gird. Additionally, the proposed algorithm is capable of providing power to the utility giving peak hours from RESs. Two algorithms were taken: ACO effectively reduces carbon emission and electricity cost, whereas GA reduces PAR and MKP ensures the grid's reliability. The proposed HGACO algorithm and benchmark algorithms including RESs and ESS were compiled and run by considering three scenarios to evaluate the performance of said algorithms, where the first scenario is comprised without photovoltaic-battery systems, the second scenario only consists of PV, and the third scenario is comprised of photovoltaic-battery systems. Consider three power supply sources that are utility grid power supply, which is available 24 h a day; RES; and ESS to implement the proposed HGACO where the grid signals that are RTP, solar irradiance, and forecasted temperature is used in the proposed HGACO model, which are illustrated in Figures 4–6. The PV is considered a renewable source, and the power generated by PV system is dependent on solar irradiance and ambient temperature. The charging level of ESS and the estimated renewable energy is shown in Figures 7 and 8. The proposed and benchmark algorithm-based EMCs actively engage consumers in DR programs to schedule consumers' power usage using pricing signal (RTP), power signal from RESs and utility, and consumers preferences. The generated schedule without participating in the DR program and the proposed algorithm-based EMC that actively engages consumers in DR programs are shown in Tables 4 and 5, respectively.

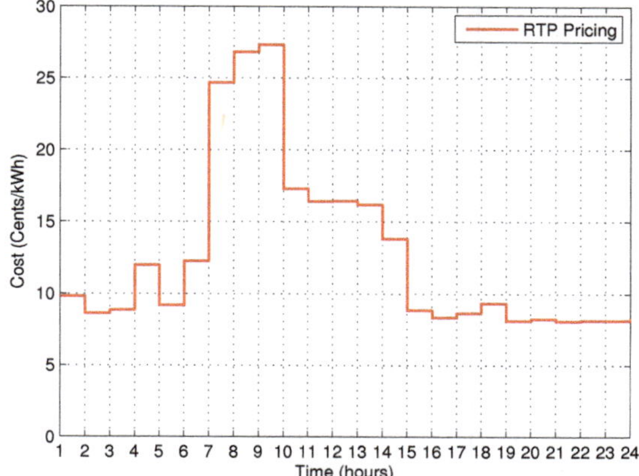

Figure 4. Real-Time Pricing (RTP) price-based Demand Response (DR) program.

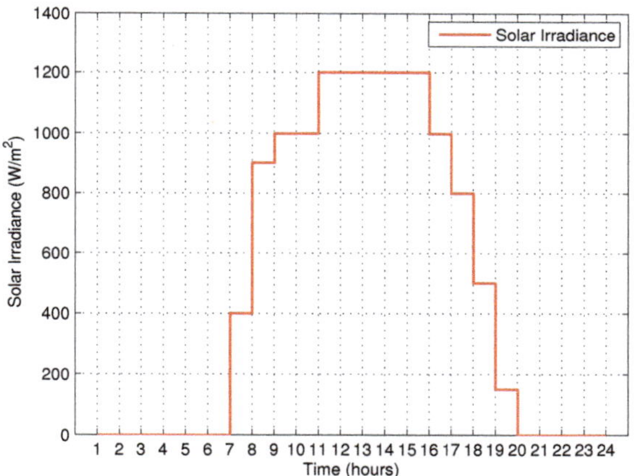

Figure 5. Solar irradiance.

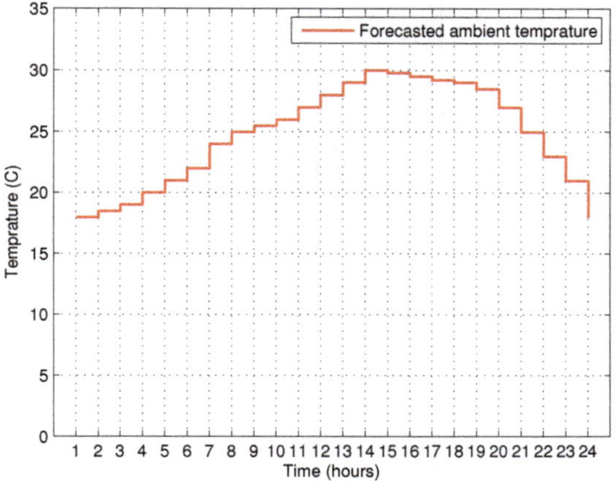

Figure 6. Day-ahead temperature profile.

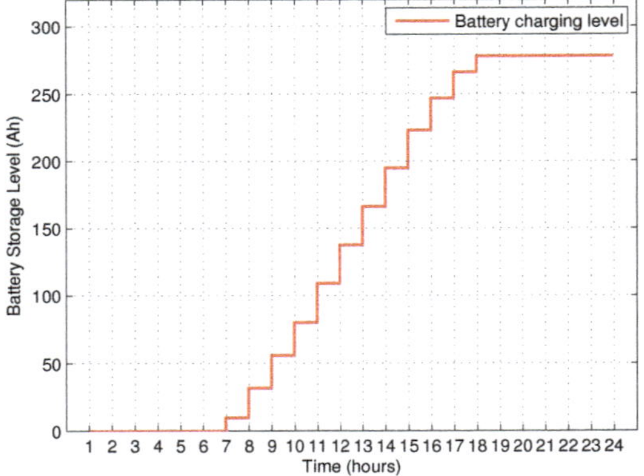

Figure 7. Charging behavior of Energy Storage System (ESS).

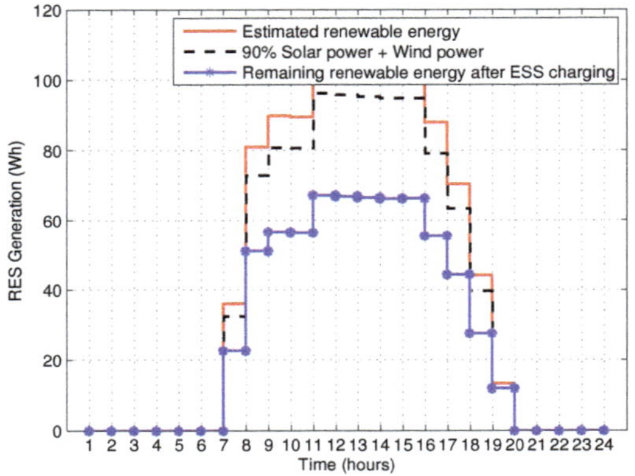

Figure 8. Estimated renewable energy.

Table 4. Consumer schedules without participating in a DR program.

Category	Appliances	1	2	3	4	5	6	7	8	9	10	11	12	13	14	15	16	17	18	19	20	21	22	23	24
SIAs	Humidifier	0	0	0	1	0	1	1	1	1	1	1	1	1	1	1	0	0	0	0	0	0	0	0	0
	WHeater	1	0	0	1	0	0	1	1	0	1	1	1	1	1	0	0	0	0	0	0	0	0	0	0
	DWasher	0	0	0	0	0	1	1	0	1	1	1	1	1	1	1	1	1	0	0	1	0	0	0	0
SUAs	WMachine	0	0	0	0	0	0	0	0	0	1	0	0	1	1	0	0	0	0	0	0	0	0	0	0
	CDryer	0	0	0	0	0	0	0	0	1	1	0	1	1	1	1	0	0	0	1	1	0	0	0	0
	EV	0	0	0	0	0	0	0	1	1	1	0	1	1	1	1	0	0	0	0	0	0	0	0	0

Table 5. Consumer schedules when participating in a DR program.

Category	Appliances	1	2	3	4	5	6	7	8	9	10	11	12	13	14	15	16	17	18	19	20	21	22	23	24
SIAs	Humidifier	0	1	1	0	1	0	0	0	0	0	0	0	0	0	1	1	0	1	1	1	0	1	1	1
	WHeater	0	1	1	0	0	0	0	0	0	0	0	0	0	0	0	1	1	0	1	1	0	1	1	1
	DWasher	1	0	1	1	1	1	0	0	0	0	0	0	0	0	1	0	1	1	0	1	1	1	1	0
SUAs	WMachine	0	0	0	0	0	0	0	0	0	0	0	0	0	0	0	0	0	0	1	0	1	0	1	0
	CDryer	0	1	0	0	0	0	0	0	0	0	0	0	0	0	0	1	0	0	1	1	1	1	1	1
	EV	0	1	0	0	0	0	0	0	0	0	0	0	0	0	0	1	1	0	0	1	1	1	0	1

From the above results listed in Tables 4 and 5, it is evident that the schedule generated by the proposed HGACO algorithm is optimal. This optimal schedule is because the proposed algorithm shifted the load to off-peak hours and mid-peak hours operating in on-peaks subjected to avoiding rebound peaks. The detailed discussion and evaluation of each scenario are presented in the subsequent sections.

3.1. Scenario I

In scenario I, the consumer's power usage schedule of the proposed algorithm and benchmark algorithms without photovoltaic-battery systems is discussed. In scenario I, the evaluation of each objective's achievement, such as electricity cost, PAR, carbon emission, and user comfort, is discussed as follows. The electricity cost of scheduled and unscheduled loads without photovoltaic-battery systems are illustrated in Figure 9 and Table 6. The maximum electricity cost of GA is 63.50 cents in time slot 7, that of PSO is 51.12 cents in time slot 4, that of ACO is 57.92 cents in time slot 16, that of HGPO is 63.76 cents in time slot 16, and that for the proposed HGACO algorithm is 48.36 cents in time slot 23. In 24 h, the unscheduled load's electricity cost is 921.19 cents compared to GA, PSO, ACO, HGPO, and HGACO, 660.86, 571.68, 512.69, 548.90, and 465.03 cents, respectively. Similarly, the total electricity cost evaluation of the proposed and existing algorithms is graphically shown in Figure 10. Thus, the proposed HGACO algorithm has a minimum electricity cost either per time slot or aggregated compared to the existing algorithms.

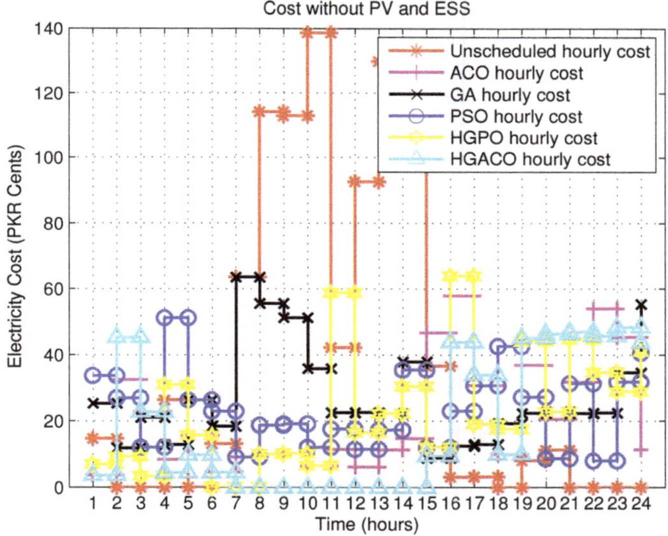

Figure 9. Electricity cost without photovoltaic-battery systems.

Table 6. Scenario I: electricity cost assessment without photovoltaic-battery systems.

Algorithms	Cost (Cents)	Difference	Reduction (%)
Unscheduled	921.1647	-	-
GA	660.8627	260.3377	28.26
ACO	512.6907	408.5057	44.34
PSO	571.6865	349.5099	37.94
HGPO	548.9073	372.2891	40.41
HGACO	465.0335	456.1628	49.51

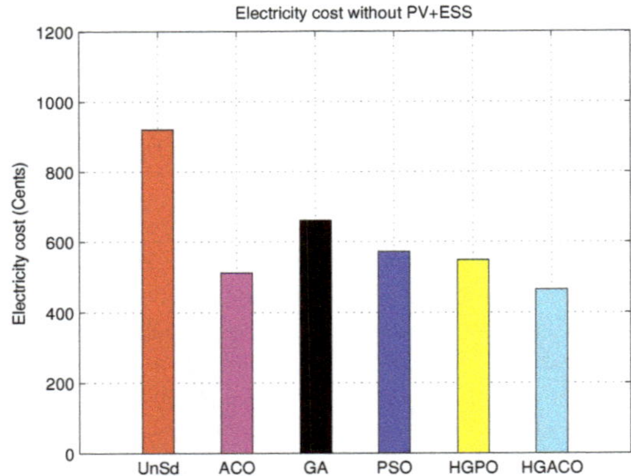

Figure 10. Aggregated electricity cost without photovoltaic-battery systems.

The assessment of PAR with and without scheduling is shown in Figure 11 and Table 7. The existing algorithms GA, PSO, ACO, and HGPO and the proposed algorithm HGACO minimized the PAR by 17.13, 13.38, 39.10, 4.64, and 25.72, respectively. The HGACO algorithm uniformly distributed the load in off-peak and mid-peak hours and achieved the desired objectives. In contrast, the benchmark algorithms generated reboud peaks while creating power usage schedules, which is dangerous for the reliability of the power grid. Thus, the proposed HGACO algorithm alleviated the PAR significantly compared to the existing algorithms.

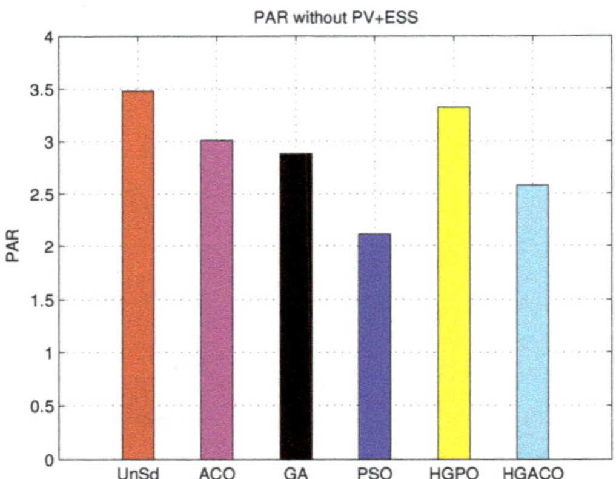

Figure 11. Peak to Average Ratio (PAR0 without photovoltaic-battery systems.

Table 7. Scenario I: PAR assessment without photovoltaic-battery systems.

Algorithms	PAR	Difference	Reduction (%)
Unscheduled	3.48	-	-
GA	2.88	0.59	17.13
ACO	3.01	0.46	13.38
PSO	2.11	1.36	39.10
HGPO	3.319	0.16	4.64
HGACO	2.58	0.89	25.72

In scenario I, the unscheduled and scheduled load's carbon emission is shown in Figure 12 and Table 8. The existing algorithms and the proposed algorithm emit less carbon compared to the without scheduling case. However, the proposed algorithm emits less carbon compared to all benchmark algorithms. The unscheduled load produces the maximum carbon emission at time slot 2, which is 6.49 pounds. In contrast, the existing algorithms GA, PSO, ACO, and HGPO emit maximum carbon at 4.65 pounds, 3.61 pounds, 4.02 pounds, and 3.86 pounds in time slot 21, respectively. Thus, all of the benchmark algorithms outperformed the without scheduling case in terms of carbon emission. However, the emission of carbon at time slot 21 of the proposed algorithm is 3.27, which is the lowest per time slot carbon emission compared to the existing algorithms. Similarly, the total carbon emission is discussed as follows: The without power usage scheduling case emits a total carbon emission of 116.78 pounds in scenario I. On the other hand, the benchmark algorithms GA, ACO, PSO, and HGPO and proposed algorithm HGACO emit total carbon at 83.77, 64.99, 72.47, 69.58, and 58.95 pounds, respectively. As compared to unscheduled carbon emission, GA reduced carbon emissions by 28.26 percent, ACO reduced carbon emissions by 44.34 percent, PSO reduced carbon emissions by 37.94 percent, HGPSO reduced carbon emissions by 40.41 percent, and HGACO reduced carbon emissions by 48.01 percent. Thus, the proposed algorithm is effective in carbon emission reduction either per time slot or in total.

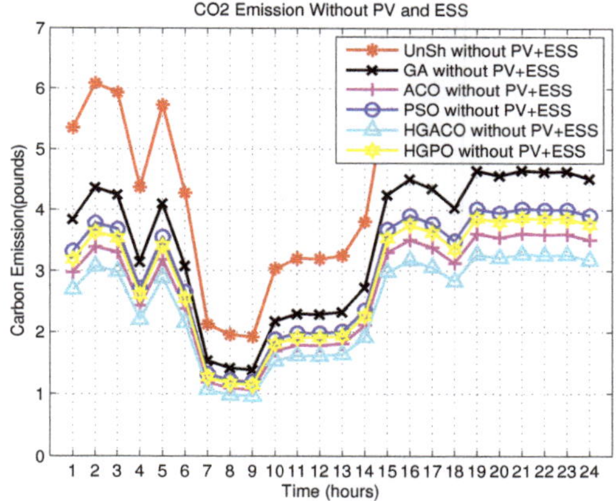

Figure 12. Scenario I: Carbon emission assessment without photovoltaic-battery systems.

Table 8. Scenario I: Carbon emission assessment without photovoltaic-battery systems.

Algorithms	Carbon Emission (Pounds)	Difference	Reduction (%)
Unscheduled	116.78	-	-
GA	83.77	33.00	28.26
ACO	64.99	51.78	44.34
PSO	72.47	44.30	37.94
HGPO	69.58	47.19	40.41
HGACO	58.95	57.82	48.01

3.2. Scenario II

In scenario II, the proposed model is compared with other benchmark algorithms via scheduling the home appliances with PV and achieving the best results, discussed as follows.

The electricity cost of scheduled and unscheduled load with PV is illustrated in Figure 13, and their numerical results are listed in Table 9. The maximum electricity cost of GA is 57.87 cents in time slot 7, that of PSO is 51.12 cents in time slot 4, that of ACO is 54.02 cents in time slot 22, that of HGPO is 59.14 cents in time slot 16, and that of HGACO algorithm is 48.36 cents in time slot 23. In 24 h, the unscheduled load's electricity cost is 816.68 cents compared to GA, PSO, ACO, HGPO, and HGACO algorithm, 556.35, 467.17, 408.18, 444.39, and 360.52 cents, respectively. Similarly, the net electricity results for the proposed and existing algorithms compared to the without scheduling case is depicted in Figure 14. Thus, from the graphical and numerical results, it is evident that the proposed algorithm outperforms the existing algorithms in terms of per time slot and aggregated electricity cost.

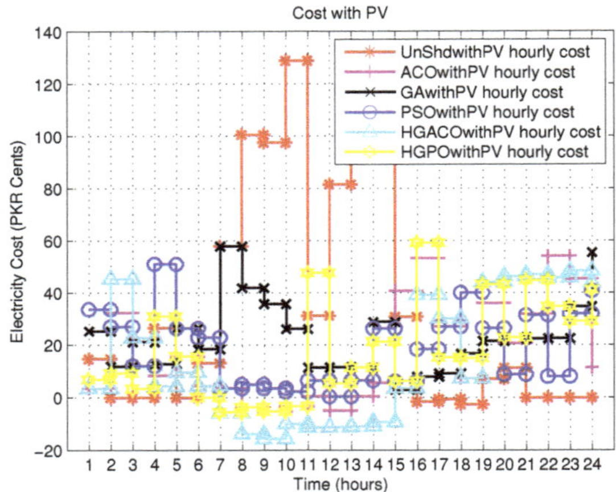

Figure 13. Electricity cost with photovoltaic systems.

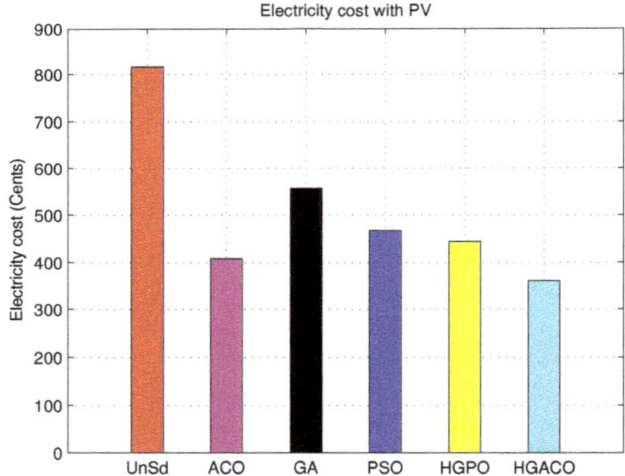

Figure 14. Aggregated electricity cost with PV.

Table 9. Scenario II: electricity cost assessment with PV.

Algorithms	Cost (Cents)	Difference	Reduction (%)
Unscheduled	816.6876	-	-
GA	556.3538	260.3337	31.8767
ACO	408.1818	408.5057	50.0198
PSO	467.1777	349.5098	42.7960
HGPO	444.3984	372.2891	45.5852
HGACO	360.5247	456.1628	55.8552

The graphical and numerical results of PAR of scheduled and unscheduled load with PV are shown in Figure 15 and Table 10, respectively. The proposed HGACO algorithm and existing algorithms GA, PSO, ACO, HGPO, and HGACO reduced PAR by 21.69, 17.56, 34.49, 8.42, 13.29, respectively. The proposed algorithm uniformly distributes the load in off-peak hours and checks the grid capacity with the help of knapsack problem formulation to balance the load and to avoid rebound peaks. In contrast, the benchmark algorithms shift the load uniformly, resulting in rebound peaks that disturb the reliability of the power grid. From the results and discussion, it is evident that the proposed algorithm effectively shifts the load from peak hours to off-peak hours and reduced PAR, which is beneficial for both consumers and utility. Thus, the performance of the proposed algorithm is outstanding while scheduling with PV.

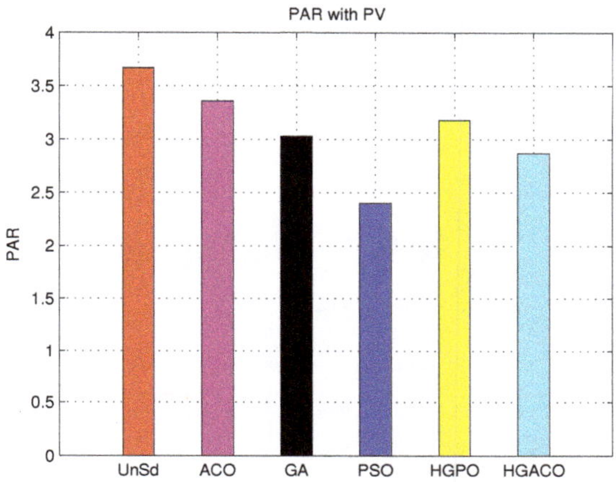

Figure 15. PAR assessment with PV.

Table 10. Scenario II: PAR assessment with PV.

Algorithms	PAR	Difference	Reduction (%)
Unscheduled	3.67	-	-
GA	3.03	0.64	17.57
ACO	3.36	0.30	8.42
PSO	2.40	1.26	34.49
HGPO	3.18	0.48	13.29
HGACO	2.87	0.80	21.69

The performance evaluation of the proposed algorithm compared with existing algorithms in terms of carbon emission with and without scheduling using PV is depicted in Figure 16 and Table 11. From the results, it is evident that scheduling carbon emission to the environment is more than scheduling based on proposed and benchmark algorithm. The proposed HGACO algorithm emit less carbon to the atmosphere compared to all benchmark algorithms. The maximum carbon emitted by the unscheduled load with PV is 5.75 pounds in time slot 21. In contrast, the existing algorithms GA, ACO, PSO, and HGPO emit maximum carbon emissions of 3.92 pounds at time slot 21, 2.87 pounds at time slot 21, 3.29 pounds at time slot 21, and 3.13 pounds at time slot 21. The maximum carbon emission at time slot 21 of the proposed HGACO algorithm is 2.54 pounds, which is the lowest of the existing algorithms. The unscheduled load with PV emits 103.53 pounds of carbon while the existing and proposed algorithms GA, ACO, PSO, HGPO, and HGACO algorithm emits 70.52, 51.74, 59.22, 56.33, and 45.70 pounds of carbon. GA reduced carbon emissions by 31.87% compared to the unscheduled load, ACO reduced carbon emissions by 50.01%, PSO reduced carbon emissions by 42.79%, and HGPO reduced carbon emissions by 45.58%. On the other hand, the proposed HGACO algorithm reduced carbon emissions by 54.22% which is the highest reduction compared to the without and with scheduling based on existing algorithms. Thus, the proposed algorithm outperforms the existing algorithms in terms of carbon emission reduction.

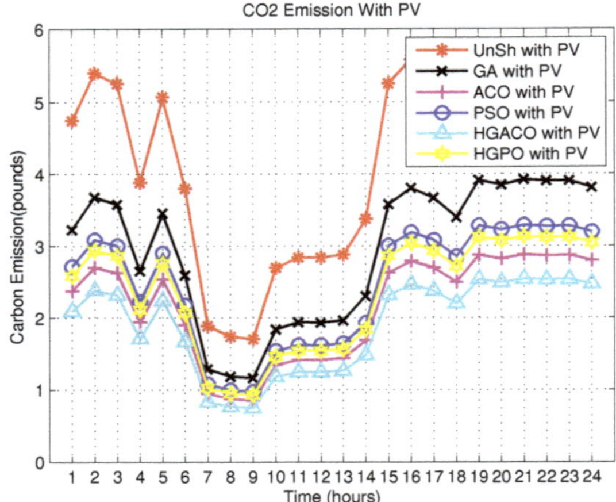

Figure 16. Carbon emission with PV.

Table 11. Scenario II: carbon emission assessment with PV.

Algorithms	Carbon Emission (Pounds)	Difference	Reduction (%)
Unscheduled	103.53	-	-
GA	70.52	33.00	31.87
ACO	51.74	51.78	50.01
PSO	59.22	44.30	42.79
HGPO	56.33	47.19	45.58
HGACO	45.70	57.82	54.22

3.3. Scenario III

In scenario III, the proposed algorithm and existing algorithms with photovoltaic-battery systems-based scheduling are performed to minimize electricity cost, to alleviate PAR, to reduce carbon emission, and to mitigate user discomfort. A detail discussion follows.

The electricity cost assessment using photovoltaic-battery systems with and without scheduling-based proposed algorithm and existing algorithms are depicted in Figure 17 and Table 12. The maximum electricity cost of GA is 57.87 cents in time slot 7, that of PSO is 52.12 cents in time slot 4, that of ACO is 54.02 cents in time slot 22, that of HGPO is 57.15 cents in time slot 16, and that of the proposed HGACO algorithm is 48.36 cents in time slot 23. The unscheduled load's electricity cost is 772.31 cents compared to GA, PSO, ACO, HGPO, and the proposed HGACO algorithm, 510.59, 422.8, 363.80, 400.02, and 316.15 cents, respectively, for the 24 h time horizon. Similarly, the results of the aggregated electricity cost is shown in Figure 18. The graphical and numerical results of the electricity cost validate that the electricity minimization of HGACO algorithm is significant compared to all other benchmark algorithms and without scheduling cases. Thus, the proposed HGACO algorithm is useful in terms of per time slot and aggregated cost reduction compared to the existing algorithms.

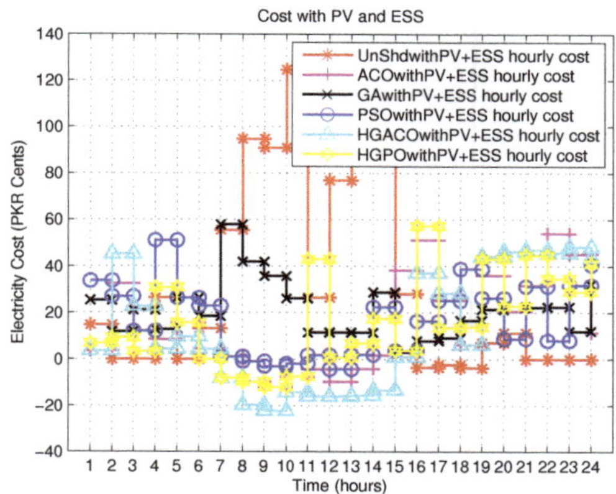

Figure 17. Scenario III: electricity cost assessment with photovoltaic-battery systems.

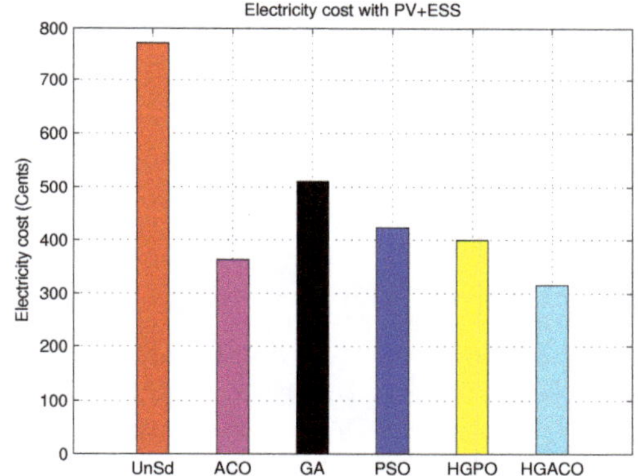

Figure 18. Aggregated electricity cost with photovoltaic-battery systems.

Table 12. Scenario III: numerical results of electricity cost with photovoltaic-battery systems.

Algorithms	Cost (Cents)	Difference	Reduction (%)
Unscheduled	772.313	-	-
GA	510.5990	261.713	33.8870
ACO	363.8072	408.5057	52.8938
PSO	422.8031	349.5098	45.2549
HGPO	400.0238	372.2891	48.2044
HGACO	316.1501	456.1628	59.0645

The PAR evaluation graphical and numerical results with and without scheduling, considering photovoltaic-battery systems is shown in Figure 19 and Table 13. The proposed algorithm HGACO algorithm minimize the PAR while adapting the behaviour of ACO and GA algorithms. The heuristic algorithms GA, PSO, ACO, and HGPO and the proposed HGACO algorithm minimize the PAR by 40.85, 32.29, 7.85, 4.96, and 17.40, respectively. The HGACO algorithm uniformly distributed the load in off-peak hours and achieved the desired objectives. Some of the benchmark algorithms create rebound peaks that disturb the reliability of the grid. From the results and discussion on PAR evaluation considering photovoltaic-battery systems, the proposed algorithm significantly reduces PAR compared to the existing algorithms, which is beneficial for both utility and consumers.

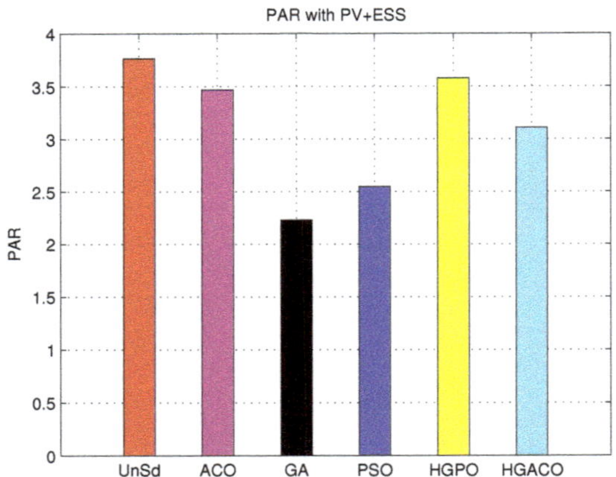

Figure 19. PAR with photovoltaic-battery systems.

Table 13. Scenario III: PAR assessment with photovoltaic-battery systems.

Algorithms	PAR	Difference	Reduction (%)
Unscheduled	3.77	-	-
GA	2.23	1.54	40.85
ACO	3.47	0.30	7.85
PSO	2.55	1.21	32.29
HGPO	3.58	0.18	4.96
HGACO	3.11	0.65	17.40

The carbon emission of unscheduled and scheduled loads considering photovoltaic-battery systems is shown in Figure 20 and in Table 14. The existing algorithms and the proposed algorithm emitted less carbon than the unscheduled load. The without scheduling maximum carbon emitted into the atmosphere is 5.44 pounds in time slot 21. Contrarily, the maximum carbon emission for GA is 3.92 pounds in time slot 21, that for ACO is 2.56 pounds at time slot 21, that for PSO is 2.97 at time slot 21, that for HGPO at time slot 21 is 2.81 pounds, and the proposed HGACO algorithm emits less carbon is 2.22 pounds at time slot 21. The unscheduled load emits 97.90 pounds of total carbon emission in scenario III, whereas GA, ACO, PSO, HGPO, and HGACO algorithm emit 70.52, 46.12, 53.59, 50.71, and 40.07 pounds of carbon emission. Compared to unscheduled carbon emission, GA reduced carbon emissions by 27.96 percent, ACO reduced carbon emissions by 52.89 percent, PSO reduced carbon emissions by 45.20 percent, HGPO reduced carbon emissions by 48.20 percent, and HGACO reduced carbon emissions by 57.42 percent. Thus, the carbon released into the atmosphere by the HGACO algorithm is less than all benchmark algorithms.

Table 14. Scenario III: carbon emission assessment with photovoltaic-battery systems.

Algorithms	Carbon Emission (Pounds)	Difference	Reduction (%)
Unscheduled	97.90	-	-
GA	70.52	27.37	27.96
ACO	46.12	51.78	52.89
PSO	53.59	44.30	45.20
HGPO	50.71	47.19	48.20
HGACO	40.07	57.82	57.42

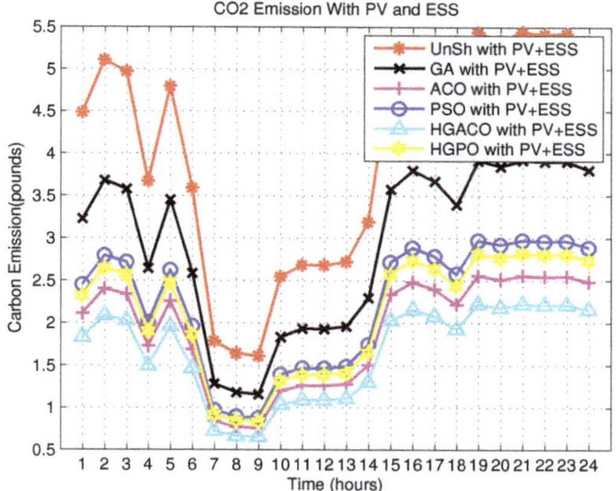

Figure 20. Scenario III: carbon emission with photovoltaic-battery systems.

The proposed HGACO algorithm-based scheduling was compared with other benchmark algorithm (GA, PSO, ACO, and HGPO)-based scheduling to evaluate the DTR that consumers face. The DTR evaluation of scheduled load using the proposed algorithm compared to benchmark algorithms are shown in Figure 21. The detailed discussion is as follows. In GA-based scheduling, average delays of 0.9, 1.6, and 0.5 h are confronted by SIAs such as humidifiers, water heaters, and dishwashers, respectively. Similarly, SUAs such as EVs, washing machines, and clothes dryers face average delays of 1.7, 2, and 0.75 h, respectively, depicted in Figure 21.

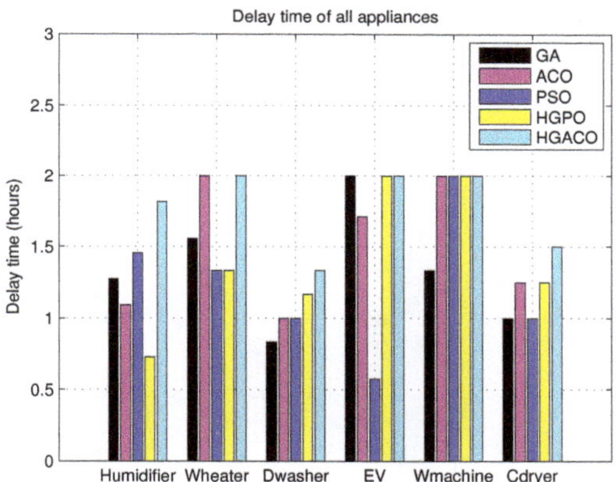

Figure 21. Delay faced by consumers while operating appliances according to power usage schedule generated by proposed and existing algorithms

The ACO algorithm-based EMC generated power usage schedule for SIAs such as humidifiers, water heaters, and dishwashers faces average delays of 1.4, 2, and 0.8 h. Likewise, the SUAs face delays of 2, 2, and 1.5 h observed for EVs, washing machines, and clothes dryers, respectively, which are illustrated in Figure 21.

The PSO-based EMC created power usage schedule faces delays of 1.95, 0.7, and 0.5 h for SIAs such as humidifiers, water heaters, and dishwashers, respectively. In a similar manner, SUAs confronted delays of 1.95, 2, and 1.75 h for EVs, washing machines, and clothes dryers, respectively, shown in Figure 21.

HGPO algorithm-based schedules introduce average delays of 1.2, 0.9, and 0.85 h for SIAs such as humidifiers, water heaters, and dishwashers, respectively. Likewise, average delays of 1.9, 2, and 1.5 h are posed to EVs, washing machines, and clothes dryers, respectively. This behavior is illustrated in Figure 21.

With HGACO-based power usage scheduling, average delays of 1.75, 2, and 1.6 h are posed to SIAs such as humidifiers, water heaters, and dishwashers, respectively. In a similar fashion, average delays of 2, 2, and 1.5 h are observed by SUAs such as EVs, washing machines, and clothes dryers, respectively. The DTR faced by consumers while using the HGACO algorithm is depicted in Figure 21. This evaluation concludes that a tradeoff exists between DTR and electricity because, when one wants to reduce their electricity cost, they may accept DTR.

3.4. Performance Tradeoff Analysis

The proposed HGACO algorithm confronted performance tradeoffs between different conflicting parameters while solving DSM problem via power usage scheduling. The first tradeoff analysis is observed between electricity cost and PAR. The proposed algorithm reduced the electricity cost significantly while the PAR increased a little bit. This tradeoff behavior is observed in all scenarios. However, this tradeoff is natural because it is observed for the proposed and existing algorithms. The second tradeoff analysis is visualized in the case of electricity cost and DTR. The proposed algorithm shifted the load from on-peak hours to off-peak hours to avoid rebound peaks and to reduce the electricity cost. Thus, the electricity cost is reduced significantly. However, while reducing the electricity cost, the DTR increased. Thus, the user will confront a little bit of discomfort. This tradeoff analysis is observed for both proposed and existing algorithms in all scenarios, which makes it evident that this tradeoff is natural and could not be completely avoided. However, the tradeoff between electricity cost and DTR is average for our proposed algorithm compared to existing algorithms. The performance tradeoff (cost and PAR, and cost and DTR) analyses for scenarios I, II, and II are listed in Tables 15–17, respectively.

Table 15. Scenario I: performance tradeoff analysis of the proposed and existing algorithms without photovoltaic-battery systems.

Algorithms	Cost (Cents)	PAR	DTR (Hours)	Carbon Emission (Pounds)
Unscheduled	921.19	3.48	-	116.781
GA	660.86	2.88	8	83.77
ACO	512.69	3.01	9.3	64.99
PSO	571.68	2.11	8	72.47
HGPO	548.90	3.31	9.25	69.58
HGACO	465.03	2.58	10.1	58.95

The tradeoff among unscheduled load, the benchmark, and proposed algorithms with PV is shown in Table 16; it shows that HGACO is best in terms of electricity cost, PAR, carbon emission, and DTR. The tradeoff among different algorithms and unscheduled load without photovoltaic-battery systems is shown in Figure 10, where the HGACO is best in terms of producing minimum electricity cost.

Table 16. Scenario II: performance tradeoff analysis of the proposed and existing algorithms with PV.

Algorithms	Cost (Cents)	PAR	DTR (Hours)	Carbon Emission (Pounds)
Unscheduled	816.68	3.67	-	103.53
GA	556.35	3.02	8	70.52
ACO	408.18	3.36	9.3	51.74
PSO	467.17	2.40	8	59.22
HGPO	444.39	3.18	9.25	56.33
HGACO	360.52	2.87	10.1	45.70

Table 17. Scenario III: performance tradeoff analysis of the proposed and existing algorithms with photovoltaic-battery systems.

Algorithms	Cost (Cents)	PAR	DTR (Hours)	Carbon Emission (Pounds)
Unscheduled	772.31	3.77	-	97.90
GA	510.59	2.23	8	70.52
ACO	363.80	3.47	9.3	46.12
PSO	422.80	2.55	8	53.59
HGPO	400.02	3.58	9.25	50.71
HGACO	316.15	3.11	10.1	40.07

4. Conclusions

This study introduces an optimization-based energy management framework to schedule consumers' power usage pattern in response to the RTP signal under the grid and photovoltaic-battery system. Then, the HGACO algorithm is proposed, which is a hybrid of GA and ACO using MKP to solve the complete scheduling problem for all three scenarios: without photovoltaic-battery system, with the photovoltaic system, and with the photovoltaic-battery system. This study aims to stimulate consumers to participate in RES generation and power usage scheduling to resolve the DSM problem by coping with the gap between demand and generation. The purpose of solving the DSM problem is to facilitate utility and end-users by reducing electricity bill, peak load demand, and carbon emission and by avoiding rebound peak creation. To endorse the applicability of the HGACO algorithm-based energy management framework, simulations are conducted in comparison with existing frameworks based on the GA, PSO, ACO, and HGPO algorithms. The results show that the proposed HGACO algorithm reduced electricity cost, carbon emission, and peak load by 49.51%, 48.01%, and 25.72% in scenario I; by 55.85%, 54.22%, and 21.69% in scenario II; and by 59.06%, 57.42%, and 17.40% in scenario III, respectively, compared to without scheduling.

Author Contributions: Conceptualization, S.A. and I.K.; methodology, I.K.; software, G.H.; validation, G.H., S.A. and S.J.; formal analysis, S.A., and G.H.; investigation, I.K.; resources, S.J.; data curation, S.A.; writing—original draft preparation, G.H.; writing—review and editing, G.H. and I.K.; visualization, I.K., and G.H.; supervision, S.J.; project administration, S.A. and I.K.; funding acquisition. All authors have read and agreed to the published version of the manuscript.

Funding: This research was funded by University of Engineering & Technology, Mardan, and University of Engineering & Technology, Peshawar.

Conflicts of Interest: The authors declare no conflict of interest.

References

1. Mesarić, P.; Krajcar, S. Home demand side management integrated with electric vehicles and renewable energy sources. *Energy Build.* **2015**, *108*, 1–9. [CrossRef]
2. Arora, S.; Singh, H.; Sharma, M.; Sharma, S.; Anand, P. A new hybrid algorithm based on grey wolf optimization and crow search algorithm for unconstrained function optimization and feature selection. *IEEE Access* **2019**, *7*, 26343–26361. [CrossRef]
3. Moser, A.; Muschick, D.; Gölles, M.; Nageler, P.; Schranzhofer, H.; Mach, T.; Tugores, C.R.; Leusbrock, I.; Stark, S.; Lackner, F.; et al. A MILP-based modular energy management system for urban multi-energy systems: Performance and sensitivity analysis. *Appl. Energy* **2020**, *261*, 114342. [CrossRef]
4. Rahmani-Andebili, M. Scheduling deferrable appliances and energy resources of a smart home applying multi-time scale stochastic model predictive control. *Sustain. Cities Soc.* **2017**, *32*, 338–347. [CrossRef]
5. Huang, Y.; Wang, W.; Hou, B. A hybrid algorithm for mixed integer nonlinear programming in residential energy management. *J. Clean. Prod.* **2019**, *226*, 940–948. [CrossRef]
6. Shirazi, E.; Jadid, S. Cost reduction and peak shaving through domestic load shifting and DERs. *Energy* **2017**, *124*, 146–159. [CrossRef]
7. Patnam, B.S.K.; Pindoriya, N.M. Demand response in consumer-centric electricity market: Mathematical models and optimization problems. *Electr. Power Syst. Res.* **2020**, *193*, 106923. [CrossRef]

8. Di Giorgio, A.; Liberati, F. Near real time load shifting control for residential electricity prosumers under designed and market indexed pricing models. *Appl. Energy* **2014**, *128*, 119–132. [CrossRef]
9. Luna, A.C.; Diaz, N.L.; Graells, M.; Vasquez, J.C.; Guerrero, J.M. Mixed-integer-linear-programming-based energy management system for hybrid PV-wind-battery microgrids: Modeling, design, and experimental verification. *IEEE Trans. Power Electron.* **2016**, *32*, 2769–2783. [CrossRef]
10. Nan, S.; Zhou, M.; Li, G. Optimal residential community demand response scheduling in smart grid. *Appl. Energy* **2018**, *210*, 1280–1289. [CrossRef]
11. Fernandez, E.; Hossain, M.J.; Nizami, M.S.H. Game-theoretic approach to demand-side energy management for a smart neighbourhood in Sydney incorporating renewable resources. *Appl. Energy* **2018**, *232*, 245–257. [CrossRef]
12. Srinivasan, D.; Rajgarhia, S.; Radhakrishnan, B.M.; Sharma, A.; Khincha, H.P. Game-Theory based dynamic pricing strategies for demand side management in smart grids. *Energy* **2017**, *126*, 132–143. [CrossRef]
13. Yu, M.; Hong, S.H. Supply–demand balancing for power management in smart grid: A Stackelberg game approach. *Appl. Energy* **2016**, *164*, 702–710. [CrossRef]
14. Alsalloum, H.; Merghem-Boulahia, L.; Rahim, R. Hierarchical system model for the energy management in the smart grid: A game theoretic approach. *Sustain. Energy Grids Netw.* **2020**, *21*, 100329. [CrossRef]
15. Kalair, A.R.; Abas, N.; Hasan, Q.U.; Seyedmahmoudian, M.; Khan, N. Demand side management in hybrid rooftop photovoltaic integrated smart nano grid. *J. Clean. Prod.* **2020**, *258*, 120747. [CrossRef]
16. Cheng, P.H.; Huang, T.H.; Chien, Y.W.; Wu, C.L.; Tai, C.S.; Fu, L.C. Demand-side management in residential community realizing sharing economy with bidirectional PEV while additionally considering commercial area. *Int. J. Electr. Power Energy Syst.* **2020**, *116*, 105512. [CrossRef]
17. Wu, Y.; Lau, V.K.; Tsang, D.H.; Qian, L.P.; Meng, L. Optimal energy scheduling for residential smart grid with centralized renewable energy source. *IEEE Syst. J.* **2013**, *8*, 562–576. [CrossRef]
18. Imran, A.; Hafeez, G.; Khan, I.; Usman, M.; Shafiq, Z.; Qazi, A.B.; Khalid, A.; Thoben, K.D. Heuristic-Based Programable Controller for Efficient Energy Management Under Renewable Energy Sources and Energy Storage System in Smart Grid. *IEEE Access* **2020**, *8*, 139587–139608. [CrossRef]
19. Hafeez, G.; Islam, N.; Ali, A.; Ahmad, S.; Usman, M.; Alimgeer, K.S. A Modular Framework for Optimal Load Scheduling under Price-Based Demand Response Scheme in Smart Grid. *Processes* **2019**, *7*, 499. [CrossRef]
20. Ayub, S.; Ayob, S.M.; Tan, C.W.; Ayub, L.; Bukar, A.L. Optimal residence energy management with time and device-based preferences using an enhanced binary grey wolf optimization algorithm. *Sustain. Energy Technol. Assess.* **2020**, *41*, 100798. [CrossRef]
21. Nawaz, A.; Hafeez, G.; Khan, I.; Jan, K.U.; Li, H.; Khan, S.A.; Wadud, Z. An Intelligent Integrated Approach for Efficient Demand Side Management with Forecaster and Advanced Metering Infrastructure Frameworks in Smart Grid. *IEEE Access* **2020**, *8*, 132551–132581. [CrossRef]
22. Waseem, M.; Lin, Z.; Liu, S.; Sajjad, I.A.; Aziz, T. Optimal GWCSO-based home appliances scheduling for demand response considering end-users comfort. *Electr. Power Syst. Res.* **2020**, *187*, 106477. [CrossRef]
23. Zhao, Z.; Lee, W.C.; Shin, Y.; Song, K.B. An optimal power scheduling method for demand response in home energy management system. *IEEE Trans. Smart Grid* **2013**, *4*, 1391–1400. [CrossRef]
24. Ullah, I.; Rasheed, M.B.; Alquthami, T.; Tayyaba, S. A Residential Load Scheduling with the Integration of On-Site PV and Energy Storage Systems in Micro-Grid. *Sustainability* **2020**, *12*, 184. [CrossRef]
25. Hafeez, G.; Alimgeer, K.S.; Wadud, Z.; Khan, I.; Usman, M.; Qazi, A.B.; Khan, F.A. An Innovative Optimization Strategy for Efficient Energy Management with Day-ahead Demand Response Signal and Energy Consumption Forecasting in Smart Grid using Artificial Neural Network. *IEEE Access* **2020**, *8*, 84415–84433. [CrossRef]
26. Jiang, X.; Xiao, C. Household energy demand management strategy based on operating power by genetic algorithm. *IEEE Access* **2019**, *7*, 96414–96423. [CrossRef]
27. Ma, K.; Yao, T.; Yang, J.; Guan, X. Residential power scheduling for demand response in smart grid. *Int. J. Electr. Power Energy Syst.* **2016**, *78*, 320–325. [CrossRef]
28. Samadi, P.; Mohsenian-Rad, A.H.; Schober, R.; Wong, V.W.; Jatskevich, J. Optimal real-time pricing algorithm based on utility maximization for smart grid. In Proceedings of the 2010 First IEEE International Conference on Smart Grid Communications, Gaithersburg, MD, USA, 4–6 October 2010; pp. 415–420.
29. Adika, C.O.; Wang, L. Smart charging and appliance scheduling approaches to demand side management. *Int. J. Electr. Power Energy Syst.* **2014**, *57*, 232–240. [CrossRef]

30. Jamil, M.; Mittal, S. Hourly load shifting approach for demand side management in smart grid using grasshopper optimisation algorithm. *IET Gener. Transm. Distrib.* **2019**, *14*, 808–815. [CrossRef]
31. Sarker, E.; Halder, P.; Seyedmahmoudian, M.; Jamei, E.; Horan, B.; Mekhilef, S.; Stojcevski, A. Progress on the demand side management in smart grid and optimization approaches. *Int. J. Energy Res.* **2021**, *45*, 36–64. [CrossRef]
32. Climate Central Solutions Brief: Battery Energy Storage. Available online: https://www.climatecentral.org/news/climate-central-solutions-brief-battery-energy-storage (accessed on 5 April 2021).
33. Nawaz, A.; Hafeez, G.; Khan, I.; Usman, M.; Jan, K.U.; Ullah, Z.; Diallo, D. Demand-side management of residential service area under price-based demand response program in smart grid. In Proceedings of the 2020 IEEE International Conference on Electrical, Communication, and Computer Engineering (ICECCE), Istanbul, Turkey, 12–13 June 2020; pp. 1–6.
34. Hussain, I.; Ullah, M.; Ullah, I.; Bibi, A.; Naeem, M.; Singh, M. Optimizing energy consumption in the home energy management system via a bio-inspired dragonfly algorithm and the genetic algorithm. *Electronics* **2020**, *9*, 406. [CrossRef]
35. Hafeez, G.; Wadud, Z.; Khan, I.U.; Khan, I.; Shafiq, Z.; Usman, M.; Khan, M.U.A. Efficient Energy Management of IoT-Enabled Smart Homes Under Price-Based Demand Response Program in Smart Grid. *Sensors* **2020**, *20*, 3155. [CrossRef] [PubMed]

Publisher's Note: MDPI stays neutral with regard to jurisdictional claims in published maps and institutional affiliations.

Article

A Combined Deep Learning and Ensemble Learning Methodology to Avoid Electricity Theft in Smart Grids

Zeeshan Aslam [1], Nadeem Javaid [1,*], Ashfaq Ahmad [2,*], Abrar Ahmed [3] and Sardar Muhammad Gulfam [3]

1 Department of Computer Science, COMSATS University Islamabad, Islamabad 44000, Pakistan; zeeshanxh@gmail.com
2 School of Electrical Engineering and Computing, The University of Newcastle, Callaghan 2308, Australia
3 Department of Electrical and Computer Engineering, COMSATS University Islamabad, Islamabad 44000, Pakistan; abrar_ahmed@comsats.edu.pk (A.A.); sardar_muhammad@comsats.edu.pk (S.M.G.)
* Correspondence: nadeemjavaid@comsats.edu.pk (N.J.); ashfaqahmad@ieee.org (A.A.)

Received: 21 September 2020; Accepted: 19 October 2020; Published: 26 October 2020

Abstract: Electricity is widely used around 80% of the world. Electricity theft has dangerous effects on utilities in terms of power efficiency and costs billions of dollars per annum. The enhancement of the traditional grids gave rise to smart grids that enable one to resolve the dilemma of electricity theft detection (ETD) using an extensive amount of data formulated by smart meters. This data are used by power utilities to examine the consumption behaviors of consumers and to decide whether the consumer is an electricity thief or benign. However, the traditional data-driven methods for ETD have poor detection performances due to the high-dimensional imbalanced data and their limited ETD capability. In this paper, we present a new class balancing mechanism based on the interquartile minority oversampling technique and a combined ETD model to overcome the shortcomings of conventional approaches. The combined ETD model is composed of long short-term memory (LSTM), UNet and adaptive boosting (Adaboost), and termed LSTM–UNet–Adaboost. In this regard, LSTM–UNet–Adaboost combines the advantages of deep learning (LSTM-UNet) along with ensemble learning (Adaboost) for ETD. Moreover, the performance of the proposed LSTM–UNet–Adaboost scheme was simulated and evaluated over the real-time smart meter dataset given by the State Grid Corporation of China. The simulations were conducted using the most appropriate performance indicators, such as area under the curve, precision, recall and F1 measure. The proposed solution obtained the highest results as compared to the existing benchmark schemes in terms of selected performance measures. More specifically, it achieved the detection rate of 0.92, which was the highest among existing benchmark schemes, such as logistic regression, support vector machine and random under-sampling boosting technique. Therefore, the simulation outcomes validate that the proposed LSTM–UNet–Adaboost model surpasses other traditional methods in terms of ETD and is more acceptable for real-time practices.

Keywords: electricity theft detection; smart grids; electricity consumption; electricity thefts; smart meter; imbalanced data

1. Introduction

The secure and efficient use of electricity represents a major aspect of the social and economic development of a country. Electricity losses happen during power generation, transmission and delivery to consumers. Essentially, power transmission and delivery have a couple of losses, namely, technical losses (TL) and nontechnical losses (NTL) [1]. TL occur due to the line losses, transformer losses and other power system elements. NTL occur due to the electricity stealing, defective meters, overdue bills and billing mistakes [1]. More generally, NTL are the difference between total losses and TL. Besides, NTL raise electricity prices, increase load-shedding, decrease revenue and decrease energy efficiency. Thus, NTL badly affect both the utilities and a country's financial state [2,3].

Electricity fraud is one of the chief reasons for the NTL, which accounts for 10–40% of cumulative electricity losses worldwide [4]. Electricity theft comprises of bypassing electricity meters, tampering with meter readings, tampering with meters themselves and cyber-attacks [5,6]. Therefore, the reduction of electricity fraud is a principal concern of the power distribution companies to secure significant amounts of the total electricity losses and revenue [7].

Electricity theft is spreading widely in many developing countries—e.g., India loses 20% of its total electricity due to electricity theft [8], and developed countries. For instance, in the U.S., the revenue loss as a result of electricity fraud is about $6 billion, while in the UK, it costs up to £175 million per annum [9,10]. Moreover, it is stated in [11] that electricity theft accounts for approximately a hundred million Canadian dollars annually. Globally, the utilities lose more than $20 billion per annum through electricity fraud [12].

The introduction of advanced metering infrastructure (AMI) in the smart grid environment provides a massive amount of power consumption users' records, which makes it easier for utilities to monitor the electricity theft [13,14]. AMI enables price and load forecasting [15,16], energy management [17,18] and consumer behavior characterization [19]. As electricity theft continues to increase, smart meters enable utilities to provide new and innovative solutions to perform electricity theft detection (ETD). Generally, electricity thieves can alter the smart meters' information physically or through cyber-attacks. Consequently, the primary way of ETD is manually examining the consumers' electricity meters and comparing the abnormal consumption readings with the previous normal ones, known as the audit and on-site inspection process. However, these methods are costly, inefficient and time-consuming.

In contrast to the manual methods, supervised machine learning solutions have gained the interest of utilities and academia for performing ETD. Studies based on the supervised learning techniques [20–25] focus on ETD using the large and imbalanced datasets obtained through the smart meters. However, the performances of these techniques are still not sufficient for the practical applications in utilities. that implies that the techniques raise misclassification scores that lead to the costly procedure of on-site inspections for the final verifications. Therefore, that exhibits the need for a new solution to solve the ETD problem using a large imbalanced dataset to determine the real assessment of the model's performance.

In ETD literature, the proposed methods are categorized into two major groups: ETD through technical methods and ETD through nontechnical methods. Nontechnical methods cover: the auditing and inspection of illegal electricity consumers, giving awareness to the electricity consumers about electricity theft as a crime and punishment, installing the smart meters that can not be easily tampered with by consumers and reducing electricity theft through the psychosocial methods, such as social support [26]. Technical methods are also broadly classified into three types: state-based solutions (also known as hardware-based solutions), game-theory-based solutions and data-driven solutions (also known as machine learning-based solutions) [12].

In hardware-based solutions, the major focus is on designing specific hardware devices and infrastructures to detect electricity theft. Hardware-based solutions consist of: smart meters with radio-frequency identification tags (RFID), anti-tampering sensors, wireless sensors and distribution transformers [5,27–29]. These solutions get high detection efficiency through specific devices, for instance, RFID. The major limitations of the state-based solutions are the high cost of design, the high operational and maintenance costs and the vulnerability to weather conditions. Particularly, due to the inefficiency and high cost of hardware-based solutions, data-driven solutions have gained the interest of researchers. In game-theory-based solutions [30,31], ETD is considered as a contest between the power distribution company and electricity fraudsters, known as the players. Both players want to maximize their utility functions. These solutions are low cost and provide reasonable ways to find the electricity theft. Hence, one of the complicated issues in the game-theory-based procedures is how to form the utility function for each player, which is a challenging and time-consuming problem.

Recently, machine learning approaches have achieved significant importance in ETD. The main purpose of these solutions is to analyze the electricity usage behavior of the consumers based on smart meters' data. These methods require no additional information about the network topology or hardware devices. Thus, machine learning solutions are further categorized into the supervised and unsupervised learning methods. Unsupervised learning methods have proposed clustering-based solutions to group the similar instances into one cluster [32,33], members of which each have a high false positive rate (FPR). In this paper, a unique supervised learning-based solution, namely, the LSTM–UNet–Adaboost, is proposed to perform the binary classification using the data from on-site inspections. Thus, we describe some recent advances made in this area.

Buzau et al. [34] presented a solution that is based on an extreme gradient boosted tree (XGBoost) for the detection of NTL in smart grids. Their main objective was to rank the list of consumers applying the smart meters' data and extract features from auxiliary databases. Punmiya et al. [35] introduced a gradient boosting theft detector (GBTD) model, which is composed of three variants of a gradient boosting classifier to perform the ETD. A theft detector is also used for feature engineering-based preprocessing through the GBTD's feature importance function. Another solution based on the ensemble bagged tree is presented in [21] for the NTL detection, in which an ensemble of individual decision trees is applied to improve the theft detection by aggregating their performances. Buzau et al. [23] submit a hybrid of LSTM and multilayer perceptron (MLP), termed LSTM-MLP, for the NTL detection in the smart grid. LSTM is employed to automate the feature extraction from sequential information, while MLP is used to deal with non-sequential information. Likewise, the authors [36] used a deep neural network for the feature extraction and meta-heuristic technique enabled XGBoost for ETD.

Nelson et al. [24] combined the maximal overlap discrete wavelet packet transform (MODWPT) with the random under-sampling boosting (RUSBoost) technique to obtain the most suitable features for the identification of NTL. Li et al. [37] used a semi-supervised technique to perform the detection of NTL. Tianyu et al. [38] made a semi-supervised deep learning model, known as the multitask feature extracting fraud detector (MFEFD), for ETD: both the supervised and unsupervised learning procedures are combined to capture important features from the labeled and unlabeled data. Maamar et al. [32] offered a hybrid model that utilizes the k-means clustering procedure and deep neural networks (DNN) for ETD in the AMI system, where k-means is utilized to gather consumers having the same electricity consumption behaviors, and DNN is used to detect anomalies in the electricity consumption behaviors of the consumers. Ghasemi et al. [26] proposed a solution comprising of a probabilistic neural network and mathematical model, named the PNN–Levenberg–Marquardt, for the identification of two types of illegal consumers using the observer's meters. PNN is applied to detect the suspicious consumers, and the Levenberg–Marquardt is practiced for classifying the fraud consumers. In [9], a theft detector is presented, which implements the support vector machine (SVM) for the detection of normal and abnormal consumers using their consumption patterns. Another work [39] proposed a combined data sampling mechanism and performs ETD through the bi-directional gated recurrent unit (GRU).

In recent years, the convolutional neural network (CNN) has achieved success in ETD because it is a deep learning technique that catches high-level features from the electricity consumption dataset. Authors in [20] presented a hybrid wide and deep convolutional neural network (WD-CNN) for ETD. The wide part is used to extract the global features from 1-D data, while the deep component is applied to derive periodicity and non-periodicity from 2-D data. Li et al. [40] introduce a hybrid of CNN and random forest (RF), known as the CNN-RF model, for ETD in smart grids, where the RF is used as the final layer to perform ETD on the extracted features. Hasan et al. [22] made a solution consisting of the hybrid of CNN and LSTM for ETD. LSTM is used to solve the binary classification problem using CNN's output. In another study [41], the authors introduced a hybrid of CNN and GRU for the detection of abnormal consumers. However, CNN only looks for "what" information is available in the data through down-sampling while ignoring "where" this information is present, which degrades the ETD's performance. Moreover, in the traditional CNN, final classification is performed through either softmax classifier or sigmoid function that leads to the degradation of generalization ability plus subjecting the model to the local optima.

The aforementioned techniques for ETD are innovative and efficient; however, their performances are inadequate for real practices. Generally, these techniques have several limitations that must be addressed as follows: a model's bias towards the majority class due to the class imbalance; a model's performance evaluated on synthetic data does not provide a realistic assessment of the theft detection; models require artificial feature extraction and have poor detection performances, such as low detection scores and high misclassification scores. Hence, this detection score is costly for the utilities, e.g., on-site inspections needed for the confirmation. Thus, the problem of ETD is not fixed completely, which implies a new solution that provides a more accurate theft detection score.

To overcome the limitations of previous studies, we propose a new combined LSTM–UNet–Adaboost solution for ETD in this paper. The prime intention behind the proposed solution is to perform the binary classification over the electricity consumption data to characterize the consumers as either benign or thieves. As compared to the above-mentioned studies, our proposed scheme distinguishes itself: by presenting a new class balancing technique IQMOT to overcome the limitations of class imbalance by implementing a UNet technique to automate the feature extraction that captures both "what" and "where" from the 2-D data rather than only "what", and by performing the joint training and classification through Adaboost using the features extracted from UNet and LSTM. In this regard, the proposed model gets the advantages of two powerful approaches known as deep learning and ensemble learning. Deep learning is applied to automate the feature extraction, while ensemble learning is used to perform the joint training and classification.

To address the above-mentioned problems, this paper proposes a new and practical solution for ETD utilizing long short-term memory (LSTM), UNet and adaptive boosting (Adaboost), named LSTM–UNet–Adaboost. Moreover, a novel class balancing technique, namely, the interquartile minority oversampling technique (IQMOT), is introduced to address the class imbalance issues. In the proposed methodology, real-world electricity theft cases are initially generated using the IQMOT. Then, it solves the ETD problem using the LSTM–UNet–Adaboost model. Essentially, LSTM is used to capture the daily temporal correlations, while a UNet model is applied to capture the abstract features from 2-D electricity consumption data. Finally, the Adaboost model is used to perform the joint training and classification over the extracted features through LSTM and UNet modules. The proposed methodology automates the concept of feature engineering, known as self-learning. Hence, the underlying intuition of this paper is to generate real-world theft cases through a novel sampling technique and to combine a deep learning technique with ensemble learning to improve the theft detection performance. Therefore, the proposed solution is more efficient and reliable as compared to the conventional approaches.

Thus, the chief contributions of this work are described as follows.

- *IQMOT:* A novel class balancing technique, named IQMOT, is presented in this paper to overcome the problems of imbalanced data. It generates more practical theft cases as compared to the traditional class balancing techniques.
- *LSTM–UNet–Adaboost:* We propose a new combined LSTM–UNet–Adaboost model for ETD. The proposed model leverages the advantages of the most recent deep learning technique, i.e., UNet, which is applied for very first time in ETD along with the ensemble learning technique, i.e., Adaboost.

- *Comprehensive simulations:* We conducted extensive simulations on the real electricity consumption dataset and analyzed our proposed solution with standard techniques. Simulation results demonstrated the superiority and effectiveness of our proposed model over existing benchmarks previously used for ETD.

The remainder of this paper is arranged as follows. Section 2 presents our proposed approach comprising of IQMOT and LSTM–UNet–Adaboost for ETD. Section 3 illustrates and examines the simulation results before concluding the paper in Section 4. Lastly, the future directions are described in Section 5.

2. Proposed Methodology

The proposed system model is shown in Figure 1; it has three major stages: (1) data preprocessing, (2) data balancing through IQMOT and (3) data analysis using the UNet-LSTM-Adaboost. The proposed hybrid model practices LSTM, UNet and Adaboost, which solves the limitations of the state-of-the-art techniques for ETD, as mentioned in Section 1. Moreover, we designed a new class balancing mechanism to handle the data imbalance issues faced by the conventional supervised learning techniques. The proposed methodology has the potential to integrate both 1-D and 2-D information obtained from the electricity consumption data. More generally, the LSTM module is used to derive the long-term dependencies from 1-D data, known as the sequential information. The UNet module acquires global features from 2-D electricity consumption data, i.e., non-sequential information. Furthermore, the Adaboost module performs the final joint training and binary classification over the outputs of LSTM and UNet modules, as shown in Figure 1. We validated the proposed model using the real electricity theft data in terms of selected performance indicators for ETD. The proposed LSTM–UNet–Adaboost model is efficient because it performs joint training on both types of inputs, known as the sequential and non-sequential information, provided by LSTM and UNet. In the following sub-sections, a detailed description of each module is given.

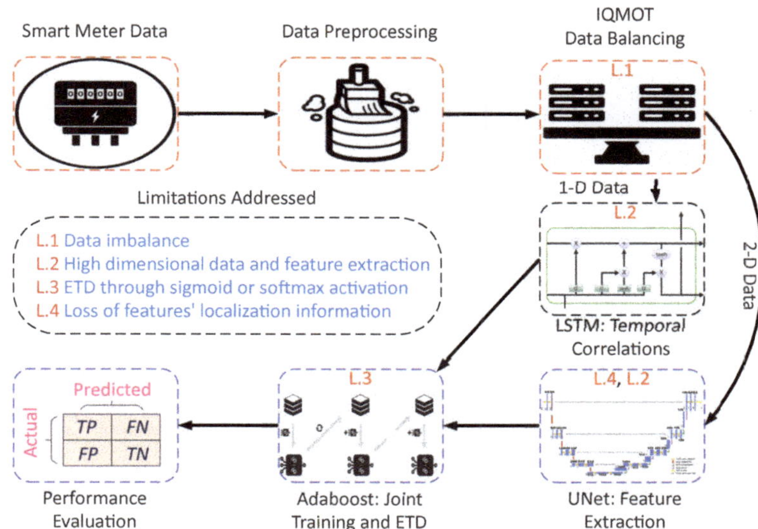

Figure 1. Overview of the proposed system model for electricity theft detection (ETD).

2.1. Data

In this paper, a real-time power consumption dataset of users is employed, which was given to us by the State Grid Corporation of China (SGCC) [42]. The metadata information of the SGCC dataset is presented in Table 1. The dataset contains the daily electricity consumption histories of residential consumers. The SGCC conducts real in-the-field inspections to verify the normal and abnormal consumers. Therefore, SGCC explicitly declares that the given dataset holds 3615 instances of electricity theft, which shows the importance of ETD in China. Additionally, it also contains the missing and erroneous values that require preprocessing, as explained in Section 2.2.

Table 1. SGCC dataset information.

Description	Numeric
Total time duration	January-2014–October-2016
Total electricity consumers	42,372
Total electricity thieves	3615
Total electricity normal consumers	38,757

2.2. Data Preprocessing

The real dataset often contains the missing and erroneous values demanded to be resolved by employing the data preprocessing techniques [43]. Thus, the electricity consumption dataset of SGCC contains the missing and erroneous values due to several reasons, such as failure of any smart meter equipment, storage issue and measurement error or unreliable transmission. Moreover, analyzing and cleansing the dataset assists one in finding and eliminating these erroneous and missing values. In this study, the concept of linear interpolation was employed to find and retrieve the missing values found in the dataset [20]. Hence, the missing values were recovered as using the Equation (1):

$$\mathrm{f}(x_i) = \begin{cases} \frac{\mathrm{z}}{2}, & x_i \in NaN, x_{i(t-1)}, x_{i(t+1)} \notin NaN, \\ 0, & x_i \in NaN, x_{i(t-1)} \text{ or } x_{i(t+1)} \in NaN, \\ x_i & x_i \notin NaN, \end{cases} \tag{1}$$

where $\mathrm{z} = x_{i(t-1)} + x_{i(t+1)}$. x_i is the current electricity consumption at a certain time t of an ith day. Likewise, $x_{i(t-1)}$ and $x_{i(t+1)}$ are the previous and next values of the current electricity consumption, respectively. Likewise, in the SGCC dataset, we have identified outliers, which skew the data, making the training process complex and have a negative impact on the final ETD performance because of overfitting. In this paper, the "three-sigma rule of thumb" [44] is practiced for detecting and recovering the outliers according to the following equation:

$$\mathrm{O}(x_{i,t}) = \begin{cases} \mathrm{w}, & if x_{i(t)} > \mathrm{w}, \\ x_{i(t)}, & otherwise, \end{cases} \tag{2}$$

where $\mathrm{w} = avg(x_{i(t)}) + 2\sigma(x_{i(t)})$. After the detection and removal of the missing and outlier values, the dataset needs to be normalized, as the deep neural networks are sensitive to the diverse data that increases the training time. Hence, the data normalization improves the training process of deep learning models by assigning the same scale to all values present in the dataset and bringing them in the range of 0 and 1. Therefore, a min-max normalization concept was applied to scale the dataset as per the following equation [40]:

$$\mathrm{N}(x_{i(t)}) = \frac{x_{i(t)} - min(x)}{max(x) - min(x)}, \tag{3}$$

where $x_{i(t)}$ is the electricity consumption at a current time t, $min(x)$ is the least electricity consumption and $max(x)$ is the highest electricity consumption.

2.3. Data Balancing

In this paper, a new data balancing technique, named IQMOT, is introduced to balance the majority and minority classes. In a real-life scenario, the number of benign consumers is always extensive as compared to the electricity thieves. Similarly, in the SGCC dataset, the benign electricity consumers are higher in number than the electricity thieves, as shown in Table 1. This imbalanced nature of the dataset adversely affects the performances of the supervised learning techniques because of the biasn towards the majority class. Hence, to reduce the class inequality problem, there are two major types of techniques, known as the cost function-based and sampling-based techniques [22].

In the sampling-based techniques, there are three major approaches, including random under-sampling (RUS), random over-sampling (ROS) and oversampling based on synthetic theft instance generation, such as the synthetic minority over-sampling technique (SMOTE). In RUS, it unintelligently discards the samples from the majority class, which contains normal consumers. This method decreases the computationally beneficial dataset size. The unintelligent removal of samples from the majority class creates a loss of potentially important information, while the remainder does not provide a realistic assessment. ROS replicates the minority class instances to balance the majority and minority classes. There is no loss of potentially useful data; however, due to the unintelligent replication of minority instances, the model leads to an overfitting problem. In this regard, SMOTE is an effective synthetic instance generation technique that generates the new minority instances based on the nearest neighbors [45].

Synthetic generation of minority instances avoids the overfitting problem, which occurs due to the ROS technique, although the synthetic forms of theft instances do not reflect the real-world electricity theft cases. Moreover, a synthetic formulation replicates the minority class instances based on the nearest neighbors, which further leads to the overfitting problem. In the light of the above-mentioned limitations, we propose a novel class balancing technique in ETD, named IQMOT, to balance the majority and minority class instances. Moreover, IQMOT-based generated instances reflect the real-word theft cases and improve the model's performance. Besides, simulation results indicate that the proposed IQMOT is superior over the existing SMOTE technique. The pseudo-code of the proposed IQMOT is described in Algorithm 1.

Algorithm 1 IQMOT algorithm.

1: **Given:** An imbalanced dataset X with majority class $y_i = 0$ and minority class $z_i = 1$,
2: $X = \{(x_{1(t)}, y_1), (x_{2(t)}, y_2), (x_{3(t)}, z_1), (x_{4(t)}, y_2), ..., (x_{n(t)}, y_n), (x_{n(t)}, z_n)\}$
where $x_{i(t)} \in \Re$, $y_i \in \{0\}$ and $z_i \in \{1\}$,
3: **Output:** Balanced dataset $X^{'}$,
4: **Initialize:** Theft consumers $x^{'}$, normal consumers x, difference between thieves and normal consumers D, 25th percentile p_1, 50th percentile p_2 and 75th percentile p_3,
5: Get the total number of thieves and normal consumers,
6: Calculate p_1, p_2, p_3 of theft consumers,
7: Calculate percentage of values falling in each percentile,
8: Get numbers: a, b, c to represent the values fall in each group,
9: **For** n = 1,2,. . . ,D **do**
10: Create $x^{'}$ by selecting values from each group with respect to a, b, c,
where $x^{'} \in X^{'}$,
11: **End for**

In the real dataset comprising of electricity theft consumers, not all theft cases fall in the median of the Gaussian distribution. The electricity theft cases exhibit irregular electricity consumption behaviors, whose consumption values fall outside the median of a normal distribution; they are usually treated as outliers. Therefore, interquartile range is a good statistical tool to indicate such non-normal instances.

In this paper, we got inspiration from the outlier detection method, named interquartile range (IQR) [46], to devise the new class balancing technique that generates the NTL instances closer to the realistic theft cases. We refer to the percentiles where the theft cases are distributed into 25th, 50th and 75th percentiles. In the proposed IQMOT technique, we first get these percentiles that tell us the range and type of theft values falling into each of the percentile groups. Likewise, the median or 50th percentile contains the middle or average energy consumption values of electricity thieves.

In the proposed IQMOT, percentiles are used to define the limit over the theft values and gives us the representation of electricity theft cases. After getting the percentiles, the percentage of values coming in each percentile is obtained. Based on the computed percentage, we get a number showing the values of theft consumers lying in each group of percentiles. Similarly, values from each of the percentiles are obtained and a new theft case is produced, which is quite similar to the real theft cases. This process iterates until the minority class (electricity thefts) becomes equal to the majority class (normal consumers). Hence, the newly created minority instances reflect electricity consumption that is similar to the available theft instances in the dataset. In Figures 2 and 3, we observe the resemblance between original theft cases in the dataset and IQMOT-generated theft cases in a month. Moreover, it is evident that the IQMOT-generated theft cases are more realistic regarding the real theft cases. In this way, the proposed IQMOT overcomes the limitations of the above-mentioned traditional class balancing techniques and generates more realistic theft cases.

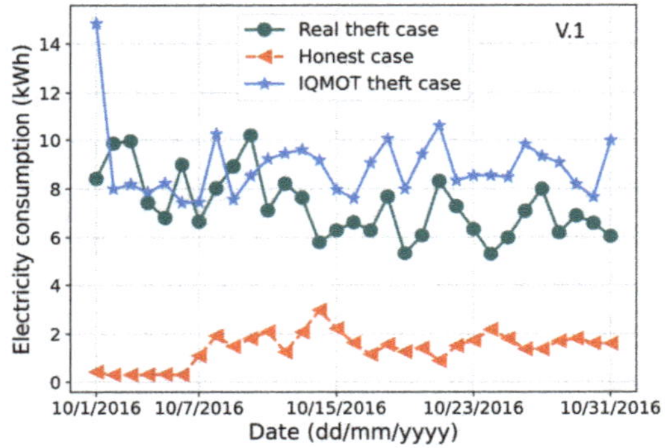

Figure 2. Example of IQMOT0based sample generation of October 2016.

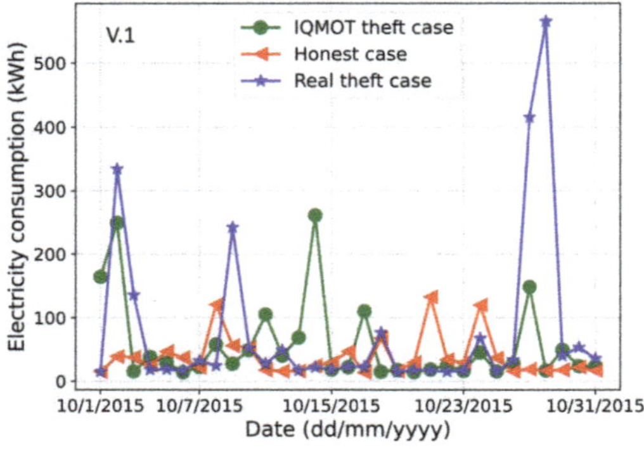

Figure 3. Example of IQMOT0based sample generation of October 2015.

2.4. Data Analysis

In the data analysis stage, we extract features from the preprocessed dataset and perform ETD. In particular, LSTM and UNet are applied to extract features from the preprocessed dataset and perform joint training through the Adaboost classifier for final classification. The following sub-sections explain the comprehensive description of each module of the LSTM–UNet–Adaboost.

2.4.1. LSTM Module

In this paper, LSTM is applied to capture the long-term associations from electricity consumption data, i.e., temporal correlations from electricity consumption time series at each time step. Therefore, 1-D daily data of electricity consumption is used as input to the LSTM. The electricity consumption data recorded by the smart meters are increasing day by day, which creates a large dataset history of a single user. A simple neural network or recurrent neural network (RNN) is not sufficient to obtain and maintain the long-term dependencies in their memory to forecast the future information. These models are difficult to train over a large historical dataset while trying to extract the long-term temporal correlations, which leads to the gradient vanishing and exploding problem [47]. For this reason, in this paper, the LSTM model is employed to memorize the long-term temporal associations from the extensive historical data.

LSTM is a special class of RNN, which has the capability to retain and propagate information from the initial stage towards the final stage of the model [48]. Figure 4 displays the general structure of the LSTM model. It has three important gates, known as input gate i_t, output gate o_t and forget gate f_t. Its main component is the cell state, which maintains the long-term dependencies along the chain. Thus, the dependencies in the cell state are managed by the aforementioned gates. In Figure 4, C_t and C_{t-1} show the current and previous cell state, respectively. h_t and h_{t-1} show the outputs of the current and previous LSTM units, respectively. Furthermore, $\sigma(x) = 1/1 + e^{-(x)}$ and $tanh(x) = e^{(2x)} + 1/e^{(2x)} - 1$ represent the sigmoid and hyperbolic tangent functions, respectively. Both functions are non-linear activations in the LSTM model. W_f, W_i, W_C, W_o and b_f, b_i, b_C, b_o are the weights and biases of the LSTM model, respectively. LSTM is based on daily electricity consumption data to produce a single output, known as h_t, which is also recognized as the hidden state at the last time step. Hence, LSTM achieves its purpose by processing the following Equations (4)–(9) [48]:

$$f_t = \sigma(W_f[C_{t-1}, h_{t-1}, x_t] + b_f), \tag{4}$$

$$i_t = \sigma(W_i[C_{t-1}, h_{t-1}, x_t] + b_i), \tag{5}$$

$$\tilde{C}_t = tanh(W_c[h_{t-1}, x_t] + b_c), \tag{6}$$

$$C_t = f_t.C_{t-1} + i_t.\tilde{C}_t, \tag{7}$$

$$o_t = \sigma(W_o[C_{t-1}, h_{t-1}, x_t] + b_o), \tag{8}$$

$$h_t = o_t.tanh(C_t). \tag{9}$$

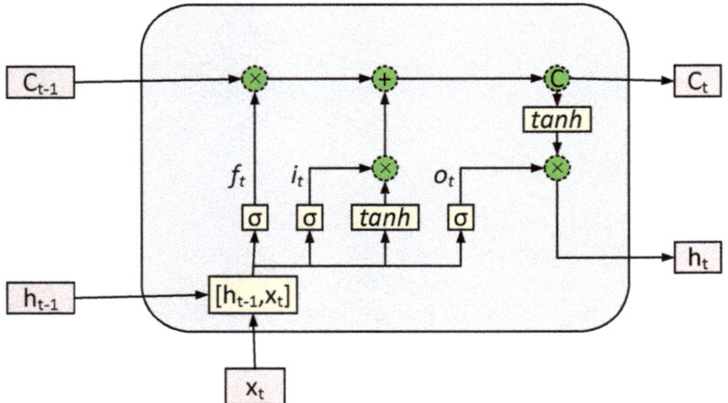

Figure 4. Architecture of the LSTM model.

In this paper, the deep LSTM model is used with a stack of recurrent LSTM layers, since a single layer LSTM model often fails to capture the complete dependencies. The output from each LSTM unit serves as an input to the batch normalization layer, which normalizes the previous layer output at runtime and forwards it to the next layers, where the batch normalization enhances the model convergence, extends the model stability and reduces both the overfitting and training time [49]. After that, the dropout layer is employed with a 0.5% probability that drops 50% of neurons randomly to inhibit the model from overfitting. Moreover, it improves the model's convergence by preventing the model from being over-dependent on a few neurons, which allows each neuron to work individually. In particular, the LSTM model comprises of three layers with batch normalization and dropout layers. It utilizes 1-D electricity consumption data using the Adam optimizer with the batch-size of 32 and binary cross-entropy as the cost function. Furthermore, a Keras callbacks concept is used during the model's training to practice the learning rate decay over five epochs and early stopping procedure over 10 epochs. Consequently, these procedures will improve the model's convergence and effectively mitigate the overfitting problem.

2.4.2. UNet Module

UNet is used in this paper to learn and derive potentially important information from 2-D electricity consumption data. As in [20], the authors explain the effect of the periodicity to illustrate how weekly data can better obtain the periodicity from consumption patterns. For this reason, the energy consumption data are transformed into 2-D weekly data and serve as an input to the UNet model. Furthermore, the authors in [20,22,40] have used a traditional CNN technique to derive high-level features from electricity consumption data. However, if we use the regular convolution network with the pooling and dense operations, the model will only extract high-level features of "what," but not their localization information, "where." As a result, in this paper, UNet is used, which derives both the high-level features and their localization information through the down-sampling and up-sampling strategies, respectively.

The UNet model was originally proposed for biomedical image segmentation [50]. The chief concept behind the semantic image segmentation is to attach the corresponding label of each pixel of the image [51]. In this way, the model predicts each pixel within the image, also known as the dense prediction. Therefore, semantic segmentation problems are considered as classification problems where each feature of a time-series is labeled with its corresponding class. In this paper, we get inspiration from such a semantic segmentation approach where the UNet model extracts the high-level features from the 2-D electricity consumption data and then labels them to their corresponding classes.

The name UNet is used because of its symmetric U-shape architecture, as shown in Figure 5. Its architecture mainly consists of two paths: the contraction path (also called down-sampling or encoder) and the expansion path (also called up-sampling or decoder) [50]. The contraction path performs down-sampling by the convolution and pooling operations, which are used to extract global features from 2-D data. On the other hand, the expansion path does up-sampling over these extracted features through the inverse or transpose convolution operation. Since transposed convolution is the inverse of convolution operation used to perform up-sampling. It tells us the whereabouts of information. We refer to down-sampling because both convolution and pooling operations reduce the size of input features or parameters. Consequently, the model determines parameters through the backpropagation procedure.

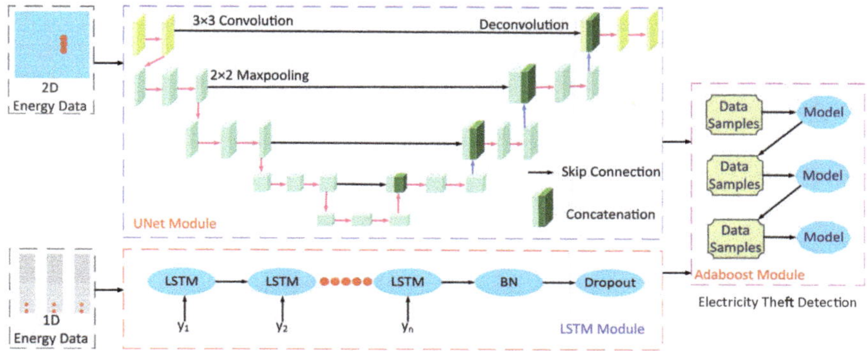

Figure 5. Overview of the proposed LSTM–UNet–Adaboost model.

UNet has both the long and short skipping connections. The short skipping connections are present in each of the major down or up-sampling blocks, while the long skipping connections are available within the contraction and expansion paths to concatenate the extracted features with their corresponding labels. In this work, the contraction path involves four major blocks where each block contains:

- Two 3 × 3 convolution layers plus *LeakyReLU* with batch normalization;
- 2 × 2 max pooling.

The feature maps are multiplied at each pooling layer, i.e., beginning from 16 feature maps in the first major block, 32 feature maps in the second block and so on. This procedure is also termed as increasing the size of depth and reducing the size of the input. Moreover, the expansion path consists of four major blocks where each block contains:

- Transpose-convolution with a stride of 2;
- Linking with regular convolution features;
- Two 3 × 3 convolution layers plus *LeakyReLU* with batch normalization.

The center of the contraction and expansion path is determined as the bottleneck; it employs a single convolution layer with batch normalization and dropout. In this paper, the UNet model is trained using an Adam optimizer with batch-size of 32 and binary cross-entropy as the cost function. Furthermore, Keras callbacks are used during the UNet training, as already described in Section 2.4.1. Simulation results validate that UNet appears to be very effective and efficient for ETD based on electricity consumption patterns.

2.4.3. Joint Training and Classification Module

For the joint training and classification mechanism, the ensemble learning boosting technique is applied where different weak classifiers are combined to build a powerful classifier, as shown in Figure 5. This is more accurate than the final joint training and classification through a single hidden layer based feed-forward neural network (FFNN)—for instance, a fully connected layer with either sigmoid activation or softmax classifier. To improve the ETD performance, we use Adaboost as a final classifier, which acts as a final layer of the LSTM-UNet to replace the single hidden layer based FFNN used in traditional models [20,22,23]. Therefore, Adaboost simply takes the outputs of LSTM-UNet modules and concatenates them to make a new input for the Adaboost model. Now, the long-term dependencies and high-level features are the inputs of the Adaboost model for the final theft detection. In this context, the proposed model derives the benefits of two powerful procedures of machine learning, known as deep learning and ensemble learning.

Adaboost is formerly designed to solve highly non-linear tasks [24]. The main focus of Adaboost is to learn from the mistakes of previous models and boost the performance of the next model. Thus, the most accurate classifier will be selected to perform the classification task. This process iterates until the training data becomes error-free or the model reaches the specified number of learners. Adaboost has several important hyperparameters, which influentially affect the model's theft detection performance. Therefore, the grid-search mechanism is applied in this paper to find the most appropriate hyperparameters of Adaboost, as described in Section 2.5.

2.5. Simulation Setting

The proposed model for ETD was implemented in python using the open-source deep learning libraries, known as Keras and Tensorflow. The proposed model was developed and simulated using the SGCC dataset, which contains a total of 42,372 consumers with 1035 days of electricity consumption history, as given in Table 1. For simulations, the dataset was first preprocessed through linear interpolation, three-sigma rule and min-max normalization. After that, the dataset was balanced through the proposed IQMOT technique. In the training procedure, the dataset was partitioned into training, validation and testing sets with a training proportion of 80%, and validation and testing proportions of 10% for each, respectively. LSTM model's configuration consisted of three layers with batch normalization and dropout layers; each LSTM layer had 60 neurons. Besides, the UNet model's configuration was the same as already defined in Section 2.4.2. For the training of LSTM, 30 iterations were run initially with a batch-size of 32 using Adam optimizer, while the training of UNet was initially performed by running 15 training iterations with the batch-size of 32. Finally, the Adaboost was executed by utilizing the outputs of LSTM and UNet modules as an input. Furthermore, to select the optimal hyperparameters of the Adaboost and other models, a grid-search algorithm was implemented. Table 2 shows the important hyperparameters selected for the Adaboost model using grid-search.

Table 2. Adaboost hyperparameter selected through grid-search.

Hyperparameter	Range of Values	Selected Value
Estimators	100, 200, 300, 400	400
Learning rate	0.1, 0.01, 0.001	0.001

2.6. Loss Function

The most widely used loss function for the classification problem is cross-entropy, to classify only two classes. In this paper, the binary cross-entropy loss function, also known as the logarithmic loss, is used to deal with the binary classification task. The predictions become more accurate as the loss function converges to zero. The binary cross-entropy loss function is calculated using the following formula [23]:

$$log\_loss = \frac{1}{N}\sum_{i=1}^{M} -(y_i log(p(y_i)) + (1 - y_i) log(1 - p(y_i)), \tag{10}$$

where N shows the accumulative consumer instances. y_i represents the actual label and $p(y_i)$ is the likelihood of the electricity theft measured by the proposed model for the ith consumer.

2.7. Performance Evaluation Metrics

In this paper, seven class imbalance metrics are employed to evaluate the performance of the proposed model, which includes area under the curve (AUC), precision, recall, Mathews correlation coefficient (MCC), F1-score, area under the precision-recall curve (PR-AUC) and accuracy. These performance evaluation metrics are determined from the confusion matrix, i.e., a matrix that describes different results in classification problems. Specifically, for the binary classification problem, the confusion matrix returns two rows and two columns, i.e., four possible outcomes. These four possible outcomes are described as follows:

- The true positive (TP) score demonstrates the number of dishonest consumers accurately predicted by the classifier;
- The true negative (TN) score shows the number of honest consumers accurately predicted by the classifier;
- The false positive (FP) score describes the number of honest consumers predicted by the model as thieves;
- The false negative (FN) score highlights the number of dishonest consumers predicted by the model as honest consumers.

The following are the performance metrics given in Equations (11)–(16), as defined in [20,21,24]:

$$Accuracy = \frac{TP + TN}{TP + TN + FP + FN}, \tag{11}$$

$$Recall = \frac{TP}{TP + FN}, \tag{12}$$

$$Precision = \frac{TP}{TP + FP}, \tag{13}$$

$$F1 = 2 \times \frac{Precison \times Recall}{Precison + Recall}, \tag{14}$$

$$AUC = \frac{\sum_{i \in positiveclass} RANK_i - \frac{P(1+P)}{2}}{P \times N}, \tag{15}$$

$$MCC = \frac{(TP \times TN - FP \times FN)}{\sqrt{(TP + FP)(TP + FN)(TN + FP)(TN + FN)}}. \tag{16}$$

where P represents the number of positive samples, N represents the number of negative samples and $RANK_i$ shows the rank value of sample i.

The accuracy of a classifier is a metric used to indicate the percentage of correct predictions. Besides, a recall is another class imbalance metric, which is also termed the detection rate (DR), sensitivity or the true positive rate (TPR) in the literature. It shows the capability of a scheme to detect electricity theft consumers. Likewise, precision is the ability of the classifier to accurately classify normal consumers. However, accuracy, precision and recall metrics using the imbalanced dataset cannot provide a realistic assessment of the model's ETD performance [7]. Hence, F1-score is another useful class imbalance measure as compared to the metrics described above for ETD's performance examination. In F1-score, we get the balance of both precision and recall, which shows its usefulness as compared to the other metrics.

Furthermore, a performance index that is more reliable and accurate for an imbalanced dataset is the AUC. It provides a more realistic assessment of the model's detection performance in terms of ETD over the imbalanced dataset. Essentially, it is the likelihood that the model ranks a positive sample higher than the negative sample. AUC is also identified as the area under the receiver operating characteristic curve (ROC-AUC). ROC-AUC is used to evaluate the model's capability to make the separation between the classes. Thus, it is a graphical representation to evaluate the ETD performance of a model by plotting the TPR against the FPR. Moreover, the area under the ROC curve is estimated between the threshold of 0 and 1. If the classifier has a ROC-AUC score higher than 0.5, then it produces a better DR against any random predictions. If the classifier has less than 0.5 ROC-AUC, then it implies that the classifier has limited classification capability.

PR-AUC is another useful class imbalance measure employed to assess the model's performance. Therefore, in this paper, we use PR-AUC, which considers the precision of the classifier and highlights the cost of on-site inspections for the utilities. PR-AUC is examined only when positive samples are on the top rather than the negative samples where the score is improved only when positive samples are on the top and negatives samples are on the bottom. Likewise, MCC is a binary classification metric used to evaluate the model's performance using the imbalanced data. Moreover, MCC is a more accurate class imbalance metric than the AUC and F1-score because MCC captures the correlation between all four possible outcomes of the confusion matrix and suggests essential evaluation metrics. The MCC score ranges from -1 to 1, where a value near to 1 shows an accurate classification. Likewise, 0 shows the result of random predictions where the model has no class separation capability and -1 dictates incorrect classification. Accordingly, a classifier is good if it achieves ETD objective effectively, i.e., a classifier with a high DR performance and low FPR. The cost of FN is pretty high and important because it shows the cost of energy stolen and not given by the theft consumers. The cost of FP is much lower than FN because it shows the cost of inspection rather than the cost of stolen energy. Hence, in ETD, more importance is given to recall than precision.

2.8. Benchmark Models

In this section, we illustrate the state-of-the-art benchmark models and basic classification techniques used for comparison with our proposed model. For a fair comparison, we implemented a grid-search algorithm to determine the most suitable hyperparameters of the benchmark models.

2.8.1. Logistic Regression (LR)

LR is the primary model for the binary classification task in ETD, which applies the notion of probability and uses the principle of neural networks. For instance, LR for binary classification is similar to the single hidden layer based neural network using the sigmoid activation function. Thus, the sigmoid score ranges between 0 and 1, where a value near to 1 is labeled as theft and near to 0 is classified as honest. Table 3 shows the hyperparameters selected for LR through grid-search.

Table 3. LR hyperparameters selected through grid-search.

Hyperparameter	Range of Values	Selected Value
C	0.1, 0.01, 0.001	0.001
R	l1 norm, l2 norm	l2 norm

2.8.2. SVM

SVM is a famous technique used to solve the ETD problem. Many previous studies, such as [52,53], have used SVM to detect the presence of electricity thieves. Moreover, SVM has important hyperparameters obtained by employing the grid-search algorithm, as shown in Table 4.

Table 4. SVM hyperparameters selected through grid-search.

Hyperparameter	Range of Values	Selected Value
C	0.1, 0.01, 0.001	0.1
γ	1, 1.5, 6, 10	10

2.8.3. RUSBoost

The RUSBoost technique is the combination of RUS and Adaboost. In [24,25], the authors used RUSBoost to perform ETD. Table 5 demonstrates the selection of the RUSBoost's hyperparameters using the grid-search technique.

Table 5. RUSBoost hyperparameters selected through grid-search.

Hyperparameter	Range of Values	Selected Value
Learning rate	1.0, 0.1, 0.01	0.1
Estimators	20, 30, 100	30

2.8.4. Bagged Tree

The authors in [21] use a bagged tree for NTL detection. A bagged tree is an ensemble learning technique, in which a number of training subsets are generated with replacements and different classifiers are trained on these subsets. Finally, a single model is selected based on the majority of votes from each model. Table 6 shows the optimal hyperparameter selection for bagged tree using grid-search.

Table 6. Bagged tree hyperparameters selected through grid-search.

Hyperparameter	Range of Values	Selected Value
Estimators	20, 30, 100	30

2.8.5. WD-CNN

The authors in [20] proposed WD-CNN to detect electricity thieves. The authors trained a wide component using 1-D data and the deep component on 2-D data. Therefore, in this paper, we practice the same WD-CNN setting as formerly proposed by the authors [20].

2.8.6. CNN-LSTM

CNN-LSTM is a hybrid deep learning model for NTL detection [22]. It consists of CNN for feature extraction and the derived features further serve as inputs to the LSTM model for classification. Hence, a similar arrangement of CNN-LSTM is used in this paper for comparison.

2.8.7. CNN-RF

CNN-RF is a composite of CNN and RF used for ETD [40]. CNN is applied to derive global features from data. After that, the derived features are delivered to RF for ETD where RF acts as a final layer of CNN. Therefore, the same model arrangement is considered in this paper as a benchmark scheme.

2.8.8. LSTM-MLP

The authors proposed a hybrid LSTM-MLP model using sequential and non-sequential data for NTL detection [23]. For a fair comparison, the same model configuration is used in this paper as that already proposed by the authors [23].

3. Simulation Results and Discussion

In this section, we describe the simulation results of our proposed model together with the performance comparison with state-of-the-art models. Moreover, to validate the LSTM–UNet–Adaboost model's performance and robustness, seven performance evaluation metrics are used to show the superiority of our proposed model over benchmark schemes for theft detection.

3.1. Performance Comparison with Benchmark Models

This section assesses the performance of our proposed LSTM–UNet–Adaboost model for ETD in the context of smart grids. To estimate the effectiveness of the proposed model, its results were analyzed against other models using the previously mentioned seven performance metrics. Table 7 presents the results of our proposed model and other existing benchmark models. It is seen that our proposed model achieved 0.94, 0.90, 0.95, 0.99, 0.92, 0.95 and 0.97 for AUC, MCC, F1-score, precision, recall, PR-AUC and accuracy, respectively. Thus, the proposed model outperformed all existing benchmark models in terms of these evaluation metrics. Likewise, CNN-RF was the second-best classifier; however, it had a low DR, i.e., 0.803, in comparison to other models. LSTM-MLP had the second-best DR, i.e., 0.889.

The principal objective of ETD is to improve theft DR and reduce FPR. Particularly, the proposed model presents the best results for each of the class imbalance metrics, as it achieved a DR of 0.92, which is the highest value among all existing benchmark models. Moreover, Table 7 also demonstrates the importance of our proposed IQMOT technique. Despite the performances of other models, such as LR, SVM, CNN-RF, WD-CNN, RUSBoost, bagged tree, LSTM-MLP and CNN-LSTM on the IQMOT-based processed data, our proposed model still outperformed existing models.

Figure 6 shows the ROC-AUC of our proposed LSTM–UNet–Adaboost model; the proposed model has achieved better ROC-AUC than other random predictions in terms of training and validation sets. Similarly to the ROC-AUC, our proposed model has also covered more PR-AUC than any random predictions using IQMOT, as shown in Figure 7. It is worth noting that our proposed scheme has covered more areas over the training and validation sets in terms of ROC-AUC and PR-AUC, which shows the superiority of our proposed scheme. Furthermore, Figure 8 shows the ROC-AUC comparison with existing benchmark models. It is evident that the proposed model has the highest ROC-AUC score and covers more area under the ROC curve. As mentioned earlier, the main goal is to maximize the DR and minimize the FPR in ETD, which has been achieved by our model as compared to the existing models.

Table 7. Proposed model's performance comparison with conventional schemes for ETD.

Model	AUC	MCC	F1-Score	Precision	Recall	PR-AUC	Accuracy
LR	0.835	0.670	0.835	0.836	0.834	0.785	0.835
SVM	0.575	0.256	0.698	0.542	0.878	0.548	0.576
CNN-RF	0.889	0.815	0.869	0.965	0.803	0.883	0.900
Bagged Tree	0.883	0.812	0.862	0.956	0.821	0.876	0.882
RUSBoost	0.881	0.689	0.848	0.825	0.824	0.787	0.844
WD-CNN	0.884	0.773	0.878	0.927	0.834	0.864	0.884
LSTM-MLP	0.866	0.732	0.869	0.849	0.889	0.818	0.866
CNN-LSTM	0.889	0.785	0.882	0.946	0.826	0.879	0.889
Proposed model	0.948	0.902	0.954	0.998	0.929	0.958	0.972

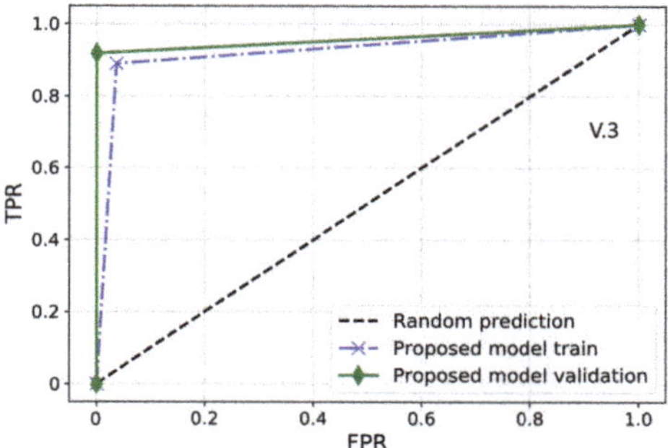

Figure 6. Proposed model's ROC-AUC-based analysis with IQMOT.

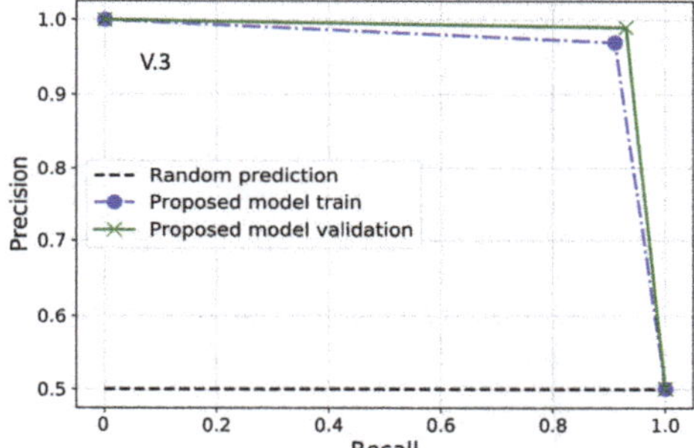

Figure 7. Proposed model's PR-AUC-based analysis with IQMOT.

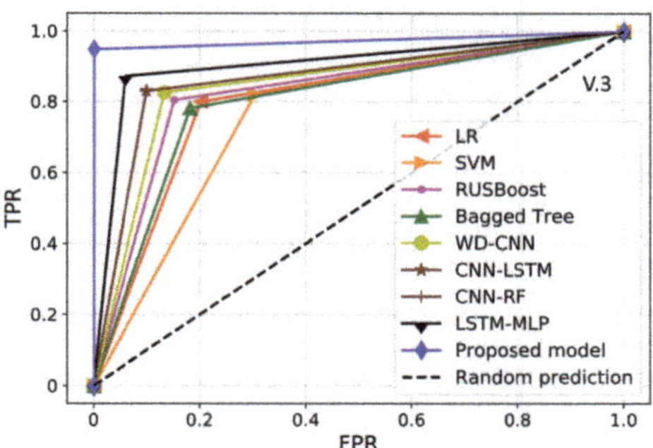

Figure 8. ROC-AUC-based performance comparison with existing benchmarks.

Similarly to ROC-AUC, the proposed model has also reported better results in terms of PR-AUC in comparison to other models, as shown in Figure 9. As we have already highlighted in Table 7, our proposed model achieved the highest PR-AUC score, higher than those of other benchmark models.

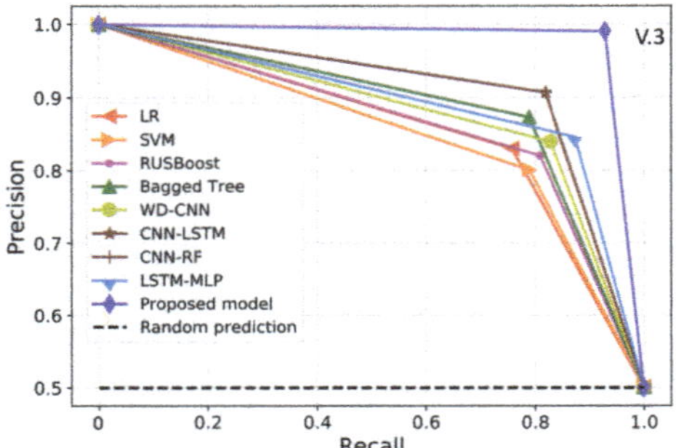

Figure 9. PR-AUC-based performance comparison with existing benchmarks.

3.2. Comparison Based on Proposed IQMOT

This section investigates the effects of imbalanced data on supervised learning techniques and the effectiveness of the novel IQMOT technique. Table 8 presents the results of the proposed model with IQMOT, SMOTE and without any class balancing technique to explain the importance of using the proposed IQMOT over the existing benchmark SMOTE technique. It is clear that the proposed IQMOT is more effective as compared to the existing SMOTE. Essentially, based on IQMOT, the proposed model achieved 0.94, 0.90, 0.95, 0.99, 0.92, 0.95 and 0.97 for the AUC, MCC, F1-score, precision, recall, PR-AUC and accuracy, respectively. Based on SMOTE, the proposed model achieved 0.90, 0.81, 0.90, 0.87, 0.90, 0.85 and 0.90 for AUC, MCC, F1, precision, recall, PR-AUC and accuracy, respectively. Consequently, this confirms the advantage of novel IQMOT over existing class balancing techniques.

Moreover, we have examined the effects of highly imbalanced data on the performance of supervised learning methods. In the third column (No Balancing) of Table 8, we express that the performance of the supervised learning model without applying any class balancing technique is worst, where the precision value of 1.00 indicates that the model misclassifies electricity theft consumers as honest consumers. In particular, this shows the significance of the class balancing mechanism and the adverse effect on the performance of supervised learning methods. Moreover, Figures 10 and 11 show that the proposed model covers less area under the PR and ROC curve as no class balancing mechanism is applied. It can be seen that without applying any class balancing technique, the model only achieves 0.63 PR-AUC and 0.60 ROC-AUC over the training and validation sets. This shows the need for a more accurate and useful class balancing technique. Furthermore, Figure 12 shows the performance analysis of the proposed model with benchmark schemes based on four important and useful class imbalance metrics, known as F1-score, MCC, PR-AUC score and ROC-AUC. It is visible that the proposed model shows excellent results compared to existing benchmark models, including LR, SVM, RUSBoost, bagged tree, WD-CNN, CNN-LSTM, CNN-RF and LSTM-MLP. This shows the superiority of the proposed methodology in terms of highly imbalanced ETD problem, which makes it acceptable of real use. Bagged tree and CNN-RF performances are quite similar to each other. Particularly, this paper focuses on raising the ETD performance without taking into account the computational cost of the proposed mechanism.

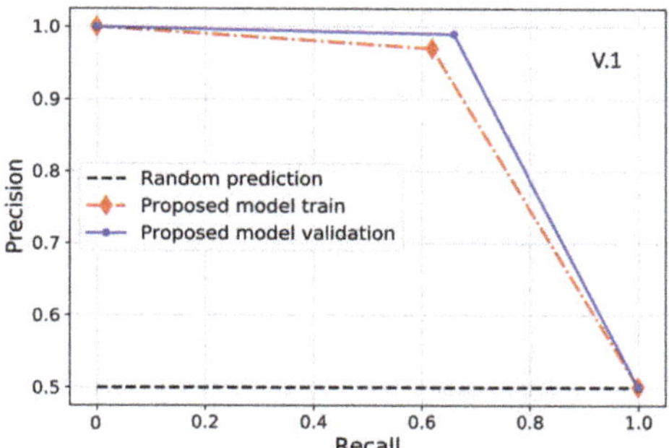

Figure 10. Proposed model's PR-AUC-based analysis without using IQMOT.

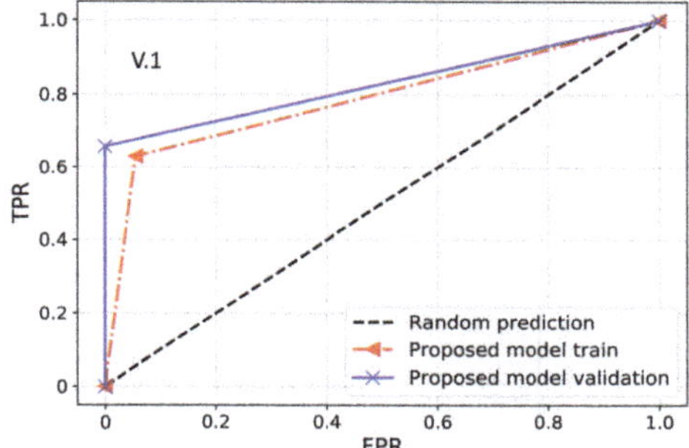

Figure 11. Proposed model's ROC-AUC-based analysis without using IQMOT.

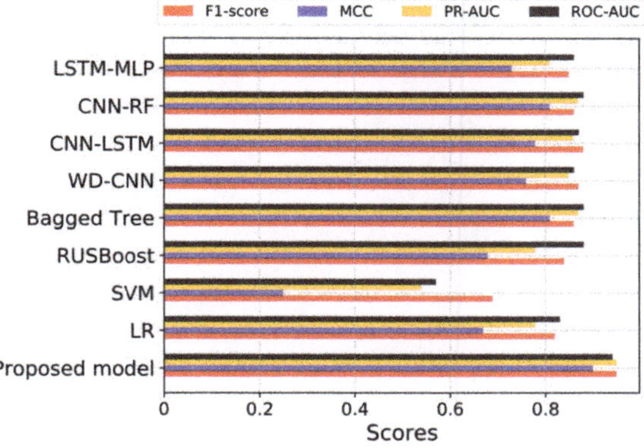

Figure 12. Performance comparison based on F1, MCC, PR-AUC and ROC-AUC.

Table 8. IQMOT-based performance comparison.

Metrics	IQMOT	No Balancing	SMOTE
AUC	0.948	0.602	0.906
MCC	0.902	0.753	0.817
F1-score	0.954	0.753	0.901
Precision	0.998	1.00	0.870
Recall	0.929	0.604	0.905
PR-AUC	0.95	0.63	0.85
Accuracy	0.972	0.745	0.906

3.3. Convergence Analysis

The simulations were done on the preprocessed data with a batch-size of 32. Epoch is a parameter that controls the model training. Figures 13 and 14 illustrate the learning process of LSTM and UNet based on the training and validation loss to select the best configuration of a particular model. Figure 13 highlights the learning curve of LSTM in terms of logarithmic loss. In the first attempt, thirty training iterations were passed to the LSTM model; it shows the smooth learning process of the LSTM model and no overfitting happened until the 28th epoch. As we can see that at 28th iteration, the LSTM model performed best and reduced the training and validation loss to 0.67. It shows the gradual convergence of training and validation loss. Finally, the best model was selected at the 28th iteration, as shown in Figure 13, to train the LSTM model.

This process explains that when we pick a small number of epochs, then the LSTM model is not well trained to capture all the temporal correlations from the electricity consumption data. On the other hand, if we choose a considerable amount of training epochs, then the model leads to the overfitting problem. Therefore, it is necessary to select the optimal number of epochs to avoid underfitting and overfitting problems. Figure 14 expresses the learning process of the UNet model based on logarithmic loss; a stable learning process of UNet is presented. For UNet training, 15 epochs were used to get the best fit model. The learning process of the UNet was also very smooth, as both the training and validation losses increasingly converged, which shows that the model gave the best fitting value at the 14th epoch. Finally, the best fit model was selected at the 14th iteration with a validation loss of 0.03 to train the UNet. Consequently, we can see the effects of smooth learning in Table 7; our proposed model outperformed all other benchmark models.

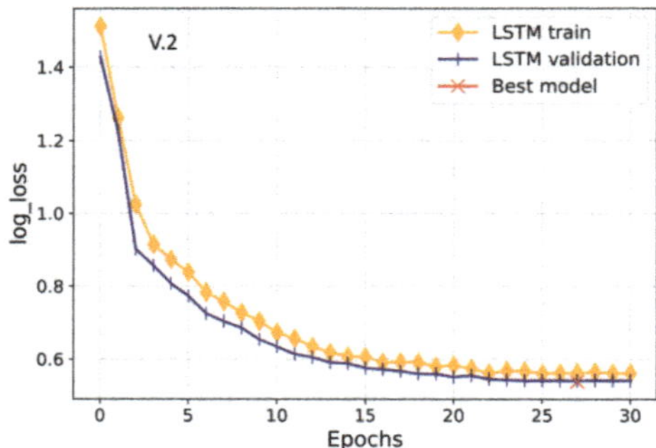

Figure 13. LSTM-based convergence analysis during training.

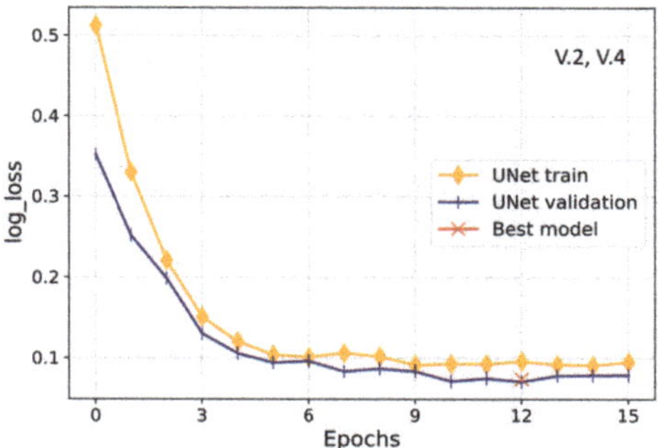

Figure 14. UNet-based convergence analysis during training.

Table 9 presents the mapping between limitations addressed, proposed solution and validation. The proposed solution addresses the limitations of traditional ETD models. It solves the problem of data imbalance through the novel IQMOT technique that generates more efficient theft samples, as depicted in Figures 2 and 3. Moreover, Figures 10 and 11 validate the significance of the proposed IQMOT in terms of no class balancing technique. Afterwards, the proposed solution utilizes the UNet and LSTM to efficiently derive the long-term dependencies and high-level features from high-dimensional data. Figures 13 and 14 show that the LSTM and UNet efficiently capture important information from high-dimensional imbalanced data. Moreover, UNet also catches features' localizations that are lost by the conventional CNN based approaches, which significantly improves the ETD results, as depicted in Figure 14. Furthermore, to make better predictions, we utilize Adaboost as a classification mechanism that utilizes the features derived by LSTM and UNet. This mechanism avoids the limitation of traditional models, which face the overfitting problem, as validated in Figures 6–9. Hence, the proposed methodology performs more better than the conventional schemes for the identification of electricity frauds.

Table 9. Mapping between the identified problems, proposed solution and validation.

Problem Identified	Proposed Solution	Validation
L.1 Model's biasness due to imbalanced data	S.1 IQMOT	V.1 The proposed IQMOT generates more realistic samples, as shown in Figures 2, 3, 10 and 11
L.2 High-dimensional data and artificial feature extraction	S.2 Deep LSTM and UNet	V.2 The deep LSTM and UNet efficiently extracts the potential features, as given in Figures 13 and 14
L.3 Final ETD through sigmoid or softmax activation based hidden layer	S.3 Adaboost based LSTM-UNet	V.3 Adaboost acts as final layer of LSTM and UNet that gives better results, as depicted in Figures 6–9
L.4 Poor ETD performance due to loss of features' localization information	S.4 UNet captures both what and whereabout of data	V.4 UNet brings significant improvement in ETD results, as shown in Figure 14

4. Conclusions

- In this work, a combined LSTM–UNet–Adaboost model and a novel class balancing mechanism IQMOT are proposed for ETD in the smart grid environment. IQMOT is introduced to solve the data imbalance concerns faced by the traditional models.
- To increase the model's theft detection performance and stability, deep learning LSTM and UNet are combined with an ensemble learning Adaboost. Deep learning automates feature extraction from 1-D and 2-D electricity consumption data, whereas ensemble learning is used for joint

training and classification. In this way, the LSTM–UNet–Adaboost model gains the benefits of most recent and powerful techniques of deep learning and ensemble learning.

- Extensive simulations were conducted using realistic electricity consumption data of SGCC. During performance evaluation, we employed the grid-search algorithm to obtain the most appropriate values for the hyperparameters of different models for a fair comparison. The proposed model, LSTM–UNet–Adaboost, achieved 0.94, 0.90, 0.95, 0.99, 0.92, 0.95 and 0.97 for AUC, MCC, F1-score, precision, recall, PR-AUC and accuracy on the test dataset, respectively. Thus, the simulation results show the superiority of the combined LSTM–UNet–Adaboost model over existing state-of-the-art methods, including LR, SVM, CNN-RF, WD-CNN, RUSBoost, bagged tree, LSTM-MLP and CNN-LSTM. Moreover, the newly proposed IQMOT was far better than the existing SMOTE by generating more real electricity theft cases. Consequently, the proposed model shows brilliance for practical terms in the smart grid and can be used in many other scenarios, for instance, anomaly detection applications.

5. Future Work

This work solely focused on enhancing the ETD performance against conventional schemes without considering the computational cost. Therefore, we did not take into account the computational time comparison, which is part of our future work. Moreover, in future, we intend to incorporate other features, such as number of appliances, geographical location and temperature, along with the electricity consumption, to improve the ETD performance.

Author Contributions: Z.A. and N.J. proposed and implemented the main idea. N.J. and A.A. (Ashfaq Ahmad) performed the mathematical modeling and wrote the simulation section. A.A. (Abrar Ahmed) and S.M.G. organized and refined the manuscript. All authors have read and agreed to the published version of the manuscript.

Funding: This research received no external funding.

Conflicts of Interest: The authors declare no conflict of interest.

References

1. Ahmad, T. Non-technical loss analysis and prevention using smart meters. *Renew. Sustain. Energy Rev.* **2017**, *72*, 573–589. [CrossRef]
2. Depuru, S.S.S.R.; Wang, L.; Devabhaktuni, V. Electricity theft: Overview, issues, prevention and a smart meter based approach to control theft. *Energy Policy* **2011**, *39*, 1007–1015. [CrossRef]
3. Khan, J.R.; Siddiqui, F.A.; Khan, R.R. Survey: NTL Detection in Electricity Energy Supply. *Int. J. Comput. Appl.* **2016**, *155*, 18–23.
4. Gaur, V.; Gupta, E. The determinants of electricity theft: An empirical analysis of Indian states. *Energy Policy* **2016**, *93*, 127–136. [CrossRef]
5. McLaughlin, S.; Holbert, B.; Fawaz, A.; Berthier, R.; Zonouz, S. A multi-sensor energy theft detection framework for advanced metering infrastructures. *IEEE J. Sel. Areas Commun.* **2013**, *31*, 1319–1330. [CrossRef]
6. Manur, A.; Venkataramanan, G.; Sehloff, D. Simple electric utility platform: A hardware/software solution for operating emergent microgrids. *Appl. Energy* **2018**, *210*, 748–763. [CrossRef]
7. Glauner, P.; Meira, J.A.; Valtchev, P.; State, R.; Bettinger, F. The challenge of non-technical loss detection using artificial intelligence: A survey. *Int. J. Comput. Intell. Syst.* **2017**, *10*, 760–775. [CrossRef]
8. Bank, T.W. *Electric Power Transmission and Distribution Losses (% of Output)*; IEA: Paris, France, 2016. Available online: https://data.worldbank.org/indicator/EG.ELC.LOSS.ZS (accessed on 10 July 2020).
9. Jokar, P.; Arianpoo, N.; Leung, V.C. Electricity theft detection in AMI using customers' consumption patterns. *IEEE Trans. Smart Grid* **2016**, *7*, 216–226. [CrossRef]
10. Lewis, F.B. Costly 'throw-ups': electricity theft and power disruptions. *Electr. J.* **2015**, *28*, 118–135. [CrossRef]
11. Smart Meters Help Reduce Electricity Theft, BC, I. Hydro, Vancouver, BC, Canada. March 2011. Available online: https://www.bchydro.com/news/conservation/2011/smart_meters_energy_theft.html (accessed on 10 July 2020).

12. Yao, D.; Wen, M.; Liang, X.; Fu, Z.; Zhang, K.; Yang, B. Energy theft detection with energy privacy preservation in the smart grid. *IEEE Internet Things J.* **2020**, *6*, 7659–7669. [CrossRef]
13. Chou, J.S.; Yutami, I.G.A.N. Smart meter adoption and deployment strategy for residential buildings in Indonesia. *Appl. Energy* **2014**, *128*, 336–349. [CrossRef]
14. Mujeeb, S.; Javaid, N. ESAENARX and DE-RELM: Novel schemes for big data predictive analytics of electricity load and price. *Sustain. Cities Soc.* **2019**, *51*, 101642. [CrossRef]
15. Wang, K.; Xu, C.; Zhang, Y.; Guo, S.; Zomaya, A.Y. Robust big data analytics for electricity price forecasting in the smart grid. *IEEE Trans. Big Data* **2019**, *5*, 34–45. [CrossRef]
16. Wang, Y.; Chen, Q.; Hong, T.; Kang, C. Review of smart meter data analytics: Applications, methodologies, and challenges. *IEEE Trans. Smart Grid* **2018**, *10*, 3125–3148. [CrossRef]
17. Chen, S.; Liu, C.C. From demand response to transactive energy: State of the art. *J. Mod. Power Syst. Clean Energy* **2017**, *5*, 10–19. [CrossRef]
18. Samuel, O.; Javaid, N.; Khalid, A.; Khan, W.Z.; Aalsalem, M.Y.; Afzal, M.K.; Kim, B.S. Towards Real-time Energy Management of Multi-microgrid using a Deep Convolution Neural Network and Cooperative Game Approach. *IEEE Access* **2020**, *8*, 161377–161395. [CrossRef]
19. Wang, Y.; Chen, Q.; Kang, C.; Xia, Q. Clustering of electricity consumption behavior dynamics toward big data applications. *IEEE Trans. Smart Grid* **2016**, *7*, 2437–2447. [CrossRef]
20. Zheng, Z.; Yang, Y.; Niu, X.; Dai, H.N.; Zhou, Y. Wide and deep convolutional neural networks for electricity-theft detection to secure smart grids. *IEEE Trans. Ind. Inform.* **2018**, *14*, 1606–1615. [CrossRef]
21. Saeed, M.S.; Mustafa, M.W.; Sheikh, U.U.; Jumani, T.A.; Mirjat, N.H. Ensemble bagged tree based classification for reducing non-technical losses in multan electric power company of Pakistan. *Electronics* **2019**, *8*, 860. [CrossRef]
22. Hasan, M.; Toma, R.N.; Nahid, A.A.; Islam, M.M.; Kim, J.M. Electricity theft detection in smart grid systems: A CNN-LSTM based approach. *Energies* **2019**, *12*, 3310. [CrossRef]
23. Buzau, M.M.; Tejedor-Aguilera, J.; Cruz-Romero, P.; Gomez-Exposito, A. Hybrid deep neural networks for detection of non-technical losses in electricity smart meters. *IEEE Trans. Power Syst.* **2020**, *35*, 1254–1263. [CrossRef]
24. Avila, N.F.; Figueroa, G.; Chu, C.C. NTL detection in electric distribution systems using the maximal overlap discrete wavelet-packet transform and random undersampling boosting. *IEEE Trans. Power Syst.* **2018**, *33*, 7171–7180. [CrossRef]
25. Adil, M.; Javaid, N.; Qasim, U.; Ullah, I.; Shafiq, M.; Choi, J.G. LSTM and Bat-Based RUSBoost Approach for Electricity Theft Detection. *Appl. Sci.* **2020**, *10*, 4378. [CrossRef]
26. Ghasemi, A.A.; Gitizadeh, M. Detection of illegal consumers using pattern classification approach combined with Levenberg-Marquardt method in smart grid. *Int. J. Electr. Power Energy Syst.* **2018**, *99*, 363–375. [CrossRef]
27. Leite, J.B.; Mantovani, J.R.S. Detecting and locating non-technical losses in modern distribution networks. *IEEE Trans. Smart Grid* **2018**, *9*, 1023–1032. [CrossRef]
28. Lo, C.H.; Ansari, N. CONSUMER: A novel hybrid intrusion detection system for distribution networks in smart grid. *IEEE Trans. Emerg. Top. Comput.* **2013**, *1*, 33–44. [CrossRef]
29. Huang, S.C.; Lo, Y.L.; Lu, C.N. Non-technical loss detection using state estimation and analysis of variance. *IEEE Trans. Power Syst.* **2013**, *28*, 2959–2966. [CrossRef]
30. Amin, S.; Schwartz, G.A.; Cárdenas, A.A.; Sastry, S.S. Game-theoretic models of electricity theft detection in smart utility networks: Providing new capabilities with advanced metering infrastructure. *IEEE Control Syst. Mag.* **2015**, *35*, 66–81.
31. Lin, C.H.; Chen, S.J.; Kuo, C.L.; Chen, J.L. Non-cooperative game model applied to an advanced metering infrastructure for non-technical loss screening in micro-distribution systems. *IEEE Trans. Smart Grid* **2014**, *5*, 2468–2469. [CrossRef]
32. Maamar, A.; Benahmed, K. A hybrid model for anomalies detection in AMI system combining k-means clustering and deep neural network. *CMC-Comput. Mater. Contin* **2019**, *60*, 15–39. [CrossRef]
33. Zheng, K.; Chen, Q.; Wang, Y.; Kang, C.; Xia, Q. A novel combined data-driven approach for electricity theft detection. *IEEE Trans. Ind. Inform.* **2019**, *15*, 1809–1819. [CrossRef]
34. Buzau, M.M.; Tejedor-Aguilera, J.; Cruz-Romero, P.; Gómez-Expósito, A. Detection of non-technical losses using smart meter data and supervised learning. *IEEE Trans. Smart Grid* **2019**, *10*, 2661–2670. [CrossRef]

35. Punmiya, R.; Choe, S. Energy theft detection using gradient boosting theft detector with feature engineering-based preprocessing. *IEEE Trans. Smart Grid* **2019**, *10*, 2326–2329. [CrossRef]
36. Khan, Z.A.; Adil, M.; Javaid, N.; Saqib, M.N.; Shafiq, M.; Choi, J.G. Electricity Theft Detection Using Supervised Learning Techniques on Smart Meter Data. *Sustainability* **2020**, *12*, 8023. [CrossRef]
37. Li, J.; Wang, F. Non-Technical Loss Detection in Power Grids with Statistical Profile Images Based on Semi-Supervised Learning. *Sensors* **2020**, *20*, 236. [CrossRef]
38. Hu, T.; Guo, Q.; Shen, X.; Sun, H.; Wu, R.; Xi, H. Utilizing unlabeled data to detect electricity fraud in AMI: A semisupervised deep learning approach. *IEEE Trans. Neural Netw. Learn. Syst.* **2019**, *30*, 3287–3299. [CrossRef]
39. Gul, H.; Javaid, N.; Ullah, I.; Qamar, A.M.; Afzal, M.K.; Joshi, G.P. Detection of Non-Technical Losses using SOSTLink and Bidirectional Gated Recurrent Unit to Secure Smart Meters. *Appl. Sci.* **2020**, *10*, 3151. [CrossRef]
40. Li, S.; Han, Y.; Yao, X.; Yingchen, S.; Wang, J.; Zhao, Q. Electricity Theft Detection in Power Grids with Deep Learning and Random Forests. *J. Electr. Comput. Eng.* **2019**. [CrossRef]
41. Ullah, A.; Javaid, N.; Samuel, O.; Imran, M.; Shoaib, M. CNN and GRU based Deep Neural Network for Electricity Theft Detection to Secure Smart Grid. In Proceedings of the 2020 IEEE International Wireless Communications and Mobile Computing (IWCMC), Limassol, Cyprus, 15–19 June 2020; pp. 1598–1602.
42. State Grid Corporation of China Dataset. Available online: https://www.sgcc.com.cn/ (accessed on 15 August 2020).
43. Khalid, R.; Javaid, N. A Survey on Hyperparameters Optimization Algorithms of Forecasting Models in Smart Grid. *Sustain. Cities Soc.* **2020**, 102275. [CrossRef]
44. Chandola, V.; Banerjee, A.; Kumar, V. Anomaly detection: A survey. *ACM Comput. Surv. (CSUR)* **2009**, *41*, 1–58. [CrossRef]
45. Yang, R.; Zhang, C.; Gao, R.; Zhang, L. A novel feature extraction method with feature selection to identify Golgi-resident protein types from imbalanced data. *Int. J. Mol. Sci.* **2016**, *17*, 218. [CrossRef] [PubMed]
46. Wan, X.; Wang, W.; Liu, J.; Tong, T. Estimating the sample mean and standard deviation from the sample size, median, range and/or interquartile range. *BMC Med. Res. Methodol.* **2014**, *14*, 135. [CrossRef] [PubMed]
47. Fei, H.; Tan, F. Bidirectional grid long short-term memory (bigridlstm): A method to address context-sensitivity and vanishing gradient. *Algorithms* **2018**, *11*, 172. [CrossRef]
48. Ding, N.; Ma, H.; Gao, H.; Ma, Y.; Tan, G. Real-time anomaly detection based on long short-Term memory and Gaussian Mixture Model. *Comput. Electr. Eng.* **2019**, *79*, 106458. [CrossRef]
49. Liu, M.; Wu, W.; Gu, Z.; Yu, Z.; Qi, F.; Li, Y. Deep learning based on Batch Normalization for P300 signal detection. *Neurocomputing* **2018**, *275*, 288–297. [CrossRef]
50. Ronneberger, O.; Fischer, P.; Brox, T. U-net: Convolutional networks for biomedical image segmentation. In *International Conference on Medical Image Computing and Computer-Assisted Intervention;* Springer: Berlin/Heidelberg, Germany, 2015; pp. 234–241.
51. Chen, L.C.; Papandreou, G.; Kokkinos, I.; Murphy, K.; Yuille, A.L. Deeplab: Semantic image segmentation with deep convolutional nets, atrous convolution, and fully connected crfs. *IEEE Trans. Pattern Anal. Mach. Intell.* **2018**, *40*, 834–848. [CrossRef]
52. Nagi, J.; Yap, K.S.; Tiong, S.K.; Ahmed, S.K.; Mohamad, M. Nontechnical loss detection for metered customers in power utility using support vector machines. *IEEE Trans. Power Deliv.* **2010**, *25*, 1162–1171. [CrossRef]
53. Nagi, J.; Yap, K.S.; Tiong, S.K.; Ahmed, S.K.; Nagi, F. Improving SVM-based nontechnical loss detection in power utility using the fuzzy inference system. *IEEE Trans. Power Deliv.* **2011**, *26*, 1284–1285. [CrossRef]

Article

Investigation of Deterministic, Statistical and Parametric NB-PLC Channel Modeling Techniques for Advanced Metering Infrastructure

Bilal Masood [1], M. Arif Khan [2], Sobia Baig [3], Guobing Song [1,*], Ateeq Ur Rehman [4,5], Saif Ur Rehman [6], Rao M. Asif [6] and Muhammad Babar Rasheed [7]

1. School of Electrical Engineering, Xi'an Jiaotong University, Xi'an 710049, China; bilal.masood@mail.xjtu.edu.cn
2. School of Computing and Mathematics, Charles Sturt University, Bathurst, NSW 2678, Australia; mkhan@csu.edu.au
3. Electrical and Computer Engineering Department, COMSATS University Isb- Lahore Campus, Punjab 54000, Pakistan; drsobia@cuilahore.edu.pk
4. College of Internet of Things Engineering, Hohai University, Changzhou 213022, China; ateqrehman@gmail.com
5. Department of Electrical Engineering, Government College University, Lahore 54000, Pakistan
6. Department of Electrical Engineering, The Superior College, Lahore 54000, Pakistan; saifurrehman@superior.edu.pk (S.U.R.); rao.m.asif@superior.edu.pk (R.M.A.)
7. Department of Electronics and Electrical Systems, The University of Lahore, Lahore, Punjab 54000, Pakistan; babarmeher@gmail.com

* Correspondence: song.gb@mail.xjtu.edu.cn

Received: 18 May 2020; Accepted: 5 June 2020; Published: 15 June 2020

Abstract: This paper is focused on the channel modeling techniques for implementation of narrowband power line communication (NB-PLC) over medium voltage (MV) network for the purpose of advanced metering infrastructure (AMI). Three different types of models, based on deterministic method, statistical method, and network parameters based method are investigated in detail. Transmission line (TL) theory model is used to express the MV network as a two-port network to examine characteristics of sending and receiving NB-PLC signals. Multipath signal propagation model is used to incorporate the effect of multipath signals to determine the NB-PLC transfer function. A Simulink model is proposed which considers the values of MV network to examine the characteristics of NB-PLC signals. Frequency selectivity is also introduced in the impedances to compare variations and characteristics with constant impedances based MV network. A state-of-the-art mechanism for the modeling of capacitive coupling device, and impedances of low voltage (LV) and MV networks is developed. Moreover, a comparative analysis of TL theory and multipath signal propagation models with the proposed Simulink model is presented to validate the performance and accuracy of proposed model. This research work will pave the way to improve the efficiency of next-generation NB-PLC technologies.

Keywords: AMI; TL; SG; NB-PLC

1. Introduction

A huge challenge faced by the 21st century grid is a large amount of greenhouse gases such as NO_X and CO_X due to the utilization of fossil fuels in power plants. This dependency of power systems on fossil fuels needs to be reduced to meet the challenges faced by 21st century grid. The aging of power system infrastructure in most of the countries is another present-day open issue [1,2]. The next generation power grid that can cater to these issues is known as smart grid (SG).

SG can be regarded as a cyber-physical system that is fully equipped with advanced information and communication technologies (ICT) to efficiently analyze the real-time performance of power grid which will improve the monitoring system, operation, and maintenance of power generation, transmission, and distribution systems. The 21st century ICT can provide various services to SG such as demand side management, demand response, and advanced metering infrastructure (AMI), which will not only facilitate the consumers to act as prosumers but will also reduce the grid reaction time against the power line faults and grid outages. In power line communications (PLC), use of the already existing infrastructure of power systems with cost-effectiveness has gained massive attention from the research community and industry due to the avoidance to install a new costly communication infrastructure. In the present scenario, PLC is certainly an obvious choice to provide SG services [3,4]. The advances and timely updated standardization process of PLC have addressed most of the concerns raised by skeptics. AMI gives an opportunity to the customers to effectively interact with the utilities for real-time electricity pricing and better services with the help of two-way communication. In addition, AMI reduces the power losses and power theft along with remote connect/disconnect of smart meters [5,6]. Despite many advantages of PLC, it still has to go through few inevitable challenges in which the unpredictable channel characteristics of PLC are on top, which makes it very difficult to model the PLC channel. Moreover, the response of PLC channel varies with the variations in the parametric values of PLC channels such as time instant, load/access impedances, and temperature effect. However, its cost-effectiveness and already existing nationwide infrastructure compels researchers and industries to opt for the other choices of communication technologies, i.e., wireless communications [7–9].

PLC is classified into three types: ultra narrowband PLC (UNB-PLC) ranging 0.3–3 kHz, narrowband PLC (NB-PLC) with frequency range 3–500 kHz, and broadband PLC (BB-PLC) with frequency range 1.8–250 MHz. This paper uses the NB-PLC frequencies for the investigation of medium voltage (MV) channel characteristics with the help of three different types of techniques. One of the more prominent advantages of NB-PLC is presenting a low attenuation profile at narrowband frequencies (under 500 kHz) as compared to higher frequencies, especially greater than 1 MHz due to capacitive coupling induced between the earth/ground and power conductor/cable of transmission and distribution system [1,10,11]. Such lower attenuations of NB-PLC are helpful in transmitting the AMI data over longer distances, making it a more suitable choice of ICT in SG. The NB-PLC frequency bands are further segregated into four types of European standard CENELEC EN 50065 to provide a better quality of services: (a) CENELEC-A band is operated in 9–95 kHz to provide services by power utilities so that execution of sophisticated monitoring and control of power system can be done; (b) CENELEC-B band is operated in 95–125 kHz and which can be used for any kind application; (c) CENELEC-C band is operated in 125–140 kHz and is dedicated to in-home networking; and (d) CENELEC-D band is operated in 140–148.5 kHz and can be used for alarm and security services. The CENELEC B band and CENELEC D band are unrestricted bands for customers.

The most suitable technology to incorporate the NB-PLC is OFDM, for which international standards such as PRIME, G3-PLC, and IEEE1901.2 are available. PRIME was the first effort made, in 2007, towards the standardization process of next generation PLC technologies with the help of DSO European region development fund (ERDF). The G3-PLC was an initiation of an alliance of 12 companies in 2011 whose main sponsorship was given by ERDF. IEEE Communication Society has also standardized the OFDM based NB-PLC system due to high level of interest shown by industries [12–14]. NB-PLC technologies also present low data rate (LDR) and high data rate (HDR) NB-PLC services. A few kbps data rate is offered by single carrier NB-PLC techniques, whereas, for HDR services, having an ability to transmit a large amount of data up to 100s of kbps can be achieved by multicarrier modulation techniques. The NB-PLC technologies operated at LDR incorporates the standards ISO/IEC (14908-3, 14543-3-5), IEC (61334-3-1, 61334-5), CEA-600.31, etc.

1.1. Mian Contribution

This paper proposes a sophisticated channel modeling technique for the investigation of narrowband frequencies (3–500 kHz) over MV network. Three different types of channel modeling techniques are presented in this paper for the comparative analyses and to examine the performance of proposed Simulink model: (1) Transmission line (TL) theory model, which is a deterministic modeling approach, is used in which constant and frequency selective (FS) low voltage LV and MV networks are taken into account. (2) Multipath signal propagation model based on statistical approach is used to examine the characteristics of NB-PLC signals by considering the effect of multipath signals. (3) The proposed Simulink model incorporates the values of network parameters that is to be investigated. Impedances are modeled by three types of impedance modeling techniques, formulated by the combination of series and parallel resonant circuits. A comparative analysis of obtained transfer functions is presented to validate the accuracy of proposed Simulink model. This research work facilitates the electric supply companies and researchers to examine the NB-PLC performance over MV network by simply incorporating the parameters of network under evaluation, instead of carrying out extensive and time taking field measurements. The analysis of MV NB-PLC network helps to assess the feasibility for the implementation of advanced metering infrastructure (AMI) using NB-PLC technologies.

1.2. Paper Structure

The paper is divided into seven sections. Section 2 discuss the related work and literature on PLC channels and their modeling techniques. Section 3 gives an overview on the characteristics of MV NB-PLC network. Section 4 characterizes the MV NB-PLC network by focusing on TL parameters such as resistance, conductance, inductance, capacitance, and characteristics impedance. Resistance variations law and impedance modeling is examined in depth by discussing their results. Section 5 elaborates various channel modeling techniques and proposed a Simulink model to compute the transfer function of MV network to analyze the performance of NB-PLC. Section 6 discusses and compares the results obtained from TL theory and multipath signal propagation models with Simulink model. This is followed by the conclusions in Section 7.

2. Related Work

Stefano Galli, a leading researcher in the field of deterministic PLC channel modeling, has made valuable contributions in the area of indoor PLC channels [14,15]. In his model, Galli addressed indoor PLC problems, using transmission lines theory, primarily focusing on model decomposition and multiconductors. The transmission lines theory analyzed propagation interaction of two coupled circuits and dominant modes: differential and pair modes. The differential propagation of signals is investigated by the differential mode while the pair mode's excitation and propagation are studied by the companion model. In [16], the author emphasized that, on the one hand, the best way to achieve a comprehensive and efficient PLC is to use the differential and companion models as a cascaded two-port network by employing transmission matrix techniques, while, on the other hand, neglecting the mode coupling and companion circuit leads to an incomplete circuit model that is unsuitable for signal propagation of PLC. He eventually came up with a single lumped network that has the ability to replace distributed parameters of circuits. He claimed experimentally and mathematically that, regardless of its topology, the transfer functions for indoor PLC remain the same. He also provided experimental evidence to show that the resonant modes and reflections can be isolated depending on their specific topologies [14].

Time-varying properties of the power line vary due to the following two factors [17]: (1) the layout of topology; and (2) the switching of devices and appliances installed indoors. These continuous varying impedances coupled with instantaneous amplitude of the mains voltage give rise to a periodically time-varying channel response. In [18,19], the authors mainly benefited from Galli's methodology with the only exception of using a different conductor scheme. This PLC model thus employed two-conductor and multi-conductor concepts. Andrea M. Tonello employed a random topology technique using statistical bottom-up approach to study PLC channel modeling, whereby he produced transfer function by utilizing computational efficiencies [20]. First, Tonello studied European in-home network topology by applying statistical methods. Second, he determined transfer function between any pair of outlets of a given topology in a unique way. Later, to study PLC channel characteristics statistically, he also developed a simulator to configure a small set of parameters for theoretical frameworks [21,22].

The authors of [23] conducted a comprehensive field trial over MV overhead power lines in which signal passes through a MV/LV transformer bridge. A simulation model to examine the PLC characteristics was also modeled. A low voltage, high frequency PLC signal was injected into the MV bus bar to cross the transformer bridge and received at LV side of transformer. The suggested PLC model was an effective tool to simulate the MV network by focusing on MV power line channel modeling. Field measurements were performed on Favignana Island on a core–shield type conductor configuration for the PLC communication over it. The measured and simulated results were compared to validate the performance of simulated model [23,24]. A line-shielded MV network with power transformers, couplers, transmitter, and receiver system was tested for diverse range of line lengths. Zimmermann [9] was among the first researchers to incorporate the physical effects of PLC channel by considering the multipath signals, losses in cable, and time delays when the PLC signal propagates throughout the length of power line. A complex transfer function of PLC channel was statistically investigated in depth to access the feasibility of PLC systems. The proposed multipath signal propagation model uses the parameters of power line; however, it provides a complex frequency response to study the PLC channels characteristics. The model employed the frequency range between 500 kHz and 20 MHz, providing the transfer function, although it is not completely familiar with the exact values of network parameters. This PLC signal of this model is an accumulation of all propagated signals in all possible paths while traveling towards the destination [10–12]. The identification of transfer functions is necessary and based on measured values. Table 1 summarizes the contributions of authors for various types of PLC models. For the characterization of power line channels, modeling of their parameters is very important. Detailed discussion on power line parameters for the purpose of channel modeling can be found in [25–27]. In [25], the voltage–current approach is applied to determine access impedance. The shunt resistance R_{sh} is used as a parameter to measure the current. Impedance modeling is expressed as,

$$Z_{a,k} = Z_{m,k} - (R + j\omega L + \frac{1}{j\omega C}) \tag{1}$$

where 1ϕ impedance at reference point B is denoted by $Z_{a,k}$, which can be calculated by measured values of impedance $Z_{m,k}$, that is 1ϕ impedance at reference point A.

$$Z_{a,3-ph} = Z_{m,3-ph} - Z_{3-phase/calibration} \tag{2}$$

where the 3ϕ coupling impedance at reference point B is denoted by $Z_{a,3-ph}$ determined by measured three-phase impedance $Z_{m,3-ph}$ at A reference point. The calibration impedance that determines the coupling network impedance is denoted by $Z_{3-phase/calibration}$. The calculation to obtain the access impedance theoretically can be achieved by parallel combination of 1ϕ impedances located at each phase by,

$$Z_{a,3-ph-theoratical} = \frac{1}{\frac{1}{Z_{a,1}} + \frac{1}{Z_{a,2}} + \frac{1}{Z_{a,3}}} \tag{3}$$

Canete et al. [27] proposed an indoor PLC channel model. Three different types of load impedances i.e., FS, constant and time varying were taken into account. The FS impedance are modeled by,

$$Z_{\omega} = \frac{R}{1 + jQ(\frac{\omega}{\omega_0} - \frac{\omega_0}{\omega})} \tag{4}$$

where R, Q, and ω_0 represent the resonant resistance, quality factor, and resonance angular frequency, respectively.

The time-varying impedances are modeled by,

$$Z_{\omega,t} = Z_A(\omega) + Z_B(\omega)|sin(\frac{2\pi}{T_0}t + \phi)|; 0 \leq t \leq T_0 \tag{5}$$

where Z_A, Z_B, and ϕ are offset impedance, amplitude variation, and phase, respectively.

Table 1. Summary of research work contributions in various types of PLC models.

Type of PLC Model	Main Contributors	Key Features of Proposed Work	Ref.
Deterministic PLC Channel Models	Stefano Galli et al.	• Multiconductor transmission line theory approach for coupled circuits • Analyzed the behavior of PLC for differential and pair mode circuits • Proposed a cascaded two port network model technique for efficient PLC	[14–16]
	Justinian Anatory et al.	• Modeled the transfer function of PLC channel • Derived the single phase PLC channel with interconnection by incorporating the various loads at different branches • Proposed model is validated with Transients Program–Electromagnetic Transients Program (ATP–EMTP)	[17–19]
Statistical PLC Channel Models	Zimmermann et al.	• Multipath model • Caters the attenuation caused due to reflections and power line • Incorporated delays due to length of line	[9,10]
	Andrea M. Tonello et al.	• Proposed a bottom-up PLC channel simulator • Derived in-home PLC channel model for Europe • Sophisticated computation method for channel transfer function • ABCD matrix based method is also proposed	[20–22]
Measured PLC Channel Models	Antonio Cataliotti, et al.	• Catered the issues involved in NB-PLC channel measurements • Proposed the suitable procedural techniques for modeling and characterization of power system components within the frequency of interest	[23,28–30]
Parametric PLC Channel Models	Canete et al.	• Impedance modeling for low voltage indoor PLC network • Proposed a simulator	[27]
	Bilal Masood et al.	• Frequency selectivity is added in resistance, conductance and impedances • Compared the transfer functions obtained from constant impedances with frequency selective impedances • Simulation and measurements for the channel modeling of LV NB-PLC system	[6,13,14,31,32]

3. Channel Modeling of NB-PLC for AMI in Medium Voltage Network

One of the challenges faced by NB-PLC-based AMI system is the inimical characteristics of MV channel [28–30,33,34]. A model of NB-PLC system for the purpose of AMI is depicted in Figure 1. The performance of NB-PLC-based AMI system is poorly influenced by various factors such as multipaths, time dispersion, reflections, propagation and time delay, and FS. When the signal is propagated in MV channel, it has to go through issues such as phase shift replication, delay, and attenuation. The root mean square technique can be used to cater the time dispersion issues during measurement process. Few problems such as dispersion and time selectivity are common in both types of channels, i.e., wireless and NB-PLC. Furthermore, consideration of additive white Gaussian noise cannot be taken as a reference to completely understand the attributes of NB-PLC channel [16,35,36]. There is an utmost need to devise a criterion for the modeling of MV NB-PLC channels that can serve to examine the NB-PLC system for various types of noises. PLC systems are usually affected by two types of noises, i.e., background noise and impulsive noise, as shown in Figure 2 [10]. Their further classifications are: (a) impulsive noise is a combination of apriodic impulsive noise with asynchronous or synchronous impulsive noise; and (b) background noise consists of colored noise and narrowband noise. In this paper, a Simulink model is proposed for the characterization and modeling of MV NB-PLC channels for the purpose of AMI. The real-time overhead MV transmission lines are used in the proposed model Simulink model. A CCD helps to inject the NB-PLC signal in the MV power line, the working principle of which is discussed in Section 5.3.1. A comparison of transfer functions computed from Simulink model are compared with TL theory (constant and FS) and multipath signal propagation model to validate the accuracy of proposed Simulink model. Characterization of MV NB-PLC network is discussed in the subsequent section.

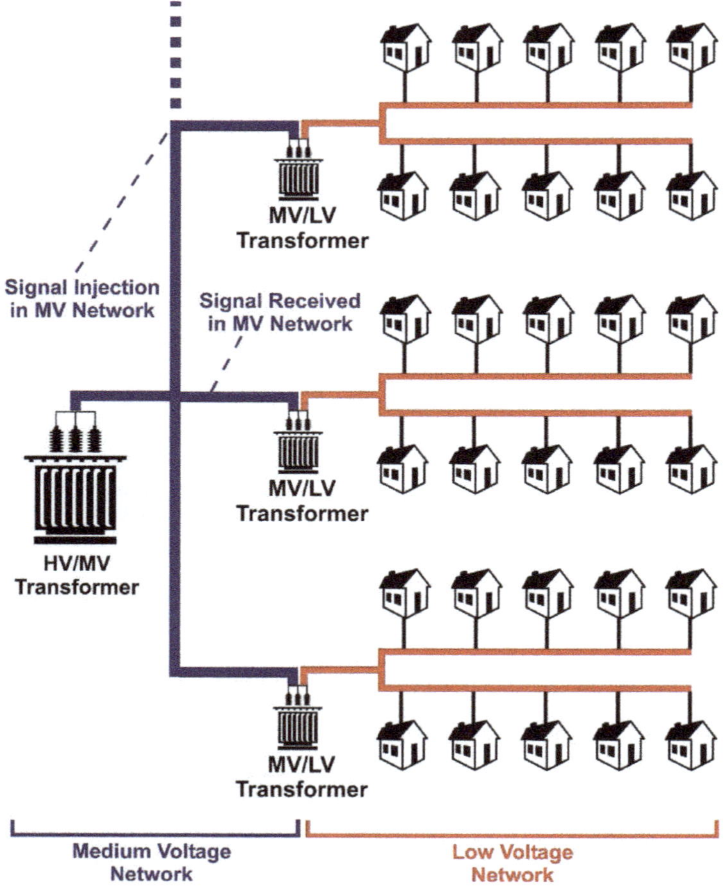

Figure 1. NB-PLC model for MV network.

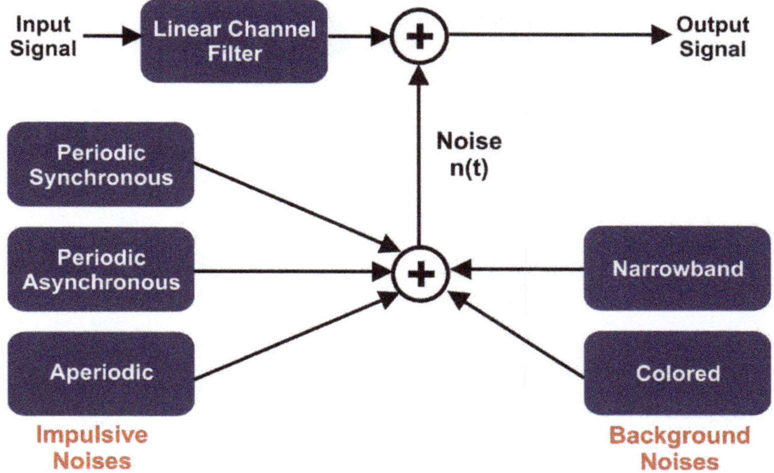

Figure 2. Noise scenario in power line communication system.

4. Characterization of Medium Voltage NB-PLC Network

This section deals with the modeling of MV transmission line. The electrical parameters of actual transmission lines are tabulated in Table 2. The conductor types Ant BS 215 and Lynex BS 215 are used for LV and MV network with ampacity up to 175 and 384 A, respectively. Ant has an all aluminium conductor (AAC) type, whereas Lynex has aluminium conductors steel reinforced (ACSR) type with a steel reinforcement to increase the tensile strength due to its weight. The characteristics of any transmission line can be analyzed by its distributed parameters such as resistance R, conductance G, capacitance C, and inductance L. These parameters in the case of overhead transmission lines can be determined as [23],

$$R = \frac{1}{\pi a}\sqrt{\frac{\pi f \mu}{\sigma}}[\Omega] \tag{6}$$

$$G = 2\pi f C tan\delta[S] \tag{7}$$

$$C = \frac{\pi\varepsilon}{\cosh^{-1}(\frac{D}{2a})}[F/m] \tag{8}$$

$$L = \frac{\mu}{\pi}\cosh^{-1}(\frac{D}{2a})[H/m] \tag{9}$$

where ε, μ, δ, D, and σ are permittivity of free space, permeability of free space, depth factor, diameter of conductor, and conductivity of material, respectively. The frequency of inters in this paper ranges 3–500 kHz. However, the characteristic impedance Z_C and propagation constant γ of transmission line can be calculated as,

$$Z_C = \sqrt{(R + j\omega L)/(G + j\omega C)} \tag{10}$$

$$\gamma = \alpha + j\beta = \sqrt{(R + j\omega L)(G + j\omega C)} \tag{11}$$

where attenuation constant is denoted by α, whereas β denotes the phase factor and ω represents the angular frequency.

Table 2. Electrical parameters of conductors.

Conductor Type/Standard	Nominal/ Section Area	No./Nominal Diameter of Wires		Approximate Overall Diameter	Nominal DC Resistance at 20 C	Current Rating	L	C
	mm^2	Aluminium (No./mm)	Steel (No./mm)	mm	Ω/km	Amps	µH/m	pF/m
Ant-BS 215	52.8	7/3.10	-	9.30	0.54190	175	0.93	12.6
Wolf BS 215	212.10	30/2.59	7/2.5	18.13	0.1828	351	1.22	9.45
Lynex BS 215	226.2	30/2.79	7/2.7	19.53	0.1576	384	1.20	9.58
Panther BS 215	261.5	30/3.0	7/3	21.0	0.1363	420	1.15	9.8

4.1. Resistance Variations Law

Resistance variations law is used to characterize the variations in transmission line's resistance with respect to change in frequency [23]. Frequency selectivity of resistance using resistance variations law can be calculated as,

$$R(f) = A_R f^2 + B_R f + C_R \tag{12}$$

where the values of A_R, B_R and C_R are $2.5 \times 10^{-10}\,\text{m}\Omega/(\text{m*kHz}^2)$, $1.5 \times 10^{-12}\,\text{m}\Omega/\text{m*kHz}$ and $3\,\text{m}\Omega/\text{m}$, respectively. Figure 3 compares the resistance variations with respect to increase in frequency obtained from Equation (6) and resistance variation law. The plotted results depict that variations in the values of resistance calculated from resistance variations law are close to the FS resistance values that validate the simulation results. Since the MV network under evaluation incorporates overhead transmission lines where the separation medium between two lines is free space, the conductance is assumed to be zero.

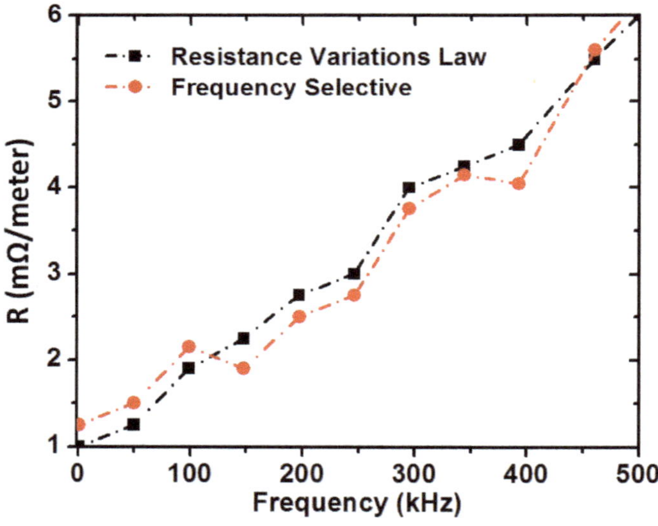

Figure 3. Comparison of FS resistance with the resistance obtained from resistance variations law.

4.2. Modeling of Impedances for Medium Voltage NB-PLC Network

Impedance modeling plays a vital role in NB-PLC systems that behave as hurdles for injected signals in power lines. Line impedances with lower values can cause a high level of attenuation to the transmitted signal, carrying high frequency with a small magnitude of injected signal. In practical situation, it becomes a challenge for field engineers to inject a signal in MV NB-PLC channels which have impedance values lower than 0.5 Ω. It is also important to note that access impedances are FS which vary with the change in frequency. In [25–27], a comprehensive discussion on the characterization and modeling can be found. Chu et al. [25] gave an overview of the LV network in the context of examining the access impedances. However, by distributing the noise over frequency range 50–500 kHz, characteristics of LV access impedances are investigated in [26].

It is worth mentioning that TL theory-based transfer functions in this paper comprise of two types:

- Constant LV and MV Networks: The constant LV and MV network includes the fixed values of transmission lines and access impedances parameters.
- FS LV and MV Network: The FS parameters includes the values of transmission lines and access impedances that varies with an increase of frequency of NB-PLC signal.

Three types of impedance modeling methods are used in this paper, as illustrated in Figure 4. The reason behind choosing the combination of parallel and series resonant circuits is critical access impedances connected to LV and MV power line channels carry the resonant behavior that can be achieved by such combination of circuits [6]. This paper incorporates Types 1, 2, and 3 for the purpose of impedance modeling in the TL theory method and Simulink model. The formulation of parallel and series combination of two resonant circuits is given as,

$$Z_S(f) = \frac{1 + j2\pi R_S C_S + (j2\pi f)^2 L_S C_S}{2\pi f C_S} \tag{13}$$

$$Z_P(f) = \frac{R_P + j2\pi f L_P}{1 + j2\pi R_P C_P + (j2\pi f)^2 L_P C_P} \tag{14}$$

where the subscript S denotes the series circuits and subscript P represents the parallel circuits.

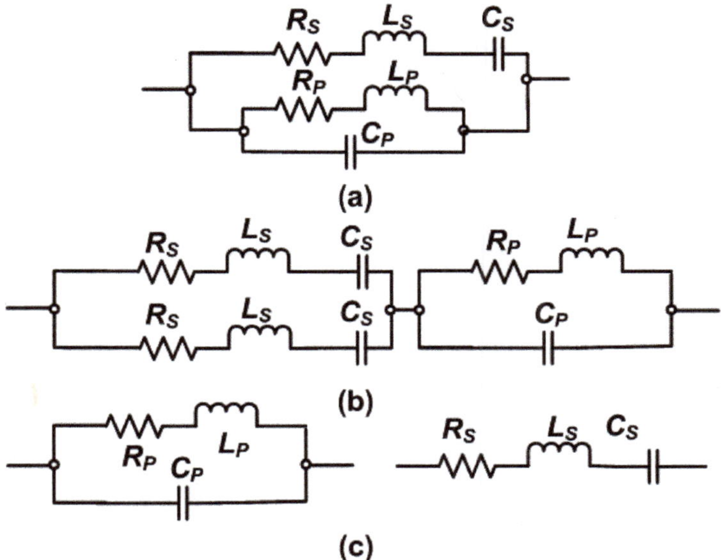

Figure 4. Modeling of access impedances by resonant circuits: (**a**) Type 1; (**b**) Type 2; and (**c**) Type 3.

4.2.1. Type 1 Circuit

By using Equations (13) and (14), Type 1 circuit is modeled by the combination of two RLC resonant circuits connected in series and parallel given as,

$$Z_{r1,r2}(f) = \frac{Z_S(f) Z_P(f)}{Z_S(f) + Z_P(f)} \tag{15}$$

4.2.2. Type 2 Circuit

The Type 2 circuit is a combination of three resonant circuits in which two series and one parallel combination of RLC elements are included given as,

$$Z_{r3}(f) = \frac{Z_{S1}(f)Z_{S2}(f)}{Z_{S1}(f) + Z_{S2}(f)} + Z_P(f) \tag{16}$$

4.2.3. Type 3 Circuit

The simplest case for analysis could be a series or parallel resonant circuit that is expressed in Type 3.

4.2.4. Input Impedance of MV Network

The Z_{in} (equivalent input impedance) of MV network is determined as,

$$Z_{in} = Z_{cbt}\frac{Z_{bt} + Z_{cbt}tanh(\gamma_{bt}l_{bt})}{Z_{cbt} + Z_{bt}tanh(\gamma_{bt}l_{bt}} \tag{17}$$

where Z_{bt} denotes the bridge tap impedance connected with LV and MV networks, Z_{cbt} is bridge tap characteristic impedance, l_{bt} represents the bridge tap length, and γ_{bt} is a propagation constant of bridge tap.

4.3. Discussion of Results on Impedance Variations of NB-PLC Network

Figure 5 elaborates the characteristics of impedance variations in NB-PLC network by incorporating both constant and FS types of impedances. The purpose of investigating both impedance types is to analyze whether the impedances vary more in the case of FS as compared to constant and then to compare their transfer functions for NB-PLC channel by TL theory model (constant and FS) and Simulink model in the next subsection.

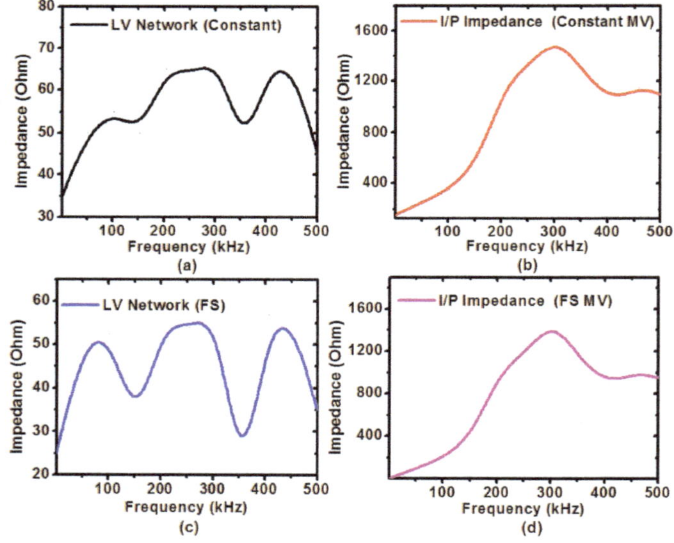

Figure 5. Impedance variations of LV and MV Networks: (**a**) constant LV network; (**b**) input impedance of constant MV network; (**c**) frequency selective LV network; and (**d**) input impedance of frequency selective MV network.

4.3.1. Discussion on Constant NB-PLC Network Impedances

Figure 5a,b illustrates the impedance variations by considering the constant impedances. The constant LV network connected with MV/LV transformer is shown in Figure 5a, whereas Figure 5b depicts the input impedance of complete MV network while considering the constant LV network. It is noteworthy that the impedance values of loads connected to main channels such as LV network connected to MV (main NB-PLC channel) vary within 35–65 Ω.

4.3.2. Discussion on Frequency Selective LV and MV Network Impedances

Figure 5c,d illustrates the impedance variations by considering the FS impedances. The FS LV network connected with MV/LV transformer is shown in Figure 5c, whereas Figure 5d depicts the input impedance of complete MV network while considering the FS LV network. When comparing the plots obtained by using constant impedances with FS impedances, it can be seen that the FS impedances-based plots show more variations in the magnitudes as compared to constant impedances based plots. In the later sections of this paper, the effect of constant and FS impedances on transfer functions is discussed as well as compared with the proposed Simulink model.

5. Methodologies to Determine NB-PLC Transfer Functions

5.1. Transmission Line Theory

The TL theory model is incorporated to determine the MV power line channel transfer function. According to TL theory, power network can be expressed by the ABCD matrix that formulates a relation between sending end current I_1 and voltage V_1 with the receiving end current I_2 and voltage V_2 given as,

$$\begin{bmatrix} V_1 \\ I_1 \end{bmatrix} = \begin{bmatrix} cosh(\gamma l) & Z_c sinh(\gamma l) \\ \frac{1}{Z_c} sinh(\gamma l) & cosh(\gamma l) \end{bmatrix} \begin{bmatrix} V_2 \\ I_2 \end{bmatrix} \tag{18}$$

The subsections T_0 (series impedance as a two port network), T_1 (power lines as a two port network), T_2 (parallel impedance as a two port network), and T_3 (power lines as a two port network) shown in Figure 6 can be given ABCD matrices form as,

$$T_0 = \begin{bmatrix} 1 & Z_S \\ 0 & 1 \end{bmatrix} \tag{19}$$

$$T_1 = \begin{bmatrix} cosh(\gamma_1 l_1) & Z_{cl1} sinh(\gamma_1 l_1) \\ \frac{1}{Z_{cl1}} sinh(\gamma_1 l_1) & cosh(\gamma_1 l_1) \end{bmatrix} \tag{20}$$

$$T_2 = \begin{bmatrix} 1 & 0 \\ \frac{1}{Z_{in}} & 1 \end{bmatrix} \tag{21}$$

$$T_3 = \begin{bmatrix} cosh(\gamma_2 l_2) & Z_{cl2} sinh(\gamma_2 l_2) \\ \frac{1}{Z_{cl2}} sinh(\gamma_2 l_2) & cosh(\gamma_2 l_2) \end{bmatrix} \tag{22}$$

where γ_1, Z_{cl1}, γ_2, and Z_{cl2} represent the propagation constants and characteristic impedances of sub-networks, whereas Z_{in} is input impedance. All of above ABCD matrices are multiplied with each other by chain rule, i.e., a generalized expression for *i* cascaded sections is given as,

$$T = \prod_{i=1}^{N} T_i \tag{23}$$

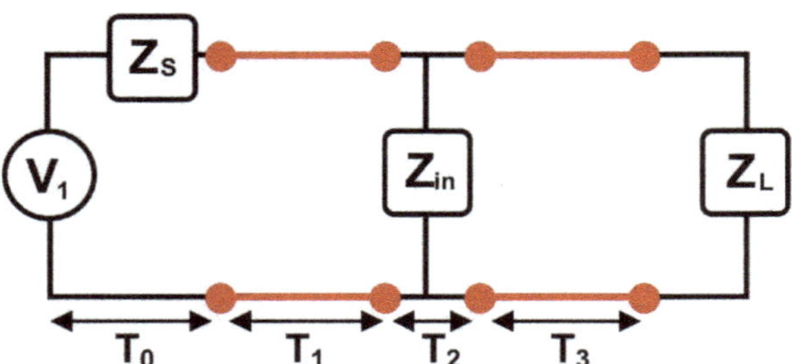

Figure 6. Equivalent transmission network of transmission line with single connection of bridge tap [1].

Finally, transfer function is determined by,

$$H = 20log_{10}\frac{Z_L}{AZ_L + B + CZ_LZ_S + DZ_S} \tag{24}$$

where Z_S and Z_L are source and load impedances.

5.2. Multipath Signal Propagation Model

The signal propagation model that is based on the signals transmitted and reflected from multiple paths uses the line of sight path for all instants of times [9]. When the signal propagates from transmitter to receiver, a few extra signals, which are known as multipaths, are superimposed and thus added, which is a leading cause of reflections and echos that need to be rectified. The resultant of such scenario is FS fading that degrades the quality of NB-PLC system. Transfer function determined in Zimmermann model is given as,

$$H(f) = \sum_{i=1}^{n} |g_i(f)|e^{\varphi g_i(f)}e^{-(a_0+a_1f^k)d_i}e^{-j2\pi f\tau_i} \tag{25}$$

where attenuation, delay, and weighting factor are denoted by $e^{-(a_0+a_1f^k)d_i}$, $e^{-j2\pi f\tau_i}$, and $|g_i(f)|e^{\varphi g_i(f)}$, respectively. It can be examined by the above-mentioned equation that, when a signal propagates, its attenuation increases with an increase of the length of conductor. Moreover, the response of system reflects the low pass characteristics on NB-PLC frequency range, i.e., 3–500 kHz. The characteristics of NB-PLC signals in regard to reflection and propagation are linked with the weighting factor g_i. Signal characteristics can be analyzed in general, frequency dependent, and complex form by multipath signal propagation model. This model suggests that N multipaths are added when the signal propagates towards the receiver side and vice versa. The simplified expression of transfer function is expressed as,

$$H(f) = \sum_{i=1}^{n} g_ie^{-(a_0+a_1f^k)d_i}e^{-j2\pi f\frac{d_i}{vp}} \tag{26}$$

The parametric values of multipaths are tabulated in Table 3.

Table 3. Parameters of multipath signal propagation model.

Parametric Values of Attenuation					
$k = 1$		$a_0 = 0$		$a_1 = 7.8 * 10^{-10}$ S/m	
Path Parameters					
i	g_i	d_i/m	i	g_i	d_i/m
1	0.70	750	3	−0.20	200
2	0.35	1000	4	0.06	225

5.3. Proposed Simulink Model for Medium Voltage NB-PLC Network

In this section, the proposed Simulink model for the channel modeling and characterization of MV NB-PLC, as shown in Figure 7, is discussed. The model is developed by incorporating the various parametric values of MV network taken from electric supply companies in Pakistan, and is mainly comprised of three MV/LV transformers supplied from three-phase source. The suggested Simulink model contains the components of power system, designed to be operated at low power system frequency, i.e., 50 Hz. However, it is significantly important to notice the role of transmitter and receiver blocks containing the CCD, as shown in Figures 7 and 8. Signal generator transmits the high frequency signal and injects in MV power line with the help of CCD thus the NB-PLC signal is imposed on 50 Hz power system. All three transformers have voltages of 11 kV at MV side and 400 V at LV side, where LV network has 230 V_{L-N} and 400 V_{L-L}. The power rating of first transformer is 150 kVA with supply to the LV network of 140 kW. The second transformer is rated at 250 kVA supplying 245 kW to LV network. The third transformer is rated at 200 kVA supplying 190 kW power to the LV network. The lengths of transmission line between source and 250 and 150 kVA transformers are 1150 m and 950 m, respectively, while there is a comparatively shorter, 110 m length between source and 200 kVA transformer. In this model, the distributed parameter lines which are commonly used by electric supply companies in Pakistan are incorporated. These lines follow British Standard 215 with conductor type Wolf, Ant, and Panther. Table 2 tabulates the various values of parameters such as diameter, ampacity, resistance, capacitance, and inductance. The power transformers used in the simulation model are three phase, two winding transformers operating on power system frequency of 50 Hz. Inductance and resistance are 0.50 H and 5 Ω for winding 1, respectively, and 0.1 H and 0.85 Ω for winding 2, respectively. The magnetization inductance L_m and resistance R_m of power transformers are 550 H and 2 MΩ, respectively, and X/R ratio is 7.

NB-PLC signal is injected and received with the help of CCD, at the same phase A of 250 kVA transformer in MV power line i.e., at the MV side of transformer. The role of CCD is explained in the next subsection. The distance between transmitted and received signal is 1150 m. It is worth noticing that NB-PLC signal can be injected and received in a similar way on the phases B and C. Furthermore, after a careful literature review, FS load of LV network is modeled as RLC load [1,26]. The ratios of active and reactive powers utilized in RLC loads of LV network, connected to MV network, are tabulated in Table 4.

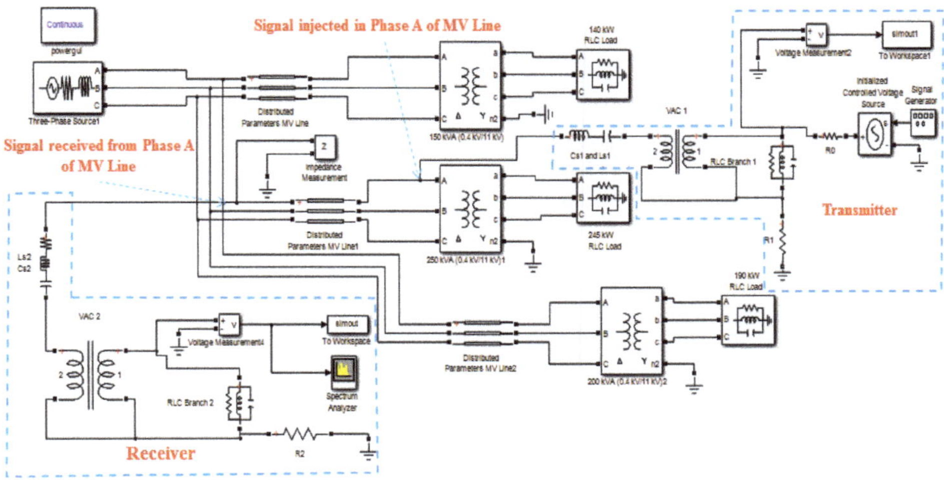

Figure 7. Proposed Simulink Model for MV NB-PLC channel modeling and characterization.

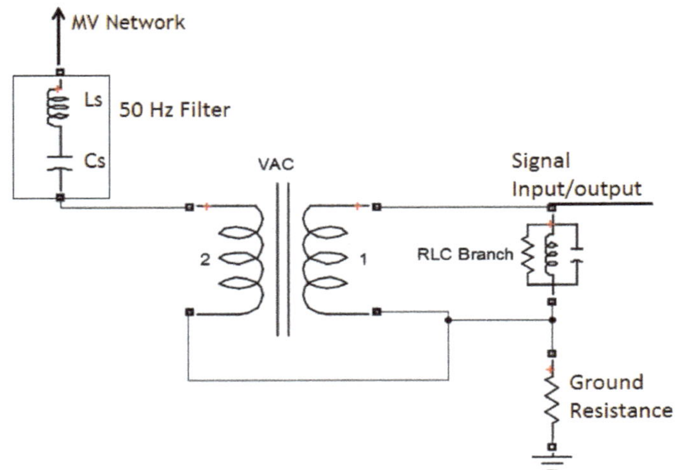

Figure 8. Circuit of MV phase to ground CCD.

Table 4. Active Power (W) and Reactive Power (VAR) ratios of LV network's RLC loads connected to MV network.

Supplying Transformers	Active and Reactive Power Ratios of Connected RLC Load of LV Network		
	Active Power (kW)	Reactive Power	
		Q_C (+VAR)	Q_L (−VAR)
150 kVA	140	25	85
200 kVA	190	30	75
250 kVA	245	20	90

5.3.1. Capacitive Coupling Device

A CCD is modeled to inject the signal in the MV power line for possible NB-PLC for AMI, without which the signal cannot be transmitted through an MV power line [36]. Figure 8 elucidates the schematic diagram of phase to ground CCD for MV network. A 1-V signal is generated with the help of a signal generator to the input of a parallel RLC circuit and further through isolation transformer tuned for the frequency range 3–500 kHz. The signal passes from a 50-Hz filter before giving input to the MV power line. CCD is grounded with a resistance of 850 Ω. The values of various parameters used in CCD are tabulated in Table 5. The same CCD is used at the receiving end of MV NB-PLC system and on the transmitter side. Figure 9 illustrates the transfer function of CCD. The plotted results of CCD depicts the variations in attenuation between 10 dB and −2 dB.

Table 5. Capacitive coupling device parameters.

Isolation Transformer			RLC Branch Parameters			MV Series Ls Cs Parameters	
Magnetization Resistance, $R_M\Omega$	Inductance, L_T [μH]	Turn Ratio	R [kΩ]	L [μH]	C [nF]	L_S [μH]	C_S nF
85	425	1:1	45	180	20	75	90

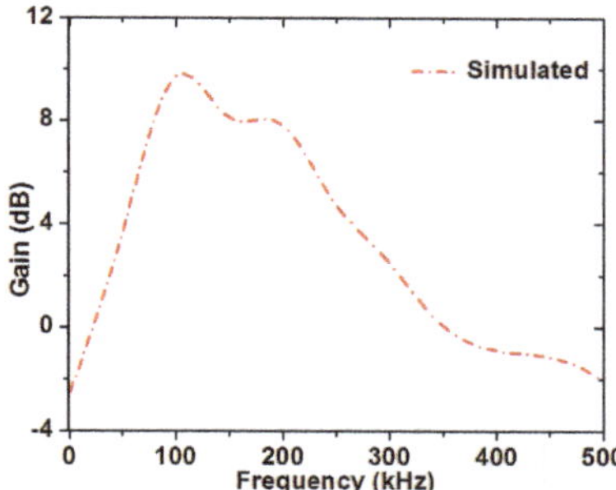

Figure 9. Simulated response of CCD.

6. Results and Discussion of MV NB-PLC Channel Transfer Functions

This paper proposes the channel modeling techniques for MV NB-PLC for AMI by comparing TL theory (with constant and FS impedances), multipath signal propagation model, and proposed Simulink model. This part of the paper discusses and compares the results of transfer functions determined from all suggested techniques.

6.1. Transmission Line Theory Based Transfer Functions

Figure 10a,b illustrates the transfer function results obtained from TL theory model by using constant and FS impedances of NB-PLC network, respectively. The attenuation profile of constant impedances-based transfer function is between −15 and −40 dB, whereas FS impedances-based transfer function varies between −11 and −48 dB. The FS impedances-based transfer function presents more peaks and dips as compared to constant impedances-based transfer function, e.g., peaks can be seen at 100, 250, and 355 kHz and dips can be noticed at 75, 200, and 275 kHz as well as a deep dip at 480 kHz.

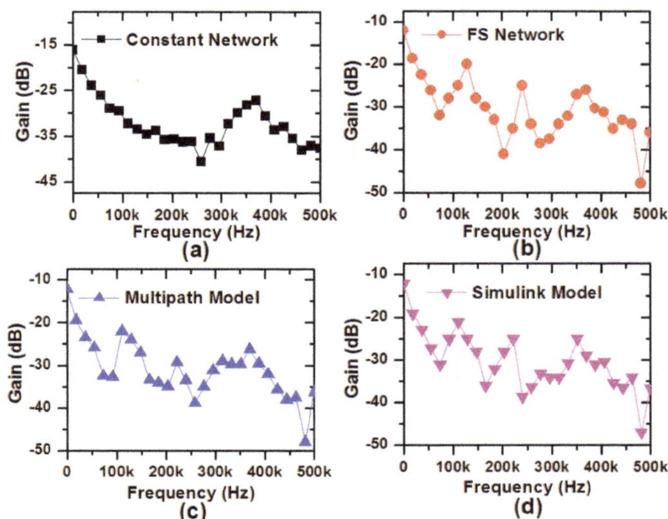

Figure 10. Transfer functions obtained from: (**a**) TL theory with constant network; (**b**) TL theory with frequency selective network; (**c**), ultipath signal propagation model; and (**d**) the proposed Simulink Model.

6.2. Multipath Signal Propagation Model Based Transfer Function

Figure 10c shows the transfer function result obtained by implementing multipath signal propagation model on NB-PLC network. The attenuation profile is varying between −11 and −44 dB, illustrating peaks at 110, 220, and 370 kHz and dips at 80–90, 200, 260, and 480 kHz.

6.3. Proposed Simulink Model Based Transfer Function

Figure 10d presents the transfer function plot obtained by using the proposed Simulink model. The attenuation profile of transfer function is varying between −11 and −45 dB, illustrating peaks at 110, 220, and 350 kHz and dips at 75, 170, 250, and 480 kHz.

It is clear from the plotted transfer function results that transfer function profile calculated from the proposed Simulink model follows the trend of transfer functions computed from FS-based TL theory and multipath signal propagation model, thus validating the performance of Simulink model. It is also noteworthy that constant impedances-based TL theory transfer function is comparatively more linear and does not give complete information about transfer function profile. However, FS-based transfer function computed by TL theory is more precise and follows the transfer function trends of multipath signal propagation and Simulink models.

6.4. Box Plot Analysis for Attenuation Profiles

A detail of attenuation profiles of transfer functions obtained from TL theory (constant and FS) model, multipath signal propagation model, and proposed Simulink model are segregated in different quartiles in the box plots shown in Figure 11. The frequency range of interest is 3–500 kHz for NB-PLC network. The box plot consists of a type of plot able to visually reveal some basic statistics of an attenuation dataset by depicting the minimum values, maximum values, and trend of attenuation gains and drops. It is clear from the box plot analysis that attenuation profiles of all techniques are in good agreement with each other, except the attenuation profile obtained from constant impedances-based TL theory model. It is particularly significant to note that Simulink model provides an exhaustive set of information with wider means and extended quartiles.

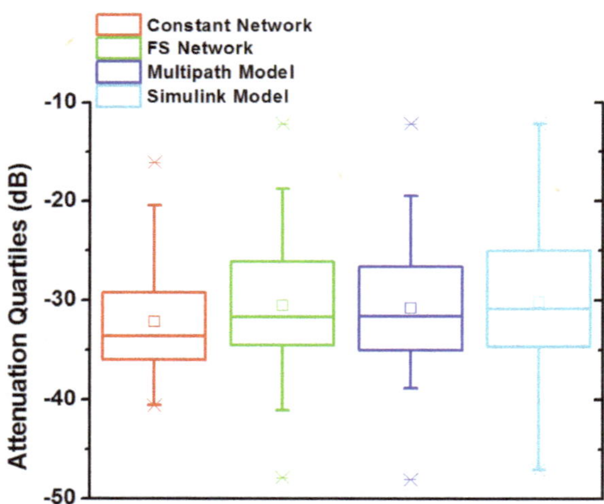

Figure 11. Summary of NB-PLC channels' transfer functions.

7. Conclusions

This paper presents the state of the art for NB-PLC channel modeling techniques for MV network by utilizing three different types of models for efficient channel modeling and characterization for

MV NB-PLC network. The first technique is based on TL theory by considering the constant and FS parameters of transmission lines and load impedances. FS is introduced in the MV power lines to get a better estimate of channel conditions, leading to a scenario that is more closer to reality. The variations in the resistance values by adding frequency selectivity in it are compared with Resistance Variations law whose results are quite similar to each other. A statistical multipath signal propagation model is used that incorporates the effect of multipaths and reflections to compute the transfer function. After a careful investigation about the behavior of LV network, impedances are modeled and their corresponding plots with respect to increase in frequency are discussed. The third technique comprises of a Simulink model, developed by using real time parameters of power network. The CCD is modeled to inject the NB-PLC signal in MV network of Simulink model. The transfer functions of all techniques such as TL theory (constant and FS impedances) model, multipath signal propagation model, and proposed Simulink model are compared with each other. It is evident from the plotted transfer function results that FS transfer function of TL theory model and multipath signal propagation model are in good agreement and have similar trends of attenuation profiles with the transfer function computed from proposed Simulink model. In practical NB-PLC systems, network parameters and impedances do not remain constant but keep on changing for different values of frequencies. Therefore, close agreement of transfer function results of Simulink model with FS-based TL theory model and multipath signal propagation model validate the accuracy of Simulink model. The presented model in this paper can prove to be a useful simulation tool to estimate any MV NB-PLC channel conditions by simply changing the parametric values as per requirements and conditions.

Author Contributions: Conceptualization, B.M.; methodology, B.M. and M.A.K.; software, B.M.; formal analysis, B.M., G.S. and A.U.R.; writing—original draft preparation, B.M. and M.A.K.; writing—review and editing, B.M., S.B., A.U.R., S.U.R. and R.M.A.; supervision, B.M., M.B.R. and M.A.K.; project administration, B.M., M.B.R. and M.A.K.; and funding acquisition, B.M. and M.A.K. All authors have read and agreed to the published version of the manuscript.

Acknowledgments: This work was supported by the National Natural Science Foundation of China (NSFC) project (51950410583) and NSFC key project (U1766209).

Conflicts of Interest: The authors declare no conflict of interest.

Abbreviations

The following abbreviations are used in this manuscript:

AAC	All aluminium conductor
ACSR	Aluminium conductors steel reinforced
AMI	Advanced metering infrastructure
CCD	Capacitive coupling device
FS	Frequency selective
HDR	High data rate
LDR	Low data rate
LV	low voltage
MV	Medium voltage
NB-PLC	Narrowband power line communications
PLC	Power line communication
SG	Smart grid
TL	Transmission line

References

1. Masood, B.; Baig, S. Standardization and deployment scenario of next generation NB-PLC technologies. *Renew. Sustain. Energy Rev.* **2016**, *65*, 1033–1047. [CrossRef]
2. Althaher, S.; Mancarella, P.; Mutale, J. Automated demand response from home energy management system under dynamic pricing and power and comfort constraints. *IEEE Trans. Smart Grid* **2015**, *6*, 1874–1883. [CrossRef]

3. Goldfisher, S.; Tanabe, S. IEEE 1901 access system: An overview of its uniqueness and motivation. *IEEE Commun. Mag.* **2010**, *48*, 150–157. [CrossRef]
4. Smith, R.; Meng, K.; Dong, Z.; Simpson, R. Demand response: A strategy to address residential air-conditioning peak load in Australia. *J. Mod. Power Syst. Clean Energy* **2013**, *1*, 223–230. [CrossRef]
5. Yan, B.L.H.; Chen, S.; Zhong, M.; Li, D.; Jiang, L.; He, G. Future evolution of automated demand response system in smart grid for low-carbon economy. *J. Mod. Power Syst. Clean Energy* **2015**, *3*, 72–81. [CrossRef]
6. Masood, B.; Usman, M.; Gul, M.U.; Khan, W.A. Measurements and characterization of power transformer and low voltage access network for NB-PLC. *Int. J. Commun. Syst.* **2017**, *30*, e3344. [CrossRef]
7. Sharma, S.K.; Chandra, A.; Saad, M.; Lefebvre, S.; Asber, D.; Lenoir, L. Voltage flicker mitigation employing smart loads with high penetration of renewable energy in distribution systems. *IEEE Trans. Sustain. Energy* **2017**, *8*, 414–424. [CrossRef]
8. Gatsis, N.; Giannakis, G.B. Residential load control: Distributed scheduling and convergence with lost AMI messages. *IEEE Trans. Smart Grid* **2012**, *3*, 770–786. [CrossRef]
9. Zimmermann, M.; Dostert, K. A Multipath Model for the Powerline Channel. *IEEE Trans. Commun.* **2002**, *50*, 553–559. [CrossRef]
10. Zimmermann, M.; Dostert, K. Analysis and modeling of impulsive noise in broad-band powerline communications. *IEEE Trans. Electromagn. Compat.* **2002**, *44*, 249–258. [CrossRef]
11. Philipps, H. Modelling of Powerline Communication Channels. In Proceedings of the 3rd International Symposium on Powerline Communications and Its Applications, Lancaster, UK, 30 March–1 April 1999; pp. 14–21.
12. Matthias, G.; Rapp, M.; Dostert, K. Power line channel characteristics and their effect on communication system design. *IEEE Commun. Mag.* **2004**, *42*, 78–86.
13. Masood, B.; Usman, M.; Din, F.U.; Haider, A. Effect of transient and non-transient models on the performance of PLC. *Telecommun. Syst.* **2016**, *65*, 55–64. [CrossRef]
14. Masood, B.; Baig, S. Channel Modeling of NB-PLC for Smart Grid. In Proceedings of the 2015 IEEE Symposium on Computers and Communications (ISCC), Larnaca, Cyprus, 6–9 July 2015.
15. Banwell, T.; Galli, S. A Novel Approach to the Modeling of the Indoor Power Line Channel Part I: Circuit analysis and Companion Model. *IEEE Trans. Power Deliv.* **2005**, *20*, 655–663. [CrossRef]
16. Galli, S.; Banwell, T. A novel approach to the modeling of the indoor power line channel part II: Transfer function and its properties. *IEEE Trans. Power Deliv.* **2005**, *20*, 1869–1878. [CrossRef]
17. Sung, T.E.; Scaglione, A.; Galli, S. Time-Varying Power Line Block Transmission Models over Doubly Selective Channels. In Proceedings of the 2008 IEEE International Symposium on Power Line Communications and Its Applications, Jeju City, Korea, 2–4 April 2008.
18. Anatory, J.; Theethayi, N.; Thottappillil, R. Power-Line Communication Channel Model for Interconnected Networks–Part I: Two-Conductor System. *IEEE Trans. Power Deliv.* **2009**, *24*, 118–123. [CrossRef]
19. Anatory, J.; Theethayi, N.; Thottappillil, R. Power-Line Communication Channel Model for Interconnected Networks—Part II: Multiconductor System. *IEEE Trans. Power Deliv.* **2009**, *24*, 124–128. [CrossRef]
20. Tonello, A.M.; Versolatto, F. Bottom-Up Statistical PLC Channel Modeling—Part I: Random Topology Model and Efficient Transfer Function Computation. *IEEE Trans. Power Deliv.* **2011**, *26*, 891–898. [CrossRef]
21. Tonello, A.M.; Versolatto, F. Bottom-Up Statistical PLC Channel Modeling—Part II: Inferring the Statistics. *IEEE Trans. Power Deliv.* **2010**, *25*, 2356–2363. [CrossRef]
22. Tonello, A.M.; Versolatto, F. New Results on Top-down and Bottom-up Statistical PLC Channel Modeling. In Proceedings of the 2009 Third Workshop on Power Line Communications, Udine, Italy, 1–2 October 2009.
23. Cataliotti, A.; Cara, D.D.; Fiorelli, R.; Tine, G. Power-Line Communication in Medium-Voltage System: Simulation Model and Onfield Experimental Tests. *IEEE Trans. Power Deliv.* **2012**, *27*, 62–69. [CrossRef]
24. Artale, G.; Cataliotti, A.; Fiorelli, R.; Tine, G. Secondary substation power line communications for medium voltage smart grids. In Proceedings of the 2012 IEEE International Workshop on Applied Measurements for Power Systems (AMPS), Aachen, Germany, 26–28 September 2012.
25. Chu, G.; Li, J.; Liu, W. Narrow Band Power Line Channel Characteristics for Low Voltage Access Network in China. In Proceedings of the 2013 IEEE 17th International Symposium on Power Line Communications and Its Applications, Johannesburg, South Africa, 24–27 March 2013.

26. Bausch, J.; Kistner, T.; Babic, M.; Dostert, K. Characteristics of indoor power line channels in the frequency range 50–500 kHz. In Proceedings of the 2006 IEEE International Symposium on Power Line Communications and Its Applications, Orlando, FL, USA, 26–29 March 2006.
27. Canete, F.J.; Cortes, J.A.; Diez, L.; Entrambasaguas, J.T. A channel model proposal for indoor Power Line Communications. *IEEE Commun. Mag.* **2011**, *49*, 166–174. [CrossRef]
28. Dong, L.; BaoHui, Z.; DongWen, N.; Bo, Z.Q.; Klimek, A. Design and implement of adaptive BFSK PLC system for medium voltage powerline communication. In Proceedings of the 45th International Universities Power Engineering Conference UPEC2010, Cardiff, Wales, UK, 31 August–3 September 2010.
29. Cataliotti, A.; Cosentino, V.; Cara, D.D.; Tine, G. Oil-Filled MV/LV Power-Transformer Behavior in Narrow-Band Power-Line Communication Systems. *IEEE Trans. Instrum. Meas.* **2012**, *61*, 2642–2652. [CrossRef]
30. Cataliotti, A.; Cosentino, V.; Cara, D.D.; Tinè, G. Simulation and Laboratory Experimental Tests of a Line to Shield Medium-Voltage Power-Line Communication System. *IEEE Trans. Power Deliv.* **2011**, *26*, 2011. [CrossRef]
31. Masood, B.; Haider, A.; Baig, S. Modeling and Characterization of Low Voltage Access Network for Narrowband Powerline Communications. *J. Electr. Eng. Technol.* **2017**, *12*, 443–450. [CrossRef]
32. Masood, B.; Ellahi, M.; Khan, W.A.; Akram, W.; Usman, M.; Gul, M.T. Characterization and Field Measurements of NB-PLC for LV Network. *J. Electr. Eng. Technol.* **2018**, *13*, 521–531.
33. Cataliotti, A.; Daidone, A.; Tinè, G. A Medium-Voltage Cables Model for Power-Line Communication. *IEEE Trans. Power Deliv.* **2009**, *24*, 129–135. [CrossRef]
34. Benato, R.; Caldon, R. Frequency Characteristics Measurement of Overhead High-Voltage Power-Line in Low Radio-Frequency Range. *IEEE Trans. Power Deliv.* **2007**, *22*, 575–583. [CrossRef]
35. Kim, I.H.; Kim, W.; Park, B. Channel measurements and field tests of Narrowband Power Line Communication over Korean underground LV Power Lines. In Proceedings of the 18th IEEE International Symposium on Power Line Communications and Its Applications, Glasgow, UK, 30 March–2 April 2014.
36. Cataliotti, A.; Cosentino, V.; Cara, D.D.; Tine, G. Measurement issues for the characterization of medium voltage grids communications. *IEEE Trans. Instrum. Meas.* **2013**, *62*, 2185–2196. [CrossRef]

Article

Cost Efficient Real Time Electricity Management Services for Green Community Using Fog †

Rasool Bukhsh [1], Muhammad Umar Javed [1], Aisha Fatima [1], Nadeem Javaid [1,*], Muhammad Shafiq [2] and Jin-Ghoo Choi [2,*]

1 Department of Computer Science, COMSATS University Islamabad, Islamabad 44000, Pakistan; rasoolbax.rb@gmail.com (R.B.); umarkhokhar1091@gmail.com (M.U.J.); aishafatima10@gmail.com (A.F.)

2 Department of Information and Communication Engineering, Yeungnam University, Gyeongsan, Gyeongbuk 38541, Korea; shafiq.pu@gmail.com

* Correspondence: nadeemjavaidqau@gmail.com (N.J.); jchoi@yu.ac.kr (J.-G.C.)

† This paper is an extended version of our paper published in Green Fog: Cost Efficient Real Time Power Management Service for Green Community. In Proceedings of the 14th International Conference on Complex, Intelligent, and Software Intensive Systems (CISIS-2020), Lodz, Poland, 1–3 July 2020.

Received: 12 May 2020; Accepted: 14 June 2020; Published: 18 June 2020

Abstract: The computing devices in data centers of cloud and fog remain in continues running cycle to provide services. The long execution state of large number of computing devices consumes a significant amount of power, which emits an equivalent amount of heat in the environment. The performance of the devices is compromised in heating environment. The high powered cooling systems are installed to cool the data centers. Accordingly, data centers demand high electricity for computing devices and cooling systems. Moreover, in Smart Grid (SG) managing energy consumption to reduce the electricity cost for consumers and minimum rely on fossil fuel based power supply (utility) is an interesting domain for researchers. The SG applications are time-sensitive. In this paper, fog based model is proposed for a community to ensure real-time energy management service provision. Three scenarios are implemented to analyze cost efficient energy management for power-users. In first scenario, community's and fog's power demand is fulfilled from the utility. In second scenario, community's Renewable Energy Resources (RES) based Microgrid (MG) is integrated with the utility to meet the demand. In third scenario, the demand is fulfilled by integrating fog's MG, community's MG and the utility. In the scenarios, the energy demand of fog is evaluated with proposed mechanism. The required amount of energy to run computing devices against number of requests and amount of power require cooling down the devices are calculated to find energy demand by fog's data center. The simulations of case studies show that the energy cost to meet the demand of the community and fog's data center in third scenario is 15.09% and 1.2% more efficient as compared to first and second scenarios, respectively. In this paper, an energy contract is also proposed that ensures the participation of all power generating stakeholders. The results advocate the cost efficiency of proposed contract as compared to third scenario. The integration of RES reduce the energy cost and reduce emission of CO_2. The simulations for energy management and plots of results are performed in Matlab. The simulation for fog's resource management, measuring processing, and response time are performed in CloudAnalyst.

Keywords: fog computing; green community; resource allocation; processing time; response time; green data center; microgrid; renewable energy; energy trade contract; real time power management

1. Introduction

Electricity is categorized as a basic right or basic need for people in the world [1,2]. The increase in population and technological advancements has increased the global electricity demand, especially in last decade. The conventional power generators run on fossil fuels to fulfill the energy demand; however, a huge amount of CO_2 is released in the environment due to combustion of fossil fuels. The CO_2 pollutes the environment and causes greenhouse effect. The fulfillment of human need of energy without polluting the environment is challenging. The researchers have proposed variety of solutions for environment friendly power production, smart consumption of electricity, and intelligent mutual cooperation between supply and demand sides to optimize energy utilization.

Renewable Energy Sources (RES) help to reduce dependency on conventional power generators. Various countries have set plans to integrate RES with existing system in order to reduce load demand from conventional power generators [3,4]. The RES are preferred over conventional power systems due to cheap energy generation for a long time. The researchers proposed RES as sole power generation for commercial and residential sectors for geographic regions where energy infrastructure is not laid down [5–7]. RES are also integrated with existing system for economical benefits [8]. The RES on demand side reduce the dependency on conventional fossil fuel based power generators on supply side, hence reducing the emission of CO_2. Currently, around 14% of world's power demand is being fulfilled with RES [3].

The RES are a not suitable permanent alternative of conventional power generators due to their intermittent nature [9]. However, the integration of RES based Microgrids (MGs) with existing system fulfills economical and environment friendly power demand. The Information and Communication Technologies (ICT) enables the supply and demand sides to aware of the situation to intelligently optimize the power utilization. The intelligent control of power generation, transmission, integration of distributed power generators and maintainability of power grid with ICT defines the Smart Grid (SG) [10]. Hence, a conventional power grid with the introduction of ICT is defined as SG and MGs are small scale distributed power generators [11,12]. The autonomic, sustainability, and scalability of SG has the potential to integrate cheap, environment friendly, and distributed RES based power generators. In view of this, around 179 countries, including Germany, U.S.A., and the Republic of Korea have set their target for the years 2017, 2020, and 2030 to fulfill full or part of energy demand with RES [13]. The governments offer subsidized RES to their public for social welfare to reduce the dependency on conventional power generators. However, willingness to payback was not considered consequently, in 2012 Spain stopped subsidizing on PhotoVoltaics (PV) power generators and European countries partially stopped such schemes [14,15]. Hence, energy trading strategies are inevitable to integrate RES in existing power system for the fulfillment of economical and environment friendly power demand. The researchers have proposed tools, platform and strategies to provide optimized energy management on the demand-side by evaluating and analyzing data for academic research in SG [16–18].

The optimized power utilization demands efficient energy management. Centralized, decentralized and distributed electricity managements are proposed by the researchers, which optimize the power utilization for proposed scenarios [19]. The analysis of collective behavior for power utilization is very complex for very short period of time or real-time in distributed and decentralized approaches. A centralized infrastructure is crucial for efficient power management for communities. The ICT has enabled the power controlling devices to communicate and manipulate at centralized computing platform (e.g., cloud) to optimize power utilization in SG [20]. Distributed and decentralized systems require centralized computing platform for analysis of various parameters for power management services [20,21]. Hence, a centralized computing platform is inevitable to provide analysis based optimized services to masses in the SG.

The cloud based platform provides on demand economical physical and virtual resources for computing services without investing for maintenance. The scalable infrastructure of cloud allows for the addition or removal of computing resources, depending on the requirements. A variety of services can be provided from single cloud infrastructure. Resource sharing and heterogeneous services provision from single platform reduce computing cost. The cloud has efficient processing and economical as compared to on-site and customized systems. However, it also has limitations, especially for the provision of real-time services. The limitations are enlisted below,

- heterogeneous services and too many requests increase the Processing Time (PT) [22],
- the physical long distance between end-user and physical cloud infrastructure increase the Response Time (RT) [23,24],
- high computation heats the physical resources, which are cooled by high powered air-conditioning systems, which increase service cost [25], and
- economical and environmental friendly huge power generation is challenging [26], especially for increasing demand of computing devices and cooling system data center is challenging [27].

The data centers of big companies, giants like Google and Facebook, have plans to installed their cloud data centers in cold regions to avoid expensive and high power cooling systems [28,29]. The long physical distance between end-users in hot regions and the physical cloud infrastructure in cold region increases the RT due to locality issues [24]. However, power applications are time-sensitive [30,31] and the real-time services from cloud infrastructure become a challenge [32]. Hence, from the literature, it is learned that the integration of renewable energy, real-time, and economical power management services for a community are challenging.

This work is an extension of our conference paper [33] in which two case studies are proposed. In this paper, three scenarios and a contract (policy) for energy trade are proposed. The contract ensures the participation of available distributed power sources with reduced cost of energy consumption. The participation of PV based RESs during the day reduce the cost of energy. A mechanism is proposed to calculate the energy that is required to provide computing services by computing resources and cooling system of a fog. A system model is proposed in which fogs of communities store energy data on cloud. The data is usable for future projects, e.g., prediction, analysis, evaluation, and feasibility, etc. The fog equipped with green (renewable) energy for a community provides near real-time response with environment friendly power management services. However, huge computing resources installed for communities of distance locations, e.g., cloud emit heat in the environment and affect it. Hence, the proposed system model is suitable for environment friendly and near real-time response.

The hierarchy of this paper is presented in following sections. Related work is given in Section 2 and, in Section 3, Proposed System Model is elaborated. Case Studies are presented in Section 4 and the Conclusion is presented in Section 5.

2. Related Work

The SG applications are time sensitive and real-time energy management solutions are challenging, especially for masses on the demand side [30,31]. The communication between demand and supply sides in real-time ensures efficient energy utilization. Kong et al. [34], propose a radio frequency based networked communication for a community. The residents generate and store renewable energy and share with neighbors for cost efficiency. The device-to-device communication form a network, which follows the topology of SG. The optimality of channel utilization to ensure the participation of maximum mobile units with minimum transmit power of sensors is the main idea. The authors have attained significant outcomes; however, the size of maximum participants, communication delay, and their effects have not discussed in particular. In SG applications, a huge number of power users need to be updated simultaneously to change in parameters' values. For instance, pricing on the supply side is updated and the power users need to modify their usage accordingly. The huge traffic on communication medium and network nodes creates bottleneck and power users suffer from delayed information. Wang et al. [35] and Khaled et al. [36] propose 5G and 6G technologies for huge data transfer rate and high bandwidth. However, these technologies are not mature and they are under research and development phase.

Zepter et al. [37] proposed a platform in which prosumers of a community trade battery based stored energy. The batteries are charged from renewable and utility power sources. The framework allows for peer-to-peer energy trading and incentive the prosumers by giving them chance to reduce energy cost. The proposed platform allows prosumers to participate in whole sale power market. The price of prosumers' energy trade is lesser than price of the utility. The proposed peer-to-peer energy trading is implemented in a residential building in London. The study shows 20–30% reduction in electricity bills and battery storage saves almost 60%. The platform allows for each prosumers to trade the energy with peers; however, they are connected with grid power lines. The energy trade with utility and other distance power users using the power lines can increase the integration of renewable energy, efficient utilization of storage system, and maximize the incentives.

The participation of residents of a community equipped with small scale power generation units reduce the demand load on supply side. The trade of small scale energy forms self-dependent community. However, defining contract between producer and consumer is necessary. The authors in [38], proposed contract game for energy trading in SG. A direct trading contract is defined to minimize the energy cost for prosumers. A theoretical contract for deterministic environment considering short-term market is proposed, which, later extended for long-term market. In long-term, the high uncertainty of small scaled energy production is considered. The incentives for energy producers and consumers with direct contract encourage the maximum participation of producers and consumers. Qin et al. [39] also propose contract for flexible market. The objective is to have minimum communication and control responsibilities on energy operators in distribution system. The high flexible energy market is balanced with Arrow–Debreu, which encourage the traders to maximize their participation and minimize the intervention of system operators. The authors left open questions for readers by discussing the limitations of proposed framework.

The energy trading mechanism helps to integrate renewable power with existing system for a smart community. The individual renewable power production microgrids may have significant impact to reduce maximum power cost with high integration of renewable or green energy. Chen et al. [40] propose a centralized energy management for electrical vehicles. The service maximize the incentives for maximum participation by defining the optimized choice between energy consumption and trading. The cloud provides the service for communities whenconsidering electrical vehicles, energy storage systems, and distributed renewable energy sources for customized energy pricing. The authors formulate a binary linear programming model and the performance is evaluated in experimental setups. The effects of participation of storage system, electrical vehicles, and smoothness of fluctuations for demand response of electrical vehicles are discussed in detail. The PT of the service on the cloud almost doubles for every addition of 500 participants, measured in seconds. The authors claim the significant smaller PT as compared to day-head period; however, the response time for end-power-users are not discussed. Long physical distance between cloud data center and end-users requires multiple computing network nodes. The processing on each node and data transfer (from end-user to cloud and back to end-user) cause the latency and increase the response time for end-users [41]. The time sensitive, like power applications, are prone to compromising the performance.

Cloud infrastructure is scalable and it consists of large number with huge sizes (operation capabilities) of physical computing devices. Such huge infrastructure is capable of providing high computation in fraction of time, even for the analysis of huge data [42]. The power demand of such data centers is 1.5% of world's total demand with annual growth rate at 4.3% [27]. These data centers produce equivalent heat, which is controlled by cooling systems. The cooling systems consume 30–60% of total demand of data centers [43]. Researchers have proposed various solutions to reduce energy demand for data centers. Toprak et al. [43] develop a software tool to estimate the power demand for data center with optimized cooling system. The considers location, type of building, electronic equipment, and setup environment (indoor and outdoor) of the system for energy and cost estimation. Jawad et al. [44] propose an optimization of workload in data centers in coordination with multiple power sources, e.g., batteries, renewable energy generators, diesel generators, and utility to reduce the energy cost. The authors minimize the dependency on fossil fuel based power generation for data center operations. Similarly, Xu et al. [45] propose a task scheduling algorithm that is based on enhanced reinforced learning and neural network. The job scheduling and the renewable power supplies reduce the energy cost for data centers. The summary of related work wor is described in Table 1.

Table 1. Summary of Related Work.

Authors	Proposed Solution	Limitations
Kong et al. [34]	Radio frequency based device-to-device communication following the topology of grid. The residents of the community are allowed to generate and trade renewable energy in peer-to-peer fashion.	The number of participants and effects of communication delay on energy trade are essential to identify.
Zepter et al. [37]	Proposed a platform for prosumers to trade battery based energy while connected with utility power lines. Prosumers participate in whole market and peer-to-peer trade is perfromed	The platform has potential to increase the energy trade by introducing centralized computation for cost efficient energy utilization. Moreover, prosumers may trade with more suitable distant consumers.
Zhang et al. [38]	Game based energy contract for prosumers is proposed to encourage maximum participation by signing direct contract.	The contract is proposed for small scale environment.
Qin et al. [39]	Proposed energy contract for flexible market. The authors left open questions for identification of possible limitations.	The minimum intervention of system operators, controllers and reduced communication lead to security issues.
Chen et al. [40]	Proposed cloud based centralized energy management service. The integration of renewable energy from electrical vehicles and storage systems is maximized by incentivizing using power trade mechanism.	The processing time increases with the increase of participants which increase the response delay. The delayed response has negative effect on the system.
Toprak et al. [43] and Jawad et al. [44]	In [43], a software tool is proposed to estimate power demand by data center with optimized cooling system. In [44], optimization of workload in data center considering various power source, e.g., renewable power and fossil fuel based power generators	The software based tool may not include all parameters required for energy optimization, e.g., as authors did in [44]. However, method in [44] may require some software or simulator to estimate the cost.
Xu et al. [45]	Proposed reinforced learning technique to schedule the tasks on computing resource in data center. Authors proved reduced energy consumption with efficient resources allocation for computation.	The authors did not discuss the load of tasks for suitable efficiency of scheduling of job

The limitations of energy management services for communities in the literature encourage to propose a system with real time power management service for communities. In this paper, a fog based system model is proposed, which allows end-power-users to directly communicate with the fog. The communication takes place on high bandwidth (4G and 5G) wireless medium. The network latency with 5G reduced to almost zero; however, the execution time depends on the computing devices. The efficient resource utilization algorithms reduce the execution time and consume efficient energy. The fog performs energy management for the residents of a community. Three scenarios are proposed to evaluate the cost efficient energy usage for community and fog's data center. A mechanism for calculating the power required for running of computing resources to execute the service and cooling system for fog's data center is proposed. Lastly, a contract is proposed to ensure the participation of renewable power generators and the utility.

3. Proposed System Model

A systematic overview of proposed system in given in Figure 1. The residents of each community have two power supplies; RES based MG and the utility for demand of their homes. The MG is placed in the vicinity of respective community to reduce power losses. The utility is used for backup when the MG is down or to be used with other power sources under the contract signed between utility and end-power-users. The energy management services for residents of the communities are provided by the respective fogs. The physical setup of each fog is installed near or within the respective community, so that each resident has direct access (at first hop) to the computing resources. The energy management services or programs are installed in the fogs, which are run on virtual machines. The virtual resources are programs that mimic physical machines and they are used to share physical resources. The resource sharing reduces the computing cost for the services as well as efficient utilization of the physical computing resources. The residents of a community request the fog for the power management services. Each request is entertained in fog and responded back via 4G or 5G wireless and wired technologies. Each fog is connected with a cloud via Internet for data storage, which can be analyzed for future and other related to power projects.

In this paper, three scenarios are considered for the proposed system model. In first scenario, the power for the physical computing resources of fog is only supplied from the utility. The community is also dependent on utility power supply. In the second scenario, a fog has two power supplies; renewable electricity from the community and from the utility. In the third scenario, the fog owns RES based MG (FMG); it also connected with communal MG and with the utility. The cost of utility power fluctuates depending on demand; however, it is always higher when compared to RES based power. The RES generate energy from free natural sources, e.g., wind, sun, etc., while, utility generates expensive electricity by running fossil fuel based power generators. The increasing power demand increases electricity cost due to greater fuel consumption. The fuel combustion releases CO_2, which pollutes the environment. Hence, generating expensive power by damaging the environment. A contract based energy trade is also proposed to integrate all power sources, e.g., utility, MG, and FMG to fulfill power demands of the community and fog's data center.

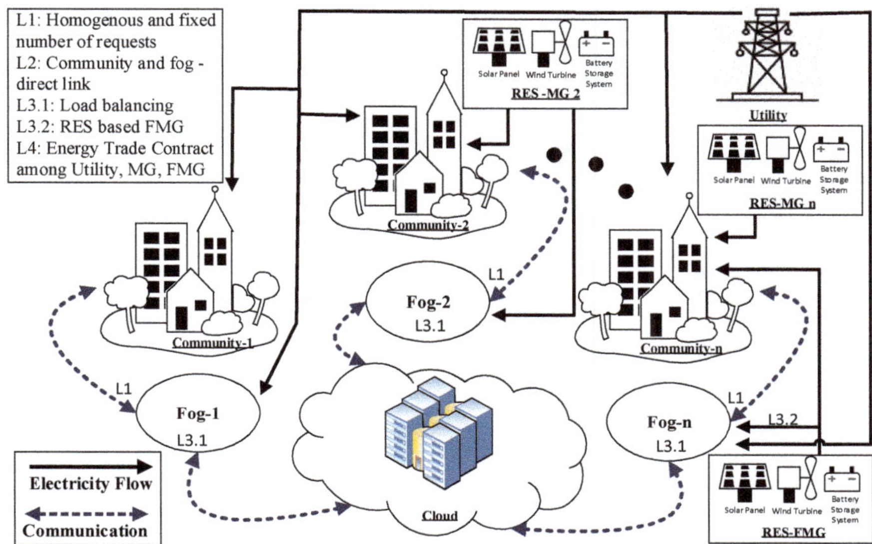

Figure 1. Proposed System Model.

The innovation of this paper is highlighted by comparing the identified problems of existing research and proposed solution in Table 2.

Table 2. Mapping of Existing Problems with Proposed Solution.

Problems Identified	Proposed Solution
Chen et al. [22], proposed cloud based energy management model. The PT increases with the increase of customers (e.g., from 500 to 1500 customers), moreover; authors do not discuss the effects of RT.	Fog based energy management model is proposed. The end-users are directly connected with the fog for power management services. The computing resources of fog are sufficient for requests of 300 SHs to process and response in near real-time.
Miodrag et al. [24], validate the efficiency of fog based monitoring and control service in SG as compared to cloud. Authors claim potential of fog based system with real-time monitoring and controlling in SG. However; prime focus is communication protocol for real-time monitoring.	Fog based energy management services, PT and RT are computing for a community of 300 SHs.
Saeed et al. [25] propose game based thermal aware resource allocation in data center (cloud) to reduce the emission of thermal energy due to high computation. Authors claim the proposed technique avoid creating hotspots as compared to counterpart strategies. The services run by the data center and possible amount of thermal energy produced are not discussed.	A mechanism to calculate the amount energy produced due to computation in fog's data center explained. Moreover, the relationship between energy produced due to computation and power required for cooling system for power management services is proposed.

Table 2. *Cont.*

Problems Identified	Proposed Solution
Authors in [46] propose VM placement technique to reduce carbon emission from cloud data centers. Green energy is beneficial to reduce carbon, however; huge platform of cloud requires an agreement between users and cloud service providers. Authors claim reduced energy cost, reduce emission of carbon and suggest integration of green energy for cloud data center. However, agreement of integration of green energy is not proposed.	A modified honey bee colony optimization technique is used to balance the load on VMs to enhance the computing efficiency. An agreement (contract) is proposed to integrate utility (fossil fuel based power generator), RESs based MG for community and FGM. The contract reduce energy cost and encourage the integration of renewable energy with incentives to the participants.
Thiago et al. in [27] conducted an intensive survey on energy efficiency and demand response for small and medium data centers. Authors claim that large data center have potential to participate in energy efficient demand response program; however, small and medium data centers are more adoptive for the program. The violation of energy policies by energy consumers also have negative impact.	Proposed system model with energy management services validate the claim of suitability of medium (fog's) data centers for energy efficiency. Moreover, the proposed energy contract runs on the fog as service, which avoid the interruption of end-power-users.
The authors in the above articles in this column discuss either computing platform or energy management. None of the author has proposed efficient solution for considering both.	In this paper, a system model is proposed for energy management service for community of 300 SHs. Energy management services are proposed considering different power sources to reduce energy cost by integration of green energy and incentive policy. The power demand for computing environment is calculated and fulfilled with multiple power sources with minimum cost.

3.1. Problem Formulation

In first scenario, the electricity demand of the community and the fog is fulfilled from utility. If the size of community is H number of Smart Homes (SHs) ($H = \{h_1, h_2, h_3, ..., h_H\}$) and the total power demand in a given time t is the sum of load of all homes ($L^t_H = \{l^t_{h1} + l^t_{h2} + l^t_{h3}, ..., l^t_{hH}\}$). The total electricity cost (T^t_{ct}) of the community is calculated with Equation (1). The cost of power consumption of a fog depends on the load demand for computing resources and cooling or air conditioning system. The energy cost of the fog is calculated with Equation (2).

$$T^t_{ct} = U^t_p \times \sum_{h=1}^{h=H} L^t_h, \tag{1}$$

$$T^t_f = U^t_p \times (L^t_{cr} + L^t_{ar}). \tag{2}$$

where, T^t_f is total cost of the fog, U^t_p is utility price, L^t_{cr} is load of computing resources, and L^t_{ar} is the load of air conditioning system at time t.

In second scenario, the surplus renewable energy of communal MG (S^t_{ct}) is bought for the fog. S^t_{ct} is the difference of energy produced by MG M^t_E and the total community demand C^t_D at time t, as shown in Equation (3).

$$S^t_{ct} = C^t_D - M^t_E. \tag{3}$$

The energy demand by a fog (F_D^t) for time t is the sum of load of L_{cr}^t and L_{ar}^t. The energy price of MG (M_p^t) is cheaper than the utility (U_p^t). The S_{ct}^t is traded with the fog or with the utility or both of these.

$$if \quad F_D^t - S_{ct}^t > 0, \tag{4}$$

then, remaining load for the fog is bought from the utility at U_p^t. Similarly,

$$if \quad F_D^t - S_{ct}^t < 0, \tag{5}$$

then, remaining power of surplus energy is sold to the utility at electricity price of the MG M_p^t.

$$if \quad F_D^t - S_{ct}^t = 0, \tag{6}$$

then, all of the surplus energy is only bought by the fog at MG's price (M_p^t).

In third scenario, the fog also owns RES based MG to ensure greener computing for fog services. The fog's computing performance varies the energy demand; moreover, the power output of FMG is also affected by natural sources, e.g., intensity of sun light, wind force, etc. The varying power demand by the fog and fluctuated power production by the FMG might have imbalanced energy generation. The imbalance is cured by trading the surplus energy $-S_f^t$ with the community or with the utility; however, deficient power (S_f^t) is bought from the community MG or from the utility. The Equation (7) shows the surplus energy of FMG, where F_{mg} is FMG's produced energy.

$$S_f^t = F_D^t - F_{mg}^t. \tag{7}$$

The relationships among prices of electricity for FMG (F_p^t), community MG, and the utility are given in Equation (8).

$$F_p^t < M_p^t < U_p^t. \tag{8}$$

The fluctuated power generation from MGs and varying energy demand by the fog constitute three situations. In the first situation, the power generation of FMG L_f^t is greater than fog demand as given in Equation (9). In this condition, surplus FMG's is traded with the community or with the utility. In the second situation (Equation (10)), FMG generates less power than demand of the fog. In this condition, the remaining energy is bought from the community MG or from the utility. However, the green power of community MG is preferred over the utility due to lesser price than electricity price of the utility. In third situation (Equation (11)), FMG's power generation is equal to the energy demand by the fog. The whole energy is utilized to fulfill fog's power demand.

$$F_D^t - L_f^t > 0, \tag{9}$$

$$F_D^t - L_f^t < 0, \tag{10}$$

$$F_D^t - L_f^t = 0. \tag{11}$$

It is stated earlier that the fog's physical resources are shared by creating virtual resources on them. The virtual resources are computer programs, which mimic the actions of a machine; hence, also known as Virtual Machine (VM). These programs continue to run on the physical resources, even in the idle condition. The energy consumed by physical resources is calculated by measuring the execution of Millions of Instructions Per Second (MIPS). The actual energy consumption is measured by observing the active and idle states of VMs. An idle VM consumes 60% energy of active state [47]. The energy required during active state is $10^{-8} \times (MIPS)^2$ Joules per Million Instructions (J/MI). The mechanism of efficient resource allocation optimizes the energy consumption by computing resources. Various techniques have been proposed to efficiently allocate the VMs [46,48–50].

The power that is required to run VMs (in idle and active states) is converted into an equal amount of heat energy [51]. The thermal heat surround the servers in data center and increase the temperature, which compromise the performance of physical computing resources. The performance of hardware affect the computing performance, which compromises the service execution. The cold air is passed through server nodes in the data center to cool down the computing devices. The phenomenon of heat circulation or thermodynamics in the environment of data center is explained in detail in [25,52].

3.2. Contract for Energy Trade

In the case of multiple power generators, one of them can have maximum participation, which affects the contribution of other generators. For example, in the third scenario, if community MG fulfills the whole power demand of the community and fog's data center then the participation of the utility is idle. In view of this, to ensure the participation of all participants, an efficient energy contract is necessary to be defined. In proposed system model, during the day, community MG generates maximum power. The part of it is traded within the community, with fog's data center and the remaining is sold to the utility. Gai et al. [53] propose approximate maximum estimate model to estimate the maximum value for tradable energy of a certain time-slot. However, in this paper, a fixed percentage of MG's energy is proposed to trade for every time-slot. The maximum threshold (α_{max}) to trade within the community is 65% of MG's total produced energy (M_E^t) for every time-slot (hour). The remaining power is sold to the utility and to the fog's data center. The amount of power from community MG to be traded with the utility and the fog (E_{Tuf}^t) is found with Equation (12).

$$E_{Tuf}^t = M_E^t - \alpha_{max}, \tag{12}$$

where,

$$\alpha_{max} = M_E^t \times 0.65. \tag{13}$$

The cost of power trading is the product of amount of energy traded and the price M_p (from Figure 2) as shown in Equation (14). The E_{Tuf}^t energy of community MG is sold to the utility and the fog's data center at M_p price. The cost (C_{Tuf}^t) of energy trade with utility and the fog is calculated with Equation (14).

$$C_{Tuf}^t = E_{Tuf}^t \times M_p. \tag{14}$$

The fog only buys energy when FMG is unable to fulfill the demand or when $S_f^t > 0$ with Equation (7). If $S_f^t > E_{Tuf}$, then E_{Tuf} is bought from the community MG and remaining ($S_f^t - E_{Tuf}$) is bought from the utility. The energy bought for fog's demand (F_D) is the sum of FMG's energy, the amount of power bought from community MG (ΔE_{Tuf}), and amount of energy bought from the utility (ΔE_u), as given in Equation (15).

$$F_D = F_{mg} + \Delta E_{Tuf} + \Delta E_u, \tag{15}$$

where,

$$\Delta E_u = S_f^t - E_{Tuf} \tag{16}$$

and ΔE_{Tuf} is amount of energy which is bought complete when $S_f^t > E_{Tuf}$ to fulfill F_D^t. Where, $F_D = L_{cr}^t + L_{ar}^t$. The cost of fog's demand C_{FD} is calculated with Equation (17),

$$C_{FD} = F_{mg} \times F_p + \Delta E_{Tuf} \times M_p + \Delta E_u \times U_p. \tag{17}$$

The total energy cost for community MG (C_{cmg}) is computed with Equation (18). However, C_{Tuf} is excluded from the power cost of the community as incentive.

$$C_{cmg} = C_{Tuf} + \alpha_{max} \times M_p. \tag{18}$$

The utility energy cost depends on the cost of the amount of energy produced from fossil fuel based generators (ΔC_{fuel}), cost of energy bought from the community MG (ΔC_{cmg}), and the cost of amount of energy bought from FMG (ΔC_{fmg}), as computed with Equation (19). The cost of fossil fuel based energy depends on amount of fuel consumed to produce desired (for demand) electricity, while considering factors, like heating and fuel cost, etc. This is calculated with quadratic function, as given in Equation (20) [54].

$$C_u = \Delta C_{fuel} + \Delta C_{cmg} + \Delta C_{fmg}. \tag{19}$$

$$C(U_{out}) = a + b.(U_{out}) + c.(U_{out})^2. \tag{20}$$

where, a, b, and c are the coefficients of heat and amount of fuel, etc., which effect the throughput (U_{out}) of fuel generator. The integration of green energy (e.g., Equations (13) and (15)) compels the utility to generate lesser power than the total demand. The utility runs fossil fuel based power generators (e.g., diesel), which emit CO_2. Lesser power generation reduces the running of the generators as a result emission of CO_2 is reduced. In this paper, the utility power pricing U_p for a day is considered, as given in Figure 2.

During the day, the fog stores surplus green energy and sells remaining (more than demand and storage). After the sun-set stored energy is utilized for computing and cooling systems. The storage capacity ST_{cap} is greater than F_D. In addition to this, the community relies on the utility power supply after the sun-sets. From the load basis of the community in Figure 3, the power demand of the community gradually reduces after the 20th and before the sixth hours. Hence, the power demand that is fulfilled from the utility during these hours is almost less than the demand during the day from the utility (after the integration of 35% of RES energy).

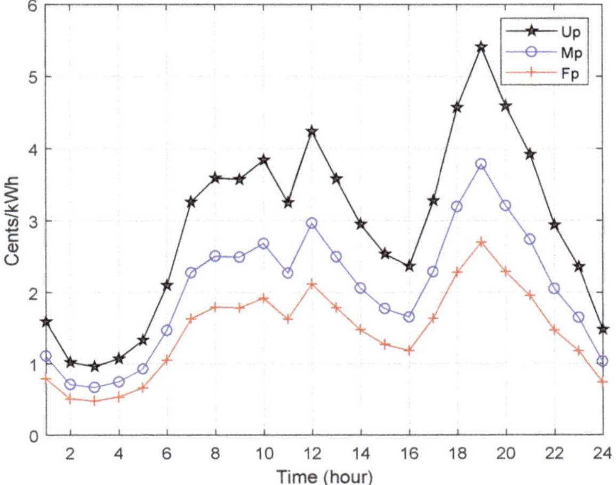

Figure 2. Energy Pricing.

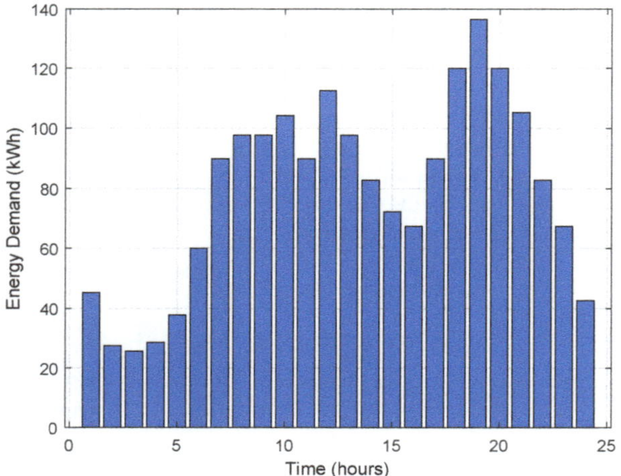

Figure 3. Demand Basis of Smart Home.

4. Case Studies

In this paper, the size of community for all three scenarios is considered 300 SHs. The power demand of each home ranges 0.5 to 1.5 of smart home demand basis given in Figure 3 [55]. In second and third scenarios, the average PV power that is required for a home is 0.4 to 1.4 of power basis given in Figure 4 [55]. The capacity of community MG is the sum of PV power generation of all SHs in the community. In the third scenario, the surplus energy of FMG is stored in energy storage system to be utilized after sunset. However, during the day the surplus energy is traded with the community or with the utility. The utility energy price for every hour of the day is taken from [55]. The community MG trades power at 70% of the utility prices and FMG trades at 50% of the utility prices as shown in Figure 2 to encourage the maximum integration of green energy with the utility. The energy management services of proposed scenarios run on the fog. The specifications of the fog are given in Table 3.

Table 3. Fog Specifications.

Parameters	Values
Operating System	Linux
Virtual Machine Manager	Xen
Architecture	X86
Physical Units	2
Processors (each unit)	4.4
VMs	5
Memory	12 GB
VM Speed	10 MIPS

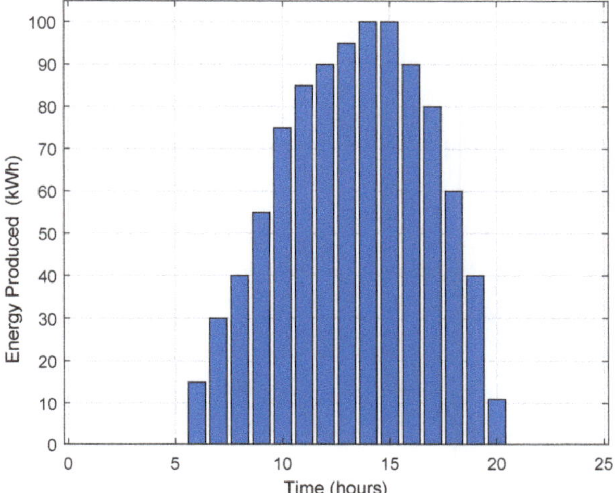

Figure 4. Basis of PV Power Generation.

In this paper, CloudAnalyst and Matlab are used for the simulation of case studies. In CloudAnalyst PT and RT for the requests generated from the community residents are measured. The fog data center and the community belong to the same region; however, Internet characteristics are defined according to the choice of generation of technology (e.g., 4G, 5G, etc.) and the system model. In proposed case studies, 4G and 5G based data transfer rates (Million Bits Per Second (Mbps) and Giga Bits Per Second (Gbps), respectively) are implemented. Matlab is used to simulate the energy use by the community and the fog for all three scenarios. Moreover, the results are also plotted using Matlab.

4.1. Discussion and Results

Figure 5 shows the flow of implementation of whole process. The SHs of the community sent requests every hour for energy management to the fog. The requests are allocated to VMs by balancing the load on them using intelligent load balancer, e.g., modified honey bee colony optimization. The load is shifted from higher loaded VMs to the lesser loaded VMs. The VMs mimic the function of a physical machine hence, each VM runs energy management program independently. The VMs enhance the computational efficiency of a single physical computing unit. The parallel execution of VMs with balanced load of requests (tasks) reduce the overall PT. In this paper, four energy management scenarios were performed in case studies. In first scenario, Equations (1) and (2) are used to calculate the energy cost of fog, e.g., computing cost and cost of cooling system, respectively, by evaluating power demand and energy pricing (of utility). In second scenario, VMs find the difference of community power demand of energy produced by the MG with Equation (3), which helps to manage energy by trading or buying or self-sufficient following the conditions of Equations (4)–(6). In the third scenario, unlike second scenario, fog owns a MG. Equation (7) finds the difference of fog demand and load of FMG. The pricing for energy trade must follow the condition given in Equation (8). The conditions given in Equations (9)–(11) help to trade power of FMG to fulfill fog demand. In contract based energy trade, the Equation (12) calculates the MG's trading amount of energy with the fog and the utility. Equation (13) finds the amount of MG's energy utilized within the community. Equation (14) calculates the cost of MG's trading energy. The fog's power demand is fulfilled with Equation (15) following Equation (16). The fog's power demand is fulfilled from different power sources; hence, the energy cost is calculated with Equation (17). The total energy cost community MG is calculated with Equation (18); however, the cost traded within the community is utilized as incentives. The proposed contract integrates the power generated by FMG and MG with its own generation and reduces the production cost calculated with Equation (19). Each VM runs services of the scenarios following the proposed equations and respond the power users (SHs).

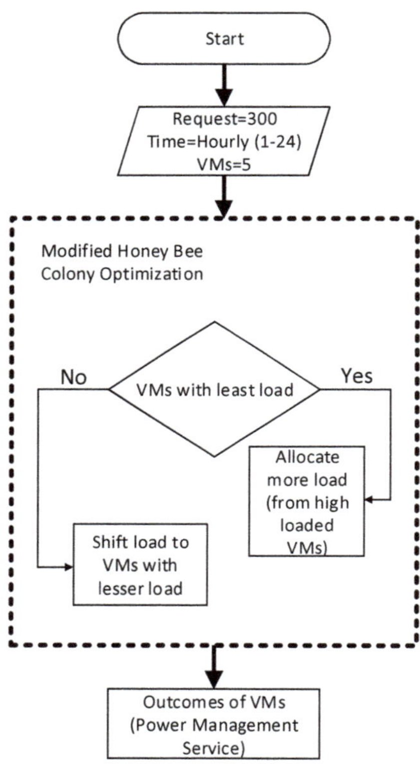

Figure 5. Implementation Flow Diagram.

In the proposed system model, unlike cloud, the physical infrastructure of fog and power users in the community are at the first hop. In cloud based infrastructure there are multiple computing nodes between end-users and cloud's data centers. Hence, cloud based infrastructure has high network latency as compared to fog's infrastructure. The requests sending time from the power users to the fog for energy management service is near real-time. The communication medium also affects the network latency. The data transfer rate of 5G is 100 times that of 4G [56,57]. Hence, requests sending with 4G is measured in milliseconds (ms), while, with 5G, it is measured in microseconds (μs). There is virtually no time delay with 5G; however, with 4G huge data may take longer time (e.g., in seconds or minutes). Once the request is sent to the fog, the processing time depends on the computing devices, architecture, and type of service. The simulation results are given in Table 4. The simulations of 7200 requests from the community in a day require an average of 0.48 ms to process the service. A modified honey bee colony optimization algorithm [58] is used to efficiently allocate the virtual resources. Hence, it is assumed that, during the processing, all VMs were in active state. The maximum time taken by a VM (T_{vm}) is the total execution time (T_{et}) divided by the total number of VMs Equation (21).

$$T_{vm} = T_{et}/VMs, \tag{21}$$

where,

$$T_{et} = 7200 \times 0.48 = 3456 \text{ ms} \tag{22}$$

and

$$T_{vm} = 3456/5 = 691.2 \text{ ms}. \tag{23}$$

Table 4. Simulations for a Community.

Parameters	Values
Total requests in a day	7200
Average PT	0.48 ms
Average RT (4G)	50.10 ms

When considering the speed of a VM given in Table 3, 6.912 Millions of Instructions (MI) are executed in 3456 ms. Using the Equation (2) of [47], each MI requires 10^6 Joules of energy when a VM is active. Hence, a total of 5×10^6 Joules is required for service execution in 3.456 minutes when VMs are active. The rest of the time VMs remain inactive and consume 60% of active state, as discussed above (1247×10^6 Joules). The total energy that is required for the fog is the sum of energy required by VMs during active and inactive states. According to the law of conservation of energy, the amount of power consumed is converted into equal amount of thermal (Joules) energy [51]. Hence, the amount of thermal Joules produced during the active and inactives states of VMs requires an equal amount of power (L_{cr}).

Various factors, e.g., indoor and outdoor temperature, power demand for cooling the data center, and execution of servers in the data centers, etc., are considered for designing the cooling system of data center [43]. A big portion of budget is spent on cooling the system of the data center [59]. In the proposed case studies, it is assumed that equal budget is allocated for cooling system, which defines the cost that is required for power consumed for execusion of requests. The cost of the cooling system is measured by the amount of required power load multiply by the energy price of the power supplier, e.g., the utility, MG, etc. The efficiency of cooling system is measured in BTU (British Thermal Unit), which is equal to 1055 Joules [60]. In this paper, it is assumed that the initial state of temperature of the data center should be maintained. Accordingly, the number of thermal Joules produced requires equal amount of power for cooling system to neutralize the change of heat in the data center. Hence, the total power that is consumed by the fog is the sum of energy required for computation (L_{cr}) and cooling or air conditioning system (L_{ar}). The energy management service is executed every hour in the day. The energy demand by the fog is also equally divided for every length of time t, e.g., $t = 1$ h.

The total power demand of community residents and the fog is given in Figure 6. In the first scenario, the community and the fog only consume power from utility. The utility prices depend on the power demand; the higher the demand, the higher the energy prices, as shown in Figures 2 and 6. In the second scenario, the community has PV based MG. The power of MG is supplied to the community at 30% lesser price as compared to the utility. The community demand is greater or lesser than MG's power generation. The power produced more than the demands of the community and the energy of FMG is sold to the utility to incentivise the community. In the proposed case studies, for the third scenario, the community fulfills the demand similar to second scenario. However, surplus renewable or green energy is first sold to the fog when FMG is unable to fulfill its demand; otherwise, it is sold to the utility. The FMG fulfills the demand of fog and surplus power is sold to the community when demand is greater than MG's generation. In second and third scenario, the fog minimizes the energy cost by consuming green energy. However, in third scenario, fog reduces energy cost by consuming power either from FMG or MG as well as combined energy from FMG, MG, and the utility. The trade of surplus energy of MG reduces the community power cost. Similarly, the trade of FMG's surplus energy reduces the power cost for the fog. The surplus energy of FMG of last two hours (e.g., 19th and 20th hours) is stored in batteries to be utilized after sun-set.

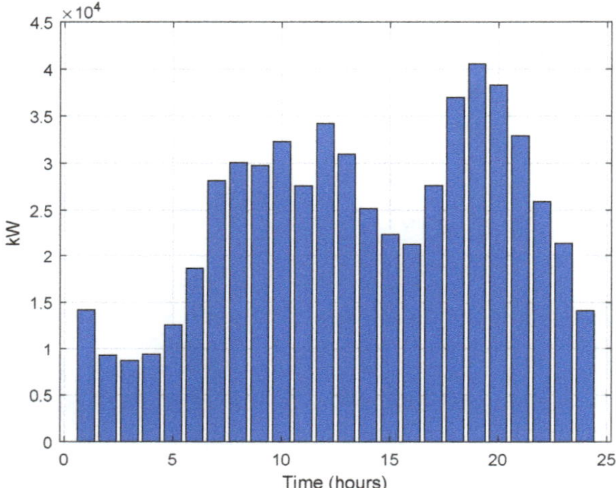

Figure 6. Total Load Demand of Community and Fog.

In Figure 7, the cost of power consumption by the utility and the fog in first scenario is the highest when compared to second and third scenarios. The cost in first and second scenarios are the same before the sixth and after the twentieth hours. The cheap energy of MG and trade of surplus power of MG with utility and the fog reduce the significant cost for the community during the day. The simulations show the lowest power cost for third scenario. The energy from MG and FMG are utilized for the community and for the fog; however, the remaining power is bought from the utility. The trade of surplus energy from MG and FMG help to reduce the extra cost as compared to first and second scenarios. In Figure 7, the cost with third scenario after 20th hour until the 23rd hour is less than other scenarios. The stored surplus energy of FMG is utilized after 20th hour (after sun-set).

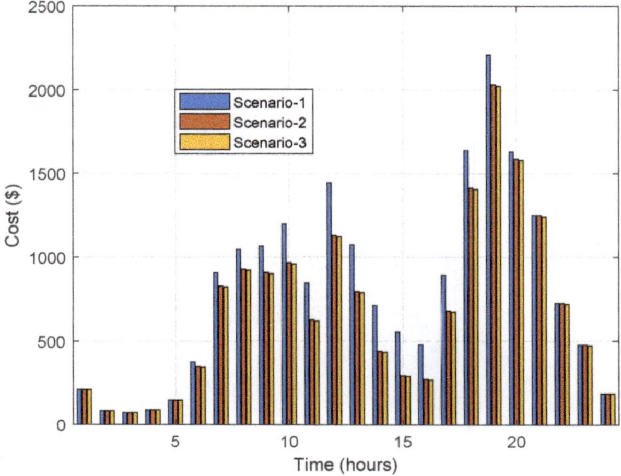

Figure 7. Energy Cost for Community and Fog.

4.2. Case Study: Energy Trade

The proposed contract is simulated for third scenario in which community and fog own their renewable MGs as well as they are connected with the utility. According to the contract, 65% of produced energy of community MG is traded within the community and the rest is sold to the utility. However, deficient power demand is bought from the utility and the FMG. The power demand for fog's data center is fulfilled with FMG and surplus power is sold to the utility. From Equation (15), power is not bought from the utility and community MG for the fog. Hence, in Figure 8, the amount of energy that is bought from the utility, FMG, community MG for community demand and amount of power sold to the utility from community MG is shown. During the day, the PV power is utilized (consumption and trade) and after sun-set the power from the utility is utilized. From Figure 6, the community's power demand is maximum during the day. Accordingly, the renewable energy trade, according to the contract, reduces the fossil fuel based power generation by the utility. The community is incentivised by reducing energy cost due to power trading of community MG.

The cost for power demand of the community with third scenario and with the proposed contract are shown in Figure 9. The renewable power is generated free of cost due to natural sources, e.g., sun light. However, the infrastructure and maintenance have cost, which are earned by providing cheaper energy. Hence, the energy sold out of the community is the incentive for the community. The contract based power utilization in the community is cost efficient as compared to the third scenario. The third scenario in Figure 7 has lesser cost for three hours when compared to first and second scenarios after 20th hour due to stored energy utilization of FMG. Similar is the case with contract cost in Figure 9; however, the contract cost is more efficient when compared to the third scenario due to energy trade incentives.

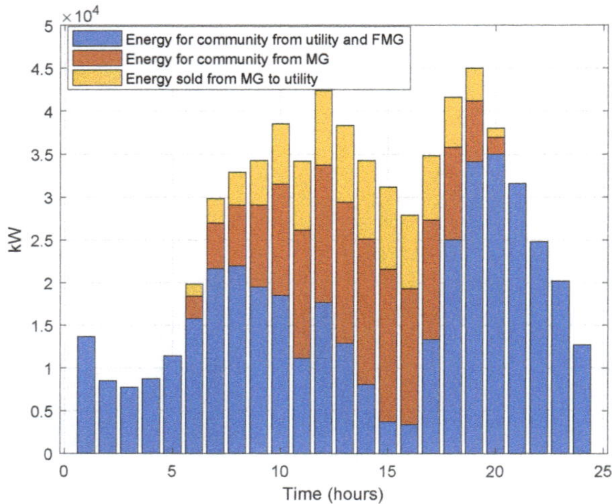

Figure 8. Contract: Power Utilization.

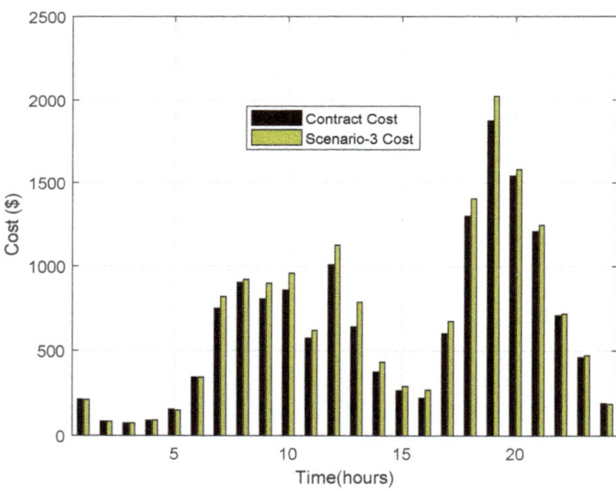

Figure 9. Contract: Power Cost for Community.

4.3. *Summary of Proposed Solution*

The cloud based energy management services suffer from latency issues. Multiple nodes between end-users and physical computing environment increase the RT. However, SG applications are time sensitive. In the proposed system model presented in Figure 1, end-power users (community) are directly connected with the fog, which reduces the network latency. The efficient resource utilization of computing resources reduces the PT. Hence, reduced network latency and PT provide a time efficient energy management service to the community. The results of efficient computing resources utilization and direct connectivity of the community with fog advocate the near real-time response of energy management. In this paper, three scenarios are proposed to fulfill the energy demand of the community and the fog's data center. The simulation results show that the integration of RES based MGs reduce the energy cost. The energy trade of surplus energy reduces more cost for end power users due to incentives. In the third scenario, apart from the utility, the fog and community also own RES based MGs. Hence, the results show that third scenario has the more cost efficient energy for the community and the fog. However, the proposed contract ensures the participation of all power sources. The results advocate the cost efficient energy is provided to the community and to the fog's data center as compared to third scenario. Hence, the proposed energy contract ensures the integration of RES with minimum cost for power users, e.g., the community and the fog's data center. The summarized mapping of limitations of exiting solutions and results validation of proposed solution is given in Table 5.

Table 5. Mapping of Problems with Validation Outcomes.

Limitation Number	Limitation	Proposed Solution	Validations
L1	Heterogeneous and too many requests on cloud increase the PT	S1	Homogeneous and fixed number of requests from community to the fog. 300 SHs directly request the fog for cost efficient power management every hour
L2	Long physical distance between end-users and computing resources increase the RT due to multiple nodes between them	S2	Direct link between end-users (community) and the fog to reduce network latency. The Table 4, shows 50.18 ms of average RT
L3.1	Computing devices heat due to high computation	S3.1	Load on computing resources are balanced intelligently (e.g., Modified Honey Bee Colony Optimization) for efficient utilization. The Table 4 shows very small average PT due to efficient resource utilization
L3.2	Increase service cost: high power demand due to computation and cooling system(s)	S3.2	Installed RES based FMG and connect with community MG for cost efficient power supply. The Figure 7 shows the cost efficient power in third scenario and in Figure 9 the contract based energy is cost efficient due to RES based MGs.
L4	Expensive fossil fuel based power supply from the utility	S4	Contract for energy trade is proposed to integrate utility, MG and FMG cost efficient and environment friendly power supply. The Figure 8 shows integration of renewable energy during day to fulfill power demand. The Figure 9 shows that contract based energy consumption is more cost efficient as compared to third scenario

5. Conclusions

In this paper, a fog based system model is proposed for the community to provide near real-time energy management service. The high bandwidth of 5G almost nullifies the data transfer delay. However, with 4G, a minute delay is observed due to 100 times lesser data transfer capacity as compared to 5G. The efficient resource allocation technique, like MHBC, minimizes the PT. The reduced data transfer delay and reduced PT minimize the RT for residents of the community. Three scenarios are proposed to analyze the cost efficiency of power utilization by the integration of green energy. In the first scenario, the community and the fog are connected with the utility only. The power generators of the utility run on expensive and environment unfriendly fossil fuel. In second scenario, the community has PV based MG along with the utility. The community's demand is fulfilled with green (MG's) energy. The surplus energy is traded with fog and the utility; however, deficient power is bought from the utility. In the third scenario, the community and fog have their respective PV based MGs along with the utility. The surplus energy of FMG is traded with the community or with the utility. The simulations of case studies show that the third scenario is more cost efficient as compared to first and second scenarios. The integration of green energy reduces power consumption cost for the community and reduce the power demand from the utility. The storage of green energy reduces the additional cost for the fog after the sun-set. The green data center of the fog and the community with green MG in the third scenario save up to 1.2% and 15.09% of energy cost as compared to second and first scenarios, respectively. In this paper, the proposed energy contract ensures the participation of all stakeholders and it is 6.54% more cost efficient for the community as compared to third scenario.

We recommend proposed distributed centralized energy management system model instead of cloud based centralized system. The heating of huge data centers of cloud in cold region affects the environment. However, distributed small data centers with green energy have a negligible impact on the environment. It is studied that the potential of integration of green energy with existing system is high when friendly energy policies are defined. For example, an individual renewable power producer can fulfill the demands of his neighbors or trade with the utility. However, in a community the trust building among producers and consumers need to be ensured by defining power trade policy. In the future, block chain based trust building mechanism among prosumers, consumers and power utility in a community shall be considered. Moreover, the power economy sharing approach should be considered to form communal MG out of existing renewable power generators among the prosumers.

Author Contributions: R.B. and N.J. proposed and implemented the main idea. M.U.J. and A.F. performed the mathematical modeling and wrote the simulation section. M.S. and J.-G.C. organized and refined the manuscript. All authors have read and agreed to the published version of the manuscript.

Acknowledgments: This research was supported by the MSIT (Ministry of Science and ICT), Korea, under the ITRC (Information Technology Research Center) support program (IITP-2020-2016-0-00313) supervised by the IITP (Institute for Information & communications Technology Planning & Evaluation, and in part by the 2019 Yeungnam University Research Grant.

Conflicts of Interest: The authors declare no conflict of interest.

References

1. Mini, G.; Debajit, P.; Rashmi, M.; Deepa, S. Gender in Electricity Policymaking in India, Nepal and Kenya. In *Energy Justice Across Borders*; Springer: Cham, Switzerland, 2020; pp. 111–135.
2. Christina, D.; Gareth, T.; Sarah, B.; Darrick, E.; Nick, P. Acceptance of energy transitions and policies: Public conceptualisations of energy as a need and basic right in the United Kingdom. *Energy Res. Soc. Sci.* **2019**, *48*, 33–45.
3. Saber, A.; Falahati, S.A.; Seyed, A.T.; Mohammad, S. Smart deregulated grid frequency control in presence of renewable energy resources by EVs charging control. *IEEE Trans. Smart Grid* **2018**, *9*, 1073–1085.
4. Jasiński, M.; Sikorski, T.; Kostyła, P.; Kaczorowska, D.; Leonowicz, Z.; Rezmer, J.; Szymańda, J.; Janik, P.; Bejmert, D.; Rybiański, M.; et al. Influence of Measurement Aggregation Algorithms on Power Quality Assessment and Correlation Analysis in Electrical Power Network with PV Power Plant. *Energies* **2019**, *12*, 3547. [CrossRef]
5. Nasir, M.; Anees, M.; Khan, H.A.; Khan, I.; Xu, Y.; Guerrero, J.M. Integration and Decentralized Control of Standalone Solar Home Systems for off-grid Community Applications. *IEEE Trans. Ind. Appl.* **2019**. [CrossRef]
6. Ryan, K.; Sini, N.; Joseph, S.; Johannes, U. Multilevel customer segmentation for off-grid solar in developing countries: Evidence from solar home systems in Rwanda and Kenya. *Energy* **2019**, *186*, 115728.
7. Fabio, R.; Emanuela, C.; Carlo, P. Towards modelling diffusion mechanisms for sustainable off-grid electricity planning. *Energy Sustain. Dev.* **2019**, *52*, 11–25.

8. Sikorski, T.; Jasiński, M.; Ropuszyńska-Surma, E.; Węglarz, M.; Kaczorowska, D.; Kostyła, P.; Leonowicz, Z.; Lis, R.; Rezmer, J.; Rojewski, W.; et al. A Case Study on Distributed Energy Resources and Energy-Storage Systems in a Virtual Power Plant Concept: Economic Aspects. *Energies* **2019**, *12*, 4447. [CrossRef]
9. Adnan, R.M.; Liang, Z.; Yuan, X.; Kisi, O.; Akhlaq, M.; Li, B. Comparison of LSSVR, M5RT, NF-GP, and NF-SC Models for Predictions of Hourly Wind Speed and Wind Power Based on Cross-Validation. *Energies* **2019**, *12*, 329. [CrossRef]
10. Mubashir, H.R.; Alan, D.; Brendan, J.; Chadi, A. Software Defined Networks based Smart Grid Communication: A Comprehensive Survey. *IEEE Commun. Surv. Tutor.* **2019**. [CrossRef]
11. Talaat, M.; Abdulaziz, S.A.; Adel, A.; Hatata, A.Y. Hybrid-cloud-based data processing for power system monitoring in smart grids. *Sustain. Cities Soc.* **2020**, *55*, 102049. [CrossRef]
12. Pasetti, M.; Ferrari, P.; Silva, D.R.C.; Silva, I.; Sisinni, E. On the Use of LoRaWAN for the Monitoring and Control of Distributed Energy Resources in a Smart Campus. *Appl. Sci.* **2020**, *10*, 320. [CrossRef]
13. Wang, B.; Wei, Y.M.; Yuan, X.C. Possible design with equity and responsibility in China's renewable portfolio standards. *Appl. Energy* **2018**, *232*, 685–694. [CrossRef]
14. Zhang, Q.; Wang, G.; Li, Y.; Li, H.; McLellan, B.; Chen, S. Substitution effect of renewable portfolio standards and renewable energy certificate trading for feed-in tariff. *Appl. Energy* **2018**, *227*, 426–435. [CrossRef]
15. Andri, P.; Angeliki, K.; Paris, A.F. The future of the Feed-in Tariff (FiT) scheme in Europe: The case of photovoltaics. *Energy Policy* **2016**, *95*, 94–102.
16. Yu, S.; Park, K.; Lee, J.; Park, Y.; Park, Y.; Lee, S.; Chung, B. Privacy-Preserving Lightweight Authentication Protocol for Demand Response Management in Smart Grid Environment. *Appl. Sci.* **2020**, *10*, 1758. [CrossRef]
17. González, I.; Calderón, A.J. Integration of open source hardware Arduino platform in automation systems applied to Smart Grids/Micro-Grids. *Sustain. Energy Technol. Assess.* **2019**, *36*, 100557. [CrossRef]
18. Vargas-Salgado, C.; Aguila-Leon, J.; Chiñas-Palacios, C.; Hurtado-Pérez, E. Low-cost web-based Supervisory Control and Data Acquisition system for a microgrid testbed: A case study in design and implementation for academic and research applications. *Heliyon* **2019**, *5*, e02474. [CrossRef]
19. Peng, Z.; Hao, L. Hierarchical and Decentralized Stochastic Energy Management for Smart Distribution Systems with High BESS Penetration. *IEEE Trans. Smart Grid* **2019**. [CrossRef]
20. Yasir, S.; Noel, C.; Mubashir, H.R.; Rebecca, C. Internet of things-aided Smart Grid: Technologies, architectures, applicat ions, prototypes, and future research directions. *IEEE Access* **2019**, *7*, 62962–63003.
21. Adia, K.; Sheraz, A.; Khursheed, A.; Syed, I.H.; Mahmood, A.; Nadeem, J. An efficient energy management approach using fog-as-a-service for sharing economy in a smart grid. *Energies* **2018**, *11*, 3500.
22. Chen, Y.W.; Chang, J. MEMaaS: Cloud-based energy management service for distributed renewable energy integration. *IEEE Trans. Smart Grid* **2015**, *6*, 2816–2824. [CrossRef]
23. Venkatraman, B.; Safa, O.; Moayad, A.; Ismaeel, A.R.; Yaser, J. Low-latency vehicular edge: A vehicular infrastructure model for 5G. *Simul. Model. Pract. Theory* **2020**, *98*, 101968.
24. Miodrag, F.; Mirjana, M. Cloud-fog-based approach for smart grid monitoring. *Simul. Model. Pract. Theory* **2020**, *101*, 101988.
25. Saeed, A.; Saif, U.R.M.; Samee, U.K.; Raymond, C.; Adeel, A.; Naveed, A. A Game-based Thermal-aware Resource Allocation Strategy for Data Centers. *IEEE Trans. Cloud Comput.* **2019**. [CrossRef]
26. Wang, S.; Zhu, X.; Song, D.; Wen, Z.; Chen, B.; Feng, K. Drivers of CO_2 emissions from power generation in China based on modified structural decomposition analysis. *J. Clean. Prod.* **2019**, *220*, 1143–1155. [CrossRef]
27. Thiago, V.L.; Pedro, M.; Aníbal, D.A. A review on energy efficiency and demand response with focus on small and medium data centers. *Energy Effic.* **2019**, *12*, 1–30.
28. Available online: http://www.google.co.uk/about/datacenters/inside/locations/hamina/ (accessed on 17 May 2020).
29. Available online: https://www.theguardian.com/environment/2011/oct/27/facebook-green-datacentre-sweden-renewables (accessed on 17 May 2020).
30. Chee, N.L.; Iromi, U.R.; Ole-Morten, M.; Lars, N. A real-time energy management system for smart grid integrated photovoltaic generation with battery storage. *Renew. Energy* **2019**, *130*, 774–785.

31. Nadeem, J.; Adnan, A.; Sohail, I.; Mahmood, A. Efficient power scheduling in smart homes using hybrid grey wolf differential evolution optimization technique with real time and critical peak pricing schemes. *Energies* **2018**, *11*, 384.
32. Wu, W.; Wang, W.; Fang, X.; Junzhou, L.; Vasilakos, A.V. Electricity Price-aware Consolidation Algorithms for Time-sensitive VM Services in Cloud Systems. *IEEE Trans. Serv. Comput.* **2019**. [CrossRef]
33. Mehmood, F.; Hamza, M.A.; Bukhsh, R.; Javaid, N.; Imran, M.I.U.; Choudri, S.; Ahmed, U. Green Fog: Cost Efficient Real Time Power Management Service for Green Community. In Proceedings of the Complex, Intelligent and Software Intensive Systems (CISIS-2020), Lodz, Poland, 1–3 July 2020.
34. Kong, P.Y.; Song, Y. Joint Consideration of Communication Network and Power Grid Topology for Communications in Community Smart Grid. *IEEE Trans. Ind. Inform.* **2019**. [CrossRef]
35. Wang, D.; Chen, D.; Song, B.; Guizani, N.; Yu, X.; Du, X. From IoT to 5G I-IoT: The Next Generation IoT-Based Intelligent Algorithms and 5G Technologies. *IEEE Commun. Mag.* **2018**. [CrossRef]
36. Letaief, K.B.; Chen, W.; Shi, Y.; Zhang, J.; Zhang, Y.J.A. The Roadmap to 6G: AI Empowered Wireless Networks. *IEEE Commun. Mag.* **2019**. [CrossRef]
37. Zepter, J.M.; Lüth, A.; del Granado, P.C.; Egging, R. Prosumer integration in wholesale electricity markets: Synergies of peer-to-peer trade and residential storage. *Energy Build.* **2019**, *184*, 163–176. [CrossRef]
38. Zhang, B.; Jiang, C.; Yu, J.L.; Han, Z. A contract game for direct energy trading in smart grid. *IEEE Trans. Smart Grid* **2016**, *9*, 2873–2884. [CrossRef]
39. Qin, J.; Rajagopal, R.; Varaiya, P. Flexible market for smart grid: Coordinated trading of contingent contracts. *IEEE Trans. Control. Netw. Syst.* **2017**, *5*, 1657–1667. [CrossRef]
40. Chen, Y.W.; Chang, J.M. Fair demand response with electric vehicles for the cloud based energy management service. *IEEE Trans. Smart Grid* **2016**, *9*, 458–468. [CrossRef]
41. Zahra, M.; Faramarz, S.E. Workflow scheduling applying adaptable and dynamic fragmentation (WSADF) based on runtime conditions in cloud computing. *Future Gener. Comput. Syst.* **2019**, *90*, 327–346.
42. Xu, F.; Zheng, H.; Jiang, H.; Shao, W.; Liu, H.; Zhou, Z. Cost-Effective Cloud Server Provisioning for Predictable Performance of Big Data Analytics. *IEEE Trans. Parallel Distrib. Syst.* **2018**, *30*, 1036–1051. [CrossRef]
43. Beyzanur, T.; Beyza, N.B.; Gül, N.G. Development of a Simulation Tool to Estimate Electricity Consumption and Determine the Optimum Cooling System for Data Centers. In Proceedings of the 2019 IEEE East-West Design & Test Symposium (EWDTS), Batumi, Georgia, 13–16 September 2019; pp. 1–6.
44. Jawad, M.; Qureshi, M.B.; Khan, U.; Ali, S.M.; Mehmood, A.; Khan, B.; Wang, X.; Khan, S.U. A robust Optimization Technique for Energy Cost Minimization of Cloud Data Centers. *IEEE Trans. Cloud Comput.* **2018**. [CrossRef]
45. Xu, C.; Wang, K.; Li, P.; Xia, R.; Guo, S.; Guo, M. Renewable energy-aware big data analytics in geo-distributed data centers with reinforcement learning. *IEEE Trans. Netw. Sci. Eng.* **2018**. [CrossRef]
46. Atefeh, K.; Lachlan, L.H.A.; Rajkumar, B. Dynamic vm placement method for minimizing energy and carbon cost in geographically distributed cloud data centers. *IEEE Trans. Sustain. Comput.* **2017**, *2*, 183–196.
47. Sambit, M.K.; Deepak, P.; Joel, J.P.C.R.; Bibhudatta, S.; Eryk, D. Sustainable Service Allocation Using a Metaheuristic Technique in a Fog Server for Industrial Applications. *IEEE Trans. Ind. Inform.* **2018**, *14*, 4497–4506.
48. Alharbi, F.; Tian, Y.C.; Tang, M.; Zhang, W.Z.; Peng, C.; Fei, M. An ant colony system for energy-efficient dynamic virtual machine placement in data centers. *Expert Syst. Appl.* **2019**, *120*, 228–238. [CrossRef]
49. Xiao, Z.; Ming, Z. A state based energy optimization framework for dynamic virtual machine placement. *Data Knowl. Eng.* **2019**, *120*, 83–99. [CrossRef]
50. Li, Z.; Yan, C.; Yu, L.; Yu, X. Energy-aware and multi-resource overload probability constraint-based virtual machine dynamic consolidation method. *Future Gener. Comput. Syst.* **2018**, *80*, 139–156. [CrossRef]
51. Chaudhry, M.T.; Ling, T.C.; Hussain, S.A.; Lu, X.Z. Thermal-aware relocation of servers in green data centers. *Front. Inf. Technol. Electron. Eng.* **2015**, *16*, 119–134. [CrossRef]
52. Rahmat, H.; Naveed, A.; Saif, U.M.; Saeed, A.; Adeel, A. Simulator for modeling, analysis, and visualizations of thermal status in data centers. *Sustain. Comput. Inform. Syst.* **2018**, *19*, 324–340.

53. Gai, K.; Wu, Y.; Zhu, L.; Qiu, M.; Shen, M. Privacy-preserving energy trading using consortium blockchain in smart grid. *IEEE Trans. Ind. Inform.* **2019**, *15*, 3548–3558. [CrossRef]
54. Bao, M.; Ding, Y.; Singh, C.; Shao, C. A Multi-State Model for Reliability Assessment of Integrated Gas and Power Systems Utilizing Universal Generating Function Techniques. *IEEE Trans. Smart Grid* **2019**, *10*, 6271–6283. [CrossRef]
55. Rasool, B.; Nadeem, J.; Raza, A.A.; Aisha, F.; Mariam, A.; Muhammad, K.A.; Farruh, I. An Efficient Fog as-a-Power-Economy-Sharing Service. *IEEE Access* **2019**. [CrossRef]
56. Eiza, M.H.; Ni, Q.; Shi, Q. Secure and privacy-aware cloud-assisted video reporting service in 5G-enabled vehicular networks. *IEEE Trans. Veh. Technol.* **2016**, *65*, 7868–7881. [CrossRef]
57. Muhammad, K.A.; Yousaf, B.Z.; Shahid, M.; Ammar, R.; Anwer, A.D.; Mohsen, G. Unlocking 5G spectrum potential for intelligent IoT: Opportunities, challenges, and solutions. *IEEE Commun. Mag.* **2018**, *56*, 92–99.
58. Rasool, B.; Nadeem, J.; Sakeena, J.; Manzoor, I.; Itrat, F. Efficient resource allocation for consumers' power requests in cloud-fog-based system. *Int. J. Web Grid Serv.* **2019**, *15*, 159–190.
59. Li, Y.; Wen, Y.; Tao, D.; Guan, K. Transforming cooling optimization for green data center via deep reinforcement learning. *IEEE Trans. Cybern.* **2019**. [CrossRef]
60. Amal, B.; Kasey, M.F. Construction waste generation estimates of institutional building projects: Leveraging waste hauling tickets. *Waste Manag.* **2019**, *87*, 301–312.

Article

Time and Cost Efficient Cloud Resource Allocation for Real-Time Data-Intensive Smart Systems

Muhammad Shuaib Qureshi [1,2], Muhammad Bilal Qureshi [3,*], Muhammad Fayaz [2], Muhammad Zakarya [4], Sheraz Aslam [5] and Asadullah Shah [1]

1 KICT, International Islamic University, Kuala Lumpur 50728, Malaysia; muhammad.qureshi@ucentralasia.org (M.S.Q.); asadullah@iium.edu.my (A.S.)
2 Department of Computer Science, School of Arts and Sciences, University of Central Asia, 310 Lenin Street, Naryn 722918, Kyrgyzstan; muhammad.fayaz@ucentralasia.org
3 Department of Computer Science, Shaheed Zulfikar Ali Bhutto Institute of Science and Technology, Islamabad 44000, Pakistan
4 Department of Computer Science, Abdul Wali Khan University, Mardan 23200, Pakistan; mohd.zakarya@awkum.edu.pk
5 Department of Electrical Engineering, Computer Engineering and Informatics, Cyprus University of Technology, Limassol 3036, Cyprus; sheraz.aslam@cut.ac.cy
* Correspondence: muhdbilal.qureshi@gmail.com

Received: 17 July 2020; Accepted: 9 September 2020; Published: 31 October 2020

Abstract: Cloud computing is the de facto platform for deploying resource- and data-intensive real-time applications due to the collaboration of large scale resources operating in cross-administrative domains. For example, real-time systems are generated by smart devices (e.g., sensors in smart homes that monitor surroundings in real-time, security cameras that produce video streams in real-time, cloud gaming, social media streams, etc.). Such low-end devices form a microgrid which has low computational and storage capacity and hence offload data unto the cloud for processing. Cloud computing still lacks mature time-oriented scheduling and resource allocation strategies which thoroughly deliberate stringent *QoS*. Traditional approaches are sufficient only when applications have real-time and data constraints, and cloud storage resources are located with computational resources where the data are locally available for task execution. Such approaches mainly focus on resource provision and latency, and are prone to missing deadlines during tasks execution due to the urgency of the tasks and limited user budget constraints. The timing and data requirements exacerbate the efficient task scheduling and resource allocation problems. To cope with the aforementioned gaps, we propose a time- and cost-efficient resource allocation strategy for smart systems that periodically offload computational and data-intensive load to the cloud. The proposed strategy minimizes the data files transfer overhead to computing resources by selecting appropriate pairs of computing and storage resources. The celebrated results show the effectiveness of the proposed technique in terms of resource selection and tasks processing within time and budget constraints when compared with the other counterparts.

Keywords: data-intensive smart application; cloud computing; resource allocation; real-time systems; smart grid

1. Introduction

With fast-growing advancements in smart systems, the real-time applications are handy candidates for utilizing the computing power in a cloud computing environment in order to maintain deadline constraints. Cloud computing is an economic-based paradigm consisting of distributed resources and providing services by collaborating in executing user applications. The cloud services are categorized into Infrastructure-as-a-Service (IaaS) that deals with providing computing such as VMs and storage resources as services, Platform-as-a-Service (PaaS) that offer deployment and development platforms as services, and Software-as-a-Service (SaaS) that facilitate users with web-based applications. The most common examples are IBM's Blue Cloud (for IaaS), Google AppEngine and Microsoft Azure (for PaaS), and EC2 (for SaaS). These services are employed by using cloud deployment models, namely public, private, and hybrid established on the basis of organization preferences. The cloud compute and storage resources are selected and allocated on the basis of nature of user application. In addition, the cloud storage resources provide facilities such as accommodating data replication to satisfy data requirements of the data-intensive real-time systems that need to access, process and transfer data files stored in distributed data repositories [1,2]. Examples of such applications are self-driving vehicles, which depend on the data and computations under a complex network of interconnected devices such as GPS, surveillance cameras, radar, laser light, odometry, etc., to perceive the surroundings [3,4]. The cloud providers such as Amazon EC2 [5] provide computing facilities (virtual machines) on a pay-as-you-go basis at the rate of 10 cents per hour. The lease prices vary depending on the virtual machines (VMs) specifications. The normal VM offers an approximate processing power of 1.2 GHz Opteron processor with a storage capacity of 160 GB disk space and 1.7 GB of memory [5–7]. Such facilities pave the way for executing time-critical and IoT applications which demand high processing and storage capabilities [8]. Stergioua et al. [9] merged cloud computing with IoT to improve the functionality of the IoT. The IoT devices offload many tasks to the cloud environment from the smart systems because these devices have very limited processing and storage capabilities. Leveraging the capabilities of virtualization technology, VMs can be scaled up and down depending on the current system workloads [10]. For executing VM code on smart platforms, smart virtual machines are proposed by Lee et al. [11]. However, there is a lack of efficient resource scheduling and allocation strategies for deploying real-time applications with stringent QoS and data requirements in cloud computing environments. The scheduling policy for data-intensive real-time systems proposed in [12] allocates HPC resources by considering that single copies of data-files are available on storage resources without taking into account the computation and transfer cost. The concept of real time scheduling is used in the deployment of smart environments [13–17].

The Global Institute [18] report on analyzing the economic impact of the IoT devices show that it will increase upto $11 trillion by the year 2025. This increase is because the IoT devices and smart home appliances ranging from small sensors to large scale biometrics offload data and computation to the cloud computing environment on a regular basis [19,20]. For this purpose, the IT companies provide solutions such as Apple's HomeKit [21], Samsung's SmartThings [22], Amazon's Alexa [5], and Google's Home [23], etc. The smart systems are considered as data-intensive systems that are different from compute-intensive or eScience applications because of the storage, access, execution and management requirements of distributed datasets and hence, require different scheduling and allocation policies. These systems basically deal with the data and transport layers for replication and access of datasets. The data-intensive smart systems can be considered as a combination of data producers and consumers geographically distributed across multiple organizations. The producers are the entities that produce data and manage its distribution over multiple locations. The consumers can be the users or their applications which need this data produced by the producers for multiple purposes. The consumers investigate efficient ways out of many to access the data for executing applications on remote compute nodes.

The driving force of cloud computing is the virtual machine manager (VMM) that creates the virtual resources of the physical machines. The basic functionality of the VMM is to separate the virtual computing environment from the underlying physical infrastructure. In this research work, we implement the rate monotonic (RM) scheduling policy to allocate cloud computing resources for the real-time data-intensive periodic tasks. The scheduling problem is divided into three parts: the processing environment (cloud virtual machines), the nature of the real-time task (fixed priority system), and the optimization criteria (time- and cost-efficient allocation). The real-time task set is a collection of multiple tasks, each of which requires data for processing. The required data files are requested from the remotely located storage resources. The intelligent selection and assignment of cloud computing resources are investigated while the data files are replicated on decoupled storage resources and accessed by utilizing networks of varying capabilities. The proposed strategy evaluates all the storage locations for the replicas of the same data file and selects the one which has minimum data access and transfer cost. It submits the application to the computing resource that is closest to the selected storage location and can complete execution within minimum possible execution time. These files are fetched to the computing resources where tasks are executed, which add transfer time to the red tasks' total execution time. In our proposed model, a user-specified budget is associated with each smart system request, and hence resources are selected not only on the basis of their high computational power, but also the cost associated with each and the storage resource are computed, as well as the ability to process jobs within tasks deadlines and scheduling preferences.

The major research contributions of our work are:

- Creating a model for selecting appropriate cloud computing and storage resources to execute real-time data-intensive systems generated by smart devices where data is replicated on multiple storage resources,
- Partitioning the task sets into groups based on common data-file demands such that the timing constraints of the original tasks set are not disturbed,
- Analysing the economic perspective of data storage and processing by scheduling real-time smart systems on distributed nodes with different storage, execution, and data transfer costs and allocating heterogeneous cloud resources,
- Allocating cloud computing resources to periodic real-time tasks such that the overall timing constraints of the smart devices remain intact,
- Analysing cloud computing and IoT usability in the context of data-intensive smart systems.

Cloud computing is considered a promising platform for executing large-scale computing-intensive IoT and smart grid applications in a cost-efficient way due to the large pool of computing resources. These resources need intelligent scheduling and allocation of the tasks such that all tasks in a batch can be processed within the stipulated time span. The research community focuses on searching mechanisms for scheduling and allocating distributed cloud resources in a systematic way which can satisfy formulated objective function such as load balancing, makespan minimization, and cost-efficiency with respect to the user-defined QoS criteria [3,24,25]. But, instead of the vast study in the cloud resource allocation domain, the existing literature is not mature in providing suitable scheduling mechanisms for real-time systems generated by smart devices due to the high deadline-miss ratio by a number of tasks in a batch [26]. The problem becomes more challenging when such systems need data from external sources. A real-time system is characterized by the deadlines of the tasks. In such systems, deadline meeting is primarily important for utilizing maximum capabilities of the cloud resources, since most of the real-time systems such as sensors in smart systems or actuators in automated systems generate periodic tasks, which are sent to the processing units after regular intervals. Such systems need proper priority-based and non priority-based strategies for scheduling. In priority-based scheduling policy, the scheduling criteria is saved for the entire duration of task execution, while in non priority-based systems, the tasks have no precedence constraints and the scheduling criteria may change with the arrival of the next job of a task. The main advantage of priority-based policies is time-saving and its simple implementation. The priority-based scheduling is also known as static priority scheduling. The well-known static priority assignment algorithm is Rate Monotonic developed by Liu and Layland in 1973. The main features of this algorithm are its simple implementation at OS kernel level, predictability in real-time behavior like in smart systems, and its easy modification for implementing task priority inheritance protocol for the purpose of synchronization [27].

In order to utilize the full potential of the cloud and IoT platforms, Suciu et al. [28] proposed a conceptual framework for the deployment of IoT based smart microgrid applications. The developed framework has the capability of integrating real-time data generated by the ubiquitous sensing devices constituting the smart home with the cloud computing environment, but their system is prone to missing deadlines because they have not evaluated their system on each time instant according to the workload of the task. The dynamic algorithm for scheduling soft real-time systems on grid resources was proposed in [29]. The proposed algorithm provides room for tasks with missed deadlines. Ye et al. [30] proposed architecture of smart home-oriented cloud. They have suggested a layered cloud design that provides efficient services for digital appliances. The authors have considered the sensors sending data continuously without specifying any real-time constraints. Caron et al. [1] consider task priorities for scheduling real-time tasks. The tasks are checked one by one. Shang et al. [1] formulated a model which elaborates grid service reliability assessment for dependable and cost-efficient applications. They have derived a cost function based on genetic and particle swarm optimization techniques that calculate service expense of each utilized resource. Isard et al. [4] considered scheduling problem of data-intensive tasks using the Hadoop structure. They have supposed coupled computing and data resources located at the same place and that the data is available for each task without transferring from remotely located resources. Therefore, they have not included the data transfer time in the task feasibility analysis. The proposed tasks are preemptable but no specific criterion is discussed for which task is to be preempted when an interruption occurs or when high priority tasks arrive. In ref. [5], the authors have focused on submitting preemptable tasks to the federated grid. They have developed a schedule which maximizes the acceptance of incoming tasks, and minimizes user-defined QoS criteria violation. The authors have not emphasized the heterogeneity of resources. Poola et al. [6] presented a mechanism for robust and fault-tolerant scheduling of scientific workflows on heterogeneous resources which concurrently optimizes makespan and execution cost. Ma et al. [7] used a hybrid approach by combining the best features of genetic and greedy approaches for QoS-aware web service composition. They have focused on minimizing cost, but they have considered non-real-time tasks. A cost optimization technique for executing data-intensive tasks on distributed resources was developed by Mansouri et al. [8]. Leveraging data storage and migration cost is addressed by using an optimal online algorithm, but their system is not suitable for executing real-time smart systems tasks. The proposed algorithm provides tasks scheduling and cloud resource allocation criteria for real-time tasks generated by smart systems considering resources heterogeneity, availability, data-intensive and timing constraints of the tasks.

The rest of the paper is structured as follows. In Section 2, we throw light on discussing task, resource, and cost model. The proposed time- and cost-efficient scheduling algorithm is explained in Section 3, while the performance of the proposed resource allocation strategy and details of the experimental setup is evaluated in Section 4. The produced results are presented in Section 5 and conclusions and future directions are provided in Section 6.

2. Task, Resource and Cost Models

In this research, we consider scheduling feasibility of real-time periodic tasks in a cloud environment. Our model as shown in Figure 1 is composed of smart devices which generate periodic tasks $(\tau_1, \tau_2, \ldots, \tau_n)$. The tasks represent data-intensive applications that generate data on a regular basis and need computing resources for processing. The represented smart devices have limited memory, storage and processing capabilities. Each task needs data stored on some remote storage locations. The tasks constitute a task set $T_i (i = 1, 2, \ldots, n)$ where the collection of task sets form a smart microgrid that sends data to a main smart hub known as Smart Grid Management System. The smart hub manages the received data and uploads the collected tasks to the cloud environment. The Cloud Resource Management System (Cloud RMS) handles task requests and manages cloud resources. The Cloud RMS has the responsibility to search suitable resources and required data files information from the Resource Files Information Directory (RFID) according to the task requirements and schedule tasks on cloud resources. The Cloud RMS also implements the scheduling policy. Our concerned cloud environment is comprised of both computational $(CR_1, CR_2, \ldots, CR_n)$ and data storage resources $(DR_1, DR_2, \ldots, DR_n)$ located remotely and connected by network links of different bandwidths. The resources are heterogeneous and characterized by power and cost constraints.

In this paper, we concentrate on two basic constraints; (a) the real-time tasks' deadlines, and (b) user-specified budget. The presented model extends the RDTA model [12] by introducing cost parameters, data files replication scenarios, and tasks' grouping criteria.

2.1. Task and Resource Model

We consider batch processing of real-time periodic tasks, each of which can generate an infinite number of jobs. In periodic tasks set $T = \{task_1, task_2, \ldots, task_n\}$, each $task_k$ is defined by the quadruple:

$$task_k = (r_k, e_k, d_k, period_k) \tag{1}$$

where r_k shows the release time of the first job, e_k the required computation time, d_k the relative deadline of $task_k$ which is the time difference between the absolute deadline and release time of a job, and $period_k$ the period which is the time difference between the two successive jobs of a $task_k$.

In the above discussed model, a job i released at time instant $r_k + (i-1).period_k$ needs to execute for e_k units before the time $r_k + (i-1).period_k + d_k$. In our task model, we concentrate on a constrained deadline model which assumes that $d_k \leq period_k, \forall k \in T$. Tasks preemption is not allowed and context switching overhead is subsumed into e_k. We also assume that $r_k = 0, \forall k \in T$, which means that feasibility of the tasks is checked when the system is most loaded.

We consider computing resource set CR such that $CR = \{CR_1, CR_2, \ldots, CR_r\}$. Each one is characterized by a computing power $CP_y (1 \leq y \leq r)$ such that $CP_y \in CP$, where $CP = \{CP_1, CP_2, \ldots, CP_r\}$, and measured in Millions of Instructions per Second (MIPS). The execution time of a $task_k$ on resource CR_y can be computed by

$$EET_{ky} = (e_k + e_{higher})/CP_y, \tag{2}$$

where e_{higher} is the execution time of higher priority tasks than $task_k$. Mathematically,

$$e_{higher} = \sum_{j=1}^{k-1} \lceil \frac{t}{period_j} \rceil e_j, \quad t \in PNP_j \tag{3}$$

where PNP set accumulates points or time instants on which task feasibility is analyzed. The PNP set is defined as follows.

Definition 1. *PNP_k is a set of positive and negative points for $task_k$ constituted by the relation $x.period_j$ such that $1 \leq j \leq k$ and $1 \leq x \leq \lfloor period_k / period_j \rfloor$, where $period_j$ represents periods of higher priority tasks than $task_k$. The point $t_P \in PNP_k$ is said to be positive if $task_k$ is declared feasible (i.e., completes its execution at or before the* *deadline) at some point t by considering all associated time and data constraints. The point* $t_N \in PNP_k$ *declares the* $task_k$ *infeasible when it misses the deadline. Each point in* PNP_k *is called rate-monotonic scheduling point. From Definition (1), it is concluded that the set* PNP *is the union of positive points set* PP *and negative points set* NP *where* $PP = PNP - NP$ *and* $NP = PNP - PP$*. In other words,* $PNP = PP \cup NP$*.*

2.2. Data Files Model

In our task model, a tasks set $T = \{task_1, task_2, \ldots, task_n\}$ consists of data-intensive real-time tasks where each task $task_k$ needs a set of data files DF_k for its execution. The set $DF_k = \{f_k1, f_k2, \ldots, f_km\} \subseteq DF$. The file $f_kx \in DF_k$ is stored on data storage resource dr_w, where $dr_w \in DR_k$ and $DR_k \subseteq DR$. The DR is the set of total storage resources in the HPC environment. In other words, files in DF_k are stored on DR_k storage resources. We assume that the data files are replicated on more than one data storage resources.

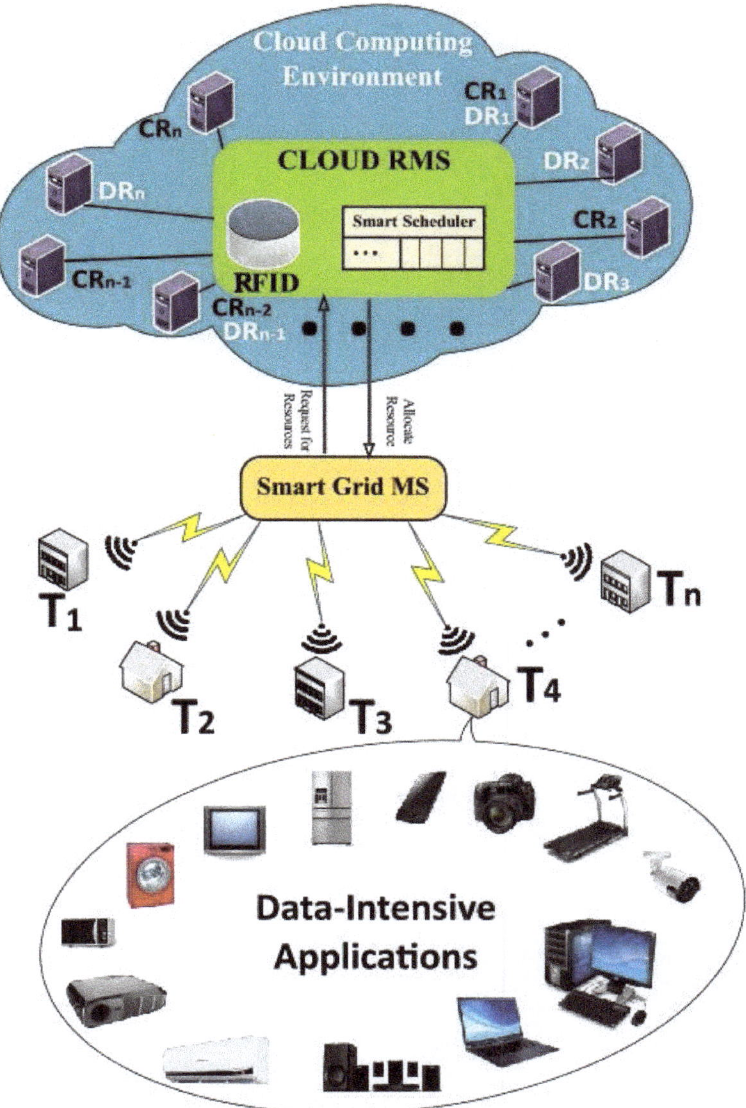

Figure 1. Model for offloading computation and data-intensive application from smart microgrid to the cloud.

The total execution time TT of a task $task_k$ is the sum of actual computation time EET of $task_k$ on computing resources CR_y and the transfer time taken by the required m data files in the set DF_k

by transferring from storage resources DR_k to the computing resource CR_k where task k is executed. Mathematically,

$$TT_k = EET_{ky} + \sum_{z=1}^{m} FT_{(f_k z)} \quad (4)$$

where $FT_{f_k z} = R_w + Size_{f_k z}/BW_{wy}$ is the transfer time of the file $f_k z$.

The R_w represents the response time of the data storage resource $dr_w \in DR_k$ where the data file $f_k z$ is stored. The response time is the time when the request to fetch the file is made at the time when the request is entertained. Algebraically, the response time of data resource dr_w is calculated as:

$$R_w = ST_{f_k z} + WT_{f_k z} \quad (5)$$

where $ST_{f_k z}$ is the service time and $WT_{f_k z}$ is the waiting time of the request respectively for accessing the file $f_k z$. Also, the $Size_{f_k z}$ denotes the size of the file $f_k z$, and BW_{wy} shows the link bandwidth between data storage resource dr_w and computing resource CR_y. The proposed model selects that storage resource for file access for which FT is minimum, i.e., $\min(FT)$.

2.3. Task Grouping

The data-intensive real-time tasks in T are grouped into x number of groups on the basis of common data files demands. The tasks in a group represent a subset of T or we can say each group is a set of tasks for easy understanding. The task grouping taxonomy is pictorially represented in Figure 2.

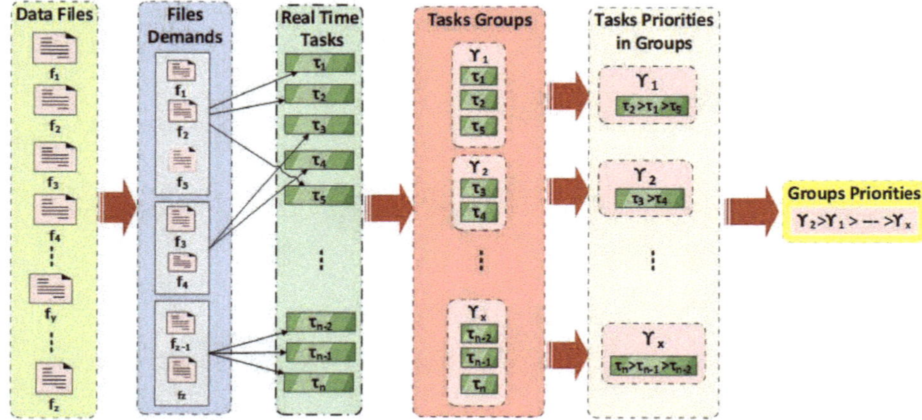

Figure 2. Task grouping taxonomy.

Based on real-time task grouping criteria, the group of tasks, its cardinality, and priority assignment is defined in the following sections.

Definition 2. *A group of real-time tasks Y_x is a subset of tasks, i.e., $Y_x \subseteq T (x \leq n)$ having common data files demands. Each group Y_x contains minimum one task which concludes that $Y_x \neq \{\}$.*

Definition 3. *The cardinality of a task group defines the total number of tasks in a group. Let there be total x number of groups, then cardinality of the original tasks set* $T = \{task_1, task_2, \ldots, task_n\}$ *can be defined as:*

$$card(T) = \sum_{l=1}^{x} card(Y_1) \tag{6}$$

The advantage of the task grouping mechanism is to reduce the total number of priority levels [25]. Additionally, $t_N \in NP$ *for a higher priority task* $task_j \in Y_x$ *remains the member of NP for all lower priority tasks than* $task_j$ *in* Y_x*, since tasks in the same group are also sorted on the basis of RM priorities. In this way, the least number of points is tested on the* PNP *set which decreases the execution time.*

From the above definitions, the following can be observed.

Observation 1. *Let each task group* Y_l *constituted from task set* $T = \{task_1, task_2, \ldots, task_n\}$ *contains a single task in a worst case scenario and x represents the total number of groups in the system, then* $x = n$ *where n represents* $card(T)$.

Each real-time task in our task model has deadline d, which demonstrates the maximum allowed time for a task to complete its execution on a single computing resource. Let the maximum and minimum deadlines of the tasks in a group Y_x are denoted by d_{max} and d_{min} respectively. Here an interesting observation can be made.

Observation 2. d_{max} *of the tasks* $\{task_1, \ldots, task_l\} \in Y_x (l \leq n)$ *sorted by RM priority assignment technique and following the implicit deadline model (period = deadline) is the period of the last task while* d_{min} *is the period of the first task in* Y_x*. The relation follows:*

$$d_{max} = period_l d_{min} = period_1 \tag{7}$$

where $period_l$ *and* $period_1$ *represent periods of the last and the first task in* Y_x*, respectively.*

Proof. The RM technique sort the tasks based on priority assignment criteria: the lesser the period of the task, the higher the priority. This means that the last task $task_l \in Y_x$ has the lowest priority and first task $task_1 \in Y_x$ has the highest priority among all tasks in Y_x. The $task_1$ is executed in the first and $task_l$ is executed in the last slot. In other words, $priority(task_1) \geq priority(task_2) \geq, \ldots, \geq priority(task_l)$ which follows that $period(task_1) \leq period(task_2) \leq, \ldots, \leq period(task_l)$. It also follows that $d_{max} = period_l$ and $d_{min} = period_1$, which completes the proof.
□

The authors in [24] discussed that RM assigns static priorities to tasks and considered an optimal scheduling algorithm among static priority assignment scheduling algorithms. By optimal they mean that RM should schedule a task, if any other static priority assignment algorithm can schedule that task. Following are the general characteristics of the RM scheduling technique which play a role in proving its optimality.

1. the system should consist of a fixed number of tasks;
2. the tasks should have execution times known in advance;
3. each task has a completion deadline equal to its period;
4. tasks should be periodic which means that instances or jobs of a task should arrive after a fixed time interval;

5. tasks should be independent;
6. all tasks should arrive at time 0. This time instant is also known as the critical instant and the system is considered as the most overloaded at this instant.

Definition 4. *The period of a task group is defined as the temporary period attached to the group of tasks which is the period of the last task in a group. In other words, $period_l$ is the group period because the tasks in a group are sorted using RM technique. For example, if a group Y_x accommodates tasks $\{task_1, task_2 \ldots, task_l\}$, then*

$$period(Y_x) = period_l. \tag{8}$$

Since the groups are sorted on the basis of RM priorities, so $period(Y_1) \leq period(Y_2) \leq, \ldots, \leq period(Y_l)$ which states that $priority(Y_1) \geq priority(Y_2) \geq, \ldots, \geq priority(Y_l)$. Equation 8 states that all tasks in Y_x must complete execution at or before $period_l$. It has been further evaluated that since the tasks in the same group Y_x are also sorted by RM priorities, so $t_N \in NP$ for a higher priority task $task_i \in Y_x$ remains the member of NP for all the lower priority tasks than $task_i$.

Definition 5. *The group capacity can be defined as the total number of tasks in a group. Tasks in a group are added on the basis of common data files demand. The group capacity can be analyzed on the basis of resource utilization by a group of tasks called group utilization (GU) which is defined as the sum of the resource utilization of each task in the group. The computing resource utilization of each task is termed as task utilization (TU). Let n denotes the total number of tasks in a group Y_x, then GU and TU can be found as follows.*

$$GU_{Y_x} = TU_1 + TU_2 + \ldots + TU_n = \sum_{i=1}^{n} TU_i \tag{9}$$

where

$$TU_i = \frac{e_i}{period_i} \tag{10}$$

Theorem 1. *Let Y_x be a group of n periodic tasks, where each task is characterized by TU. The group Y_x is said to be RM feasible if the following condition holds:*

$$GU_{Y_x} \leq n(2^{1/n} - 1) \tag{11}$$

The inequality (11) is called the Liu and Layland (LL) test reported in [24]. The expression $n(2^{1/n} - 1)$ is the threshold value of a group which means that a group Y_x can accommodate tasks as for as the GU remains lower than or equal to the threshold value. Equation (11) is checked repeatedly when a new task is added to the group. If adding a task changes the inequality to $GU \geq n(2^{1/n} - 1)$, then the incoming tasks are added to another group. The authors in [25] refer the LL test as the sufficient condition test. They claim that it is not necessary that the tasks in a group are not schedulable if Equation (11) does not hold. This means that utilization-based tests are sufficient, but not necessary. Let us explain by the following example task set taken from [26,29].

Example 1. *Consider a task group $Y = \{task_1, task_2, task_3, task_4\}$ where tasks follow RM ordering and having following characteristics and utilizations mentioned in Table 1 and Table 2 respectively.*

Table 1. Task characteristics in Y.

Tasks	C_i	$period_i$
Tak_1	2	3
Tak_2	1.5	6
Tak_3	0.5	12
Tak_4	1	24

Table 2. Task utilizations.

TU	Value
TU_1	0.666
TU_2	0.250
TU_3	0.041
TU_4	0.041

For the above tasks set Y, the $GU_Y \cong 1$ and threshold = 0.756.

It shows that the aforementioned example task group Y is not schedulable by using LL test because it does not satisfy Equation (11). However, the gantt chart in Figure 3 shows the schedulability of these tasks within deadlines under RM technique. From the above discussion, it is clear that LL test is sufficient only, so we use LL test for checking group capacity only. For analyzing task or group feasibility, we use PNP test which is necessary and sufficient conditions test.

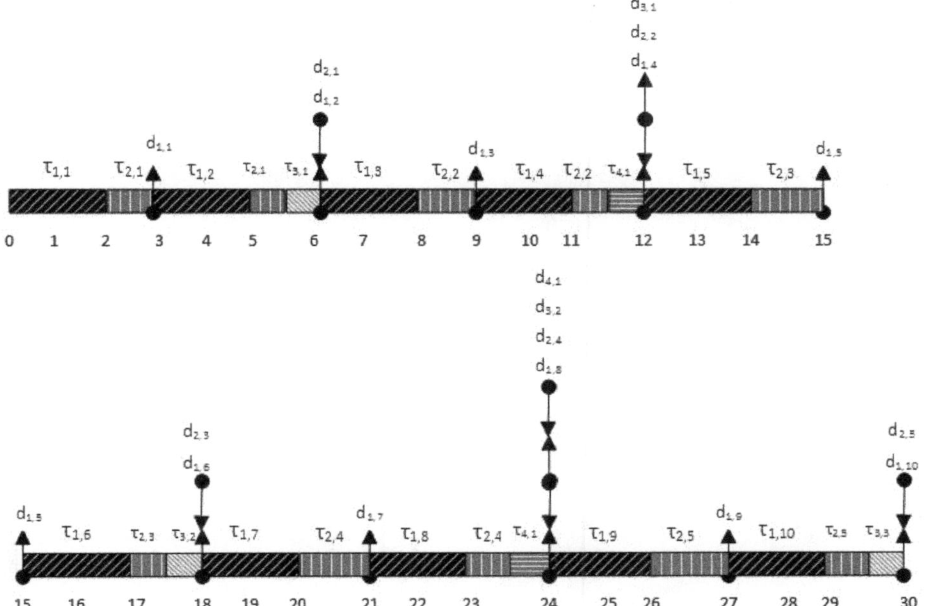

Figure 3. Gantt chart for scheduling tasks set in Example 1.

Theorem 2. *A group of real-time tasks* $Y_x = \{task_1, task_2, \ldots, task_l\}$ *is schedulable if all tasks in* Y_x *are schedulable.*

Theorem 3. *The batch of real-time tasks called periodic tasks set represented by* $T = \{task_1, task_2, \ldots, task_n\}$ *is deemed feasible if all task groups* $Y_1, Y_2, \ldots, Y_x$ *are schedulable.*

2.4. Cost Model

Scheduling decisions by integrating cost parameters change the way computational resources are selected to fulfill the user *QoS* criteria. The data-intensive real-time tasks are submitted to the broker which searches resources to process tasks within deadlines and user-specified budget constraints. The feasibility of tasks' groups on computational resources is checked by considering data transfer time, transfer costs, computational cost, tasks' deadlines, and computational power of the resources. The basic parameters considered for feasibility decisions in this research are:

(a) user-specified budget, and
(b) tasks deadlines.

Based on the above two parameters, the cloud computing and storage resources which can execute tasks within deadlines in a minimum cost by taking into account all data and processing constraints are selected.

By introducing cost model, Theorem 3 can be extended in Theorem 4 for checking schedulability of modified task set.

Theorem 4. *The batch of real-time tasks called periodic tasks set represented by* $T = \{task_1, task_2, \ldots, task_n\}$ *is deemed feasible with minimum cost if all task groups* $Y_1, Y_2, \ldots, Y_x$ *are schedulable by following all tasks constraints and holding inequality (12).*

$$cost_T \leq Budget \tag{12}$$

where $cost_T$ *is the total cost incurred by the batch of tasks, and Budget is the total user-specified budget. The cost of a resource can be expressed as execution cost per Millions of Instructions (MI), processing cost per unit time, processing cost per task, or simulation cost per unit time, etc. The cost for a single task is the sum of task execution cost and the data files transfer cost.*

3. Time- and Cost-Efficient Scheduling Algorithm

The Algorithm 1 determines the schedulability of real-time independent tasks set consisting of tasks with different data files and timing constraints. The execution procedure of the tasks involves checking task group feasibility which cumulatively constitutes tasks set. The m number of tasks in a group are checked on r number of distributed computing resources where $r \gg m$. Depending on the user budget and tasks scheduling preferences, the main objective of this algorithm is to execute distributed data-oriented applications by selecting computing and storage resources such that the tasks are processed with minimum total execution time and cost while tasks' deadlines are respected. The proposed algorithm works in three parts: (a) task initial feasibility checking which predicts a task's basic feasibility within a deadline by searching initial feasible computing resources, (b) task final feasibility and cost analysis which determines a task's schedulability after considering all the associated constraints, and finally (c) task dispatching to the best suitable resources after fulfilling all the pre-requisites. The first two parts are the matching and mapping parts which create set of time- and cost-efficient computing–storage resources pairs. The third part is the dispatching part which ensures that the selected resources can process tasks within time and budget constraints. By cost we mean the sum of a task's execution cost and data-files transfer cost. Similarly, the total execution time to be minimized is the sum of tasks actual execution time and the transfer time incurred by transferring data-files from the storage resources to the computing resources where the task is executed. The data-files are replicated on multiple storage resources and the resource which has the minimum transfer cost is selected for data-file transfer. The computing resource capability for executing a task is checked by analyzing task feasibility on PNP points. As a result, the computing resource that can execute a task by maintaining the deadlines is initially selected from the list of available computing resources. The selected resource is called an initially feasible resource. The service requests are provisioned according to the described scheduling strategy; i.e., the total execution time of the task set and the incurred cost should be minimized. To ensure the fulfilment of the aforementioned two objectives, the set of storage resources are demonstrated which accommodate data-files needed for the task $task_i$ after identifying the initially feasible resources. A single file is assumed to be replicated on more than one storage resource, so the resource which has less transfer time and cost is selected. All such computing–storage resources pairs are further checked for calculating total execution time. The total execution time is the sum of all time factors. If the total time is within the task $task_i$ deadline and the total cost is within the user-specified budget, the compute–storage resources pair is declared feasible for assigning $task_i$. After selecting all such pairs for all tasks in a group, the tasks are then dispatched to the qualified resources by the dispatcher and all required files are transferred. The tasks are scheduled and computations are carried out. In this way, if all tasks in a group Y_x are scheduled, then the group Y_x is said schedulable by the Algorithm 1. Furthermore, if all groups are scheduled, then the original task set T is declared schedulable with minimum time and cost. The resource allocation procedure completes when all the tasks are dispatched to the resources and the unmapped queue becomes empty. The pseudocode of the tasks mapping and dispatching procedures is given in Algorithm 1.

The purpose of Algorithm 2 is to find suitable compute–storage resource pairs for each data-file required by a task. For each task, sets of required data-files and initially feasible computing resources are passed from Algorithm 1 as input arguments to the file transfer time calculating function. For each data file, the storage resources are identified and the best data storage resource which qualifies the minimization criteria (transfer time and cost) are selected for retrieving data file. For each data file, all possible combinations of initially feasible compute–storage resource pairs are tried and finally, the right combination is returned with decreased transfer time and execution cost.

Algorithm 1: Time- and cost-efficient assignment of real-time data-intensive group of tasks to the HPC resources

Input: Computing resources sorted in descending order of processing capacities, and a group Y_x of unmapped real-time tasks ordered by RM priorities and having budget constraints.;
Output: Time- and cost-efficient real-time data-intensive tasks schedule on HPC resources.
Procedure
for *all* $task_i \in Y_x$ **do**
  compute $PNP_i = \{x.period_l | l = 1, \ldots, i; x = 1, \ldots, \lfloor period_i / period_l \rfloor\}$;
  // **Determining task initial feasibility for** ***all available computing resources*** $CR_r \in CR$ **do**
    for *all* $t \in PNP_i$ **do**
      calculate EET_{ir}; //gives minimum EET because resources are already sorted
      if $EET_{ir} \leq t$ **then**
        $CR_i \leftarrow CR_r$; // CR_i is set of comp resources on which $task_i$ is initially feasible
        Break; //break if t_p is found
        End
      End
    End
  if DF_i *do not locally exist* **then**
    $DF_i \leftarrow FT(CR_i, DF_i)$; //Call to Algorithm 2. CD_i is comp-storage resource pairs set for which DF_i has min transfer time and cost
    End
  calculate TT_{ir}; //TT on CD_i
  // **Determining the task final schedulability and cost analysis**
  if $TT_{ir} \leq t$ & & $cost_i \leq Budget$ **then**
    mark CD_i feasible for $task_i$;
    End
  End
// **Dispatching tasks to the feasible computing resources;**
for *schedulable tasks* $task_i$ **do**
  submit $task_i$ to CR_r;
  transfer all required files to CR_r;
  update resource information directory;
  remove $task_i$ from unmapped tasks list;
  End
initialize computing resources to maximum processing powers and update resource information directory;
End Procedure

Algorithm 2: Selection of time- and cost-efficient compute–storage resource pairs for each data-file

Specify DR_i; //storage resources on which files DF_i are stored;
for *all* $f_x \in DF_i$ **do**
  for *all* $CR_z \in CR_i$ **do**
    //compute resource z on which $task_i$ is initially feasible;
    for *all* $dr_j \in DR_i$ **do**
      $C_{zj} \leftarrow (FT_{zj}, cost_x)$ //transfer time and cost pair for file f_x in matrix C;
    End
  End
  $A_x \leftarrow (zdr, min(C))$ /pair of $comp - storage$ resources for which transfer time and cost for f_x is min;
End
Return (A); //return comp-storage resources vector on which transfer time and cost for f_x is min;

4. Performance Evaluation

This section discusses the experimental set-up, the input data, and performance metrics used to evaluate the proposed resource allocation technique.

4.1. Experimental Setup

The proposed RA technique and the existing counterparts were simulated using synthetic data sets. These experiments were carried out in MATLAB R2016a on Intel Core i5 processor, 2.50 GHz CPU and 8 GB RAM running on Microsoft Windows 10 platform. The reason for using MATLAB is that it provides a multiprocessing environment for solving complex mathematical problems demanding powerful computations. The HPC systems are difficult to implement practically due to the lack of real life experimentation environment and multiple domain administration problems which make it difficult to acquire stable configuration for evaluation. In addition, acquiring a practical HPC environment is almost impossible due to the dynamic variations in the number of users and resources at a particular moment, their characteristics, limited access, and inconsistent network conditions over the public network [1]. In addition, effective evaluation needs the study of RA technique using different user inputs and varying resource conditions. Therefore, we have created the same HPC simulated environment by managing predefined resource and network configurations such as the number of computing and storage resources connected by network links of various bandwidths randomly assigned within the range {1024, 2048} MB.

The heterogeneity in the modeled simulation environment was carried out by randomly generated resource characteristics, network bandwidths connecting computing and storage resources, file sizes, task workloads, and number of files required for each task. The data files requirements for each task were also randomly assigned in the range {x, y} showing minimum and maximum values respectively where x is assumed 1. This means that a task can demand at least 1 and at most y number of files. The files sizes are fixed at 100 MB each. To model the data files distribution, each of the data file was replicated on more than one storage resources. In our experimentation, we assume that there exist maximum 5 copies of any data file on the storage resources. The storage resources were decoupled from the nodes where computing resources are deployed. The computing resources were initially equipped with the full processing capabilities randomly chosen within the range {10,000, 40,000} Millions of Instructions per Second (MIPS). The files required by a task are either pre-fetched or transferred during execution. When the data file fetched for some higher priority task on the same computing resource is used by the lower priority task or when the required file is locally available, the file transfer time is taken as zero. This technique exploits both temporal and spatial locality of data access. This file transfer incurs communication cost and time.

As discussed above, we evaluate our proposed technique on synthetic data sets where each task (data-intensive application) gets the required computing resource within the deadlines. The applications are scheduled for getting cloud resources using the RM scheduling algorithm. The workload for the smart devices represented in Figure 1 is generated in such a way that each device generates a job periodically after the interval of $\{\alpha, \beta\}$ seconds, where $\alpha = 100$ and β = 10,000. This means that each smart device sends a request for the cloud resources after the aforementioned time interval. The execution time e_i for each application $task_i$ was generated in the range {a, b} using normal distribution function representing the best and worst-case execution times respectively, and virtual machines are allocated accordingly. We have considered worst-case execution time equal to the $period_i$ for $task_i$ in order to ensure the tasks schedulability in any changing environment. In our task model, the period of the $task_i$ is equal to its deadline, i.e., $period_i = d_i$. Initially, tasks were assigned RM priorities such that the task with high rate has higher priority, where $rate = \frac{1}{period}$. The tasks from the superset are grouped into subsets on the basis of data files demands. It is understood that when the tasks set size increases, the number of VMs needed also increases.

The task groups as well as tasks inside each group are sorted on the basis of the RM priority assignment technique. Each task in the group has respective computation requirements and is entitled to get computational resource no later than the deadline. The tasks in a group are scheduled on the HPC system and computing resources are allocated on the basis of RM priorities. Each task generates multiple jobs. Each job is generated after an interval of $\{\alpha, \beta\}$ seconds. The experimentation was carried out by considering ifferent number of computing and storage resources. The above-discussed setting is subsumed in Table 3.

Table 3. Simulation parameters settings.

Parameters	Values
Bandwidth	1024∼2048 MB
Task data files demand	$\{x, y\}$
File size	100 MB
Computing resource capacity	10,000∼40,000 MIPS
e_i	$\{a, b\}$
$period_i$	100∼10,000
Data files transfer cost	$\{1, 5\}$
Computing resource unit processing cost	$\{5, 50\}$

The computation time and cost of each job is summed into the computational requirements of each task. It is assumed that the communication cost of each job is minimal and is merged with the computation demand e_i of the task in our experimental setup. The data file is supplied to the task when it is requested from the storage resource and hence response time is zero. The transfer cost per unit size of the data file between data storage and computational resource was randomly generated between 1 and 5. In addition, the unit processing cost of the computing resources was generated between 5 and 50 depending upon the resource computing power. It is assumed that the file transfer within the same node incurs 0 transfer cost.

4.2. Performance Metrics

The proposed RA approach evaluates the HPC resource set for each task, and the overall objective is to minimize the total execution time and cost. The total execution time is the cumulative time consumed by the task set after assigning all task groups to the available computing resources. This time is also known as makespan and mathematically defined as follows.

$$Makespan = max(TT_1, TT_2, \ldots, TT_l), \tag{13}$$

where TT_j represents the total execution time of task group j. Similarly, the cost of a task set T is the overall cost incurred by all task groups. Mathematically,

$$Cost_T = max(Cost_1, Cost_2, \ldots, Cost_l), \tag{14}$$

and

$$Cost_j = \sum_{i=1}^{x} cost_i \tag{15}$$

The $cost_i$ is a combined cost incurred by a task processing on a computing resource and data files transfer taken by a $task_i$ in a group j. The time and cost for tasks context switching is negligible in our experiments and hence not included in the objective function.

5. Results and Discussion

In this section, we evaluate the performance of our proposed algorithm by comparing it with two methodologies, RDTA [12] and Greedy.

The makespan and cost minimization behavior of the proposed and the aforementioned two techniques was checked for the randomly generated task sets consisting of 100, 200, 300, 400, 500, 600, 700, 800, 900, and 1000 tasks. The plots reported in this paper are the average values of 300 runs of all the task sets. According to the task grouping criteria discussed in Section 3, the task sets are grouped into 5, 7, 7, 4, 8, 5, 7, 9, 9, and 10 groups respectively. Each group accommodates different number of tasks based on applied grouping criteria. The task grouping details are given in Table 4. The experiments were performed by checking system behavior on different number of computing resources. The number of computing resources was randomly generated within the range $\{10, 100\}$. We assume that the data storage resource gives a response immediately when a request is made by a task for data file access and hence the response time is ignored. The time delay in preparing the computing resource is also taken as zero because in our system, the computational resource is supposed to be ready for task execution as soon as the task arrives at that resource.

It was observed that for small task sets, fewer number of computing resources was involved as compared to the larger task sets. It is also understandable that choosing the proper number of computing resources can contribute to maintaining tasks' deadlines. If a lower number of computing resources is selected as compared to a large number of task sets, then it is likely that some tasks may not be RM feasible due to long waiting queues, which is crucial in real-time systems.

The main objective of this evaluation is to reduce the makespan and execution cost of the application while tasks deadlines are intact. Figure 4a,b depicts the normalized values of the makespan. The variation in magnitude depends upon the total number of tasks per task set, number of data files demands, and the computation and deadline requirements of each task. The lower the makespan value, the better the performance of the RA scheme. The other performance measurement criterion is the execution cost minimization. From Figure 5a,b, it is evident that decrease in makespan results in reduced processing cost.

It is known from Figure 4a,b, that the proposed technique continues to make scheduling decision by analyzing tasks feasibility on searching PNP sets and checking each scheduling point until some positive point t_P is found. Although, the size of PNP set for task group Y_x becomes large if the ratio between the periods of the first and the last task $\frac{period_n}{period_1}$ in Y_x is large which consumes time because large number of inequalities are tested, but this procedure enhances the chance of task feasibility because more $positive - negative$ points become available for testing tasks schedulability. Furthermore, all the initial feasible computing resources are encountered and the resource having minimum cost for the task execution is selected for task processing. The RDTA approach merely deals with executing tasks within deadlines and hence does not consider the cost parameters, which are considerably high in that case. In the case of the Greedy technique, the graph is steeply higher because a feasible resource is selected at random without taking into account the low time and cost constraints. So, the resources with high computing power are selected when termed feasible.

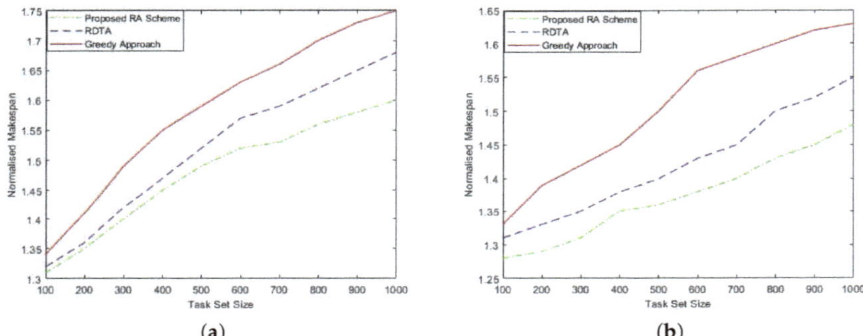

Figure 4. Average makespan for different task's requirements scenarios: (**a**) Normalised makespan for scenario 1. (**b**) Normalised makespan for scenario 2.

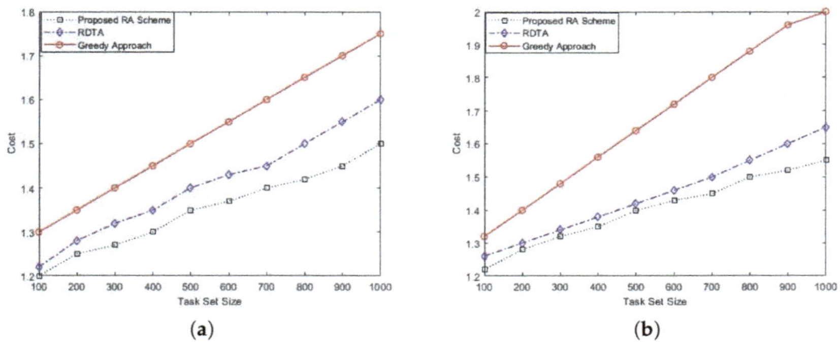

Figure 5. Average cost on two different task's requirements scenarios: (**a**) Cost for scenario 1. (**b**) Cost for scenario 2.

To further investigate the effectiveness of the proposed technique, we have conducted more experiments with different system settings. It is also noticeable from Figure 4a,b that the time taken by all tasks test also increases uniformly as the number of tasks increase because more tasks are tested. It is obvious that the makespan of some task sets is high although the computing resources were operated at full speed because they need data files from remote storage resources which increase the total completion time. The resources when operating on full capacities consume high energy, but currently energy efficiency is out of the scope of this research. The situations where makespan is low demonstrates that the data files are locally exist or perfected for some higher priority tasks and do not need re-fetching for the lower priority tasks, which adds zero file transfer time to the overall execution time. The plots show that as the task set size increases, the makespan of the Greedy and RDTA grow, as opposed to the proposed approach. This growth in case of Greedy approach is because of making a greedy selection for the data storage and computing resources among multiple choices for data files accessing and task execution. This selection does not intelligently consider the minimization criteria. The RDTA mechanism also encounters high execution time and hence cost as shown in Figure 5a,b because the data file replication is not taken into account when making a choice for data files fetch among storage resources. In the case of the proposed approach, the ratio $\frac{period_n}{period_1}$ results in a larger value which constitutes larger PNP set that provides more points for schedulability checking. This phenomenon provides more opportunities for task scheduling and hence results in large number of tasks meeting the deadlines constraints.

Table 4 shows the formation of task groups in our experimental evaluation on the basis of randomly generated data files demands. The task groups are created as for as the inequality $GU_{Y_x} \leq n(2^{1/n} - 1)$ in Theorem 1 holds.

Table 4. Task groups.

Task Set Size	Group (No. of Tasks)
100	TG1(25), TG2(10), TG3(30), TG4(8), TG5(27)
200	TG1(31), TG2(5), TG3(53), TG4(12), TG5(48), TG6(32), TG7(19)
300	TG1(4), TG2(64), TG3(20), TG4(25), TG5(40), TG6(126), TG7(21)
400	TG1(101), TG2(32), TG3(130), TG4(137)
500	TG1(10), TG2(23), TG3(116), TG4(67), TG5(50), TG6(39), TG7(120), TG8(75)
600	TG1(146), TG2(3), TG3(43), TG4(201), TG5(207)
700	TG1(2), TG2(133), TG3(108), TG4(7), TG5(211), TG6(120), TG7(119)
800	TG1(234), TG2(21), TG3(233), TG4(41), TG5(19), TG6(115), TG7(123), TG8(5), TG9(9)
900	TG1(112), TG2(21), TG3(34), TG4(321), TG5(232), TG6(18), TG7(116), TG8(29), TG9(17)
1000	TG1(12), TG2(109), TG3(120), TG4(32), TG5(19), TG6(129), TG7(127), TG8(245), TG9(21), TG10(186)

5.1. *Effect of Data Files Transfer on Performance*

One of the basic components of calculating execution time is the data files transfer time incurred by transferring data files from the remotely located storage resources to the decoupled computing resources if the required files are not locally available. In addition to the makespan and cost values, two more performance measures considered in the evaluation results are the percent share of the data transfer time and local data access.

In our experiments, the percent share of the data files transfer time in the makespan calculation is evaluated. The Figure 6a,b plot the impact of the average data transfer time for the task sets. The lower value can put significant impact on reducing the overall makespan of the task set. As it is known from the task workload, the lower priority tasks scheduled on the same computing resource can utilize the same data files retrieved for the higher priority tasks in case the data requirements of the tasks are same. In that case, the data transfer time is zero. Additionally, the transfer time is also zero if the required file resides on the same node locally where the task is being executed. In this case, the more locally accessed files decrease the impact of remote data files transfer on the performance. It is less likely that the task is scheduled on the same computing resource for which all the required data files locally exist.

The above two factors can be correlated with the makespan calculation to indicate the impact of resource selection made by the RA scheme on achieving the decided objective. It is evident from Figure 6a,b that the Greedy and RDTA schemes do not intelligently adapt for the data files locality of access procedure and hence contribute to high data transfer percentage. The percentage of locality of access rises with the increase in the task set size. In comparison to the RDTA and Greedy counterparts, there is a high chance for the lower priority tasks to reuse the pre-transferred data files by using the proposed RA scheme. In addition, it is more likely that the assigned tasks find the required data files locally. The Greedy approach exhibits degraded performance because there is a very less probability of finding appropriate computing resource for tasks assignment.

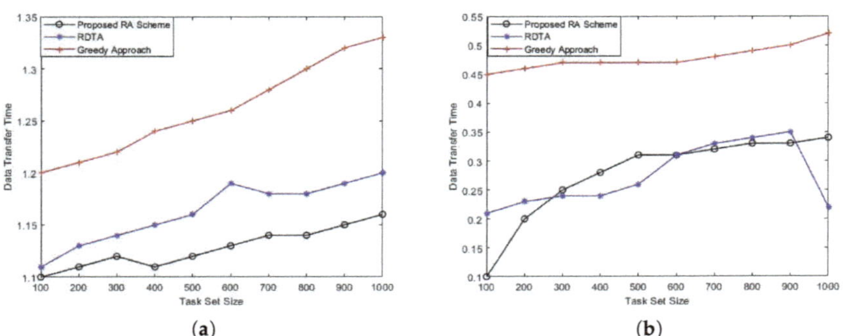

Figure 6. Effect of data transfer time on task sets in two different scenarios: (**a**) Data transfer time for scenario 1. (**b**) Data transfer time for scenario 2.

5.2. Impact on Resource Utilization

The utilization of the proposed RA scheme is measured on the basis of computing resources utilization in the HPC system. The resource utilization is directly related with the computation workload; when the task workload increases, the resource utilization also increases. The cumulative resource utilization can be calculated by the following equation.

$$Util_{cum} = \sum_{j=1}^{r} \frac{tasks\ workload\ processing\ time\ by\ a\ resource}{resource\ active\ time} \tag{16}$$

where $Util_{cum}$ represents the cumulative utilization of all computational resources spent on processing tasks workload, and r represents the total number of computing resources engaged in processing tasks sets.

It is observed from Figure 7 that the proposed RA scheme improves the resource utilization by keeping resources as busy as possible. The resource utilization is lower for the tasks sets having less number of tasks, but as soon as the number of tasks increases the resource utilization also increases. This means that the resource will be 100% utilized for the large task sets. This is an understandable phenomenon, because when the workload increases, more computational power is needed to complete tasks by their respective deadlines. If the computational power of the resources is relaxed for energy-efficient allocation, then it is very likely that some of the task groups may not be feasible. Moreover, this also will pertain to an unfair comparison. When the number of tasks increases, the proposed procedure pushes the system power to grow rapidly in order to accommodate more tasks to maintain the deadline constraints. This behavior results in high energy consumption but in this research we do not deal with the energy-efficient perspective. Figure 7 reveals that implementing the proposed approach, the minimum system utilization is between 70% and 72% for small task sets but touches 100% when the task computational demands increase. The maximum system utilization approaches 85% by the other counterparts.

5.3. Failure Ratio

Failure ratio determines the ratio of tasks missed their deadlines, or the ratio of tasks which finished their execution after the deadlines. Mathematically,

$$Failure - ratio = \frac{Number\ of\ tasks\ missing\ deadlines}{Total\ number\ of\ tasks} \tag{17}$$

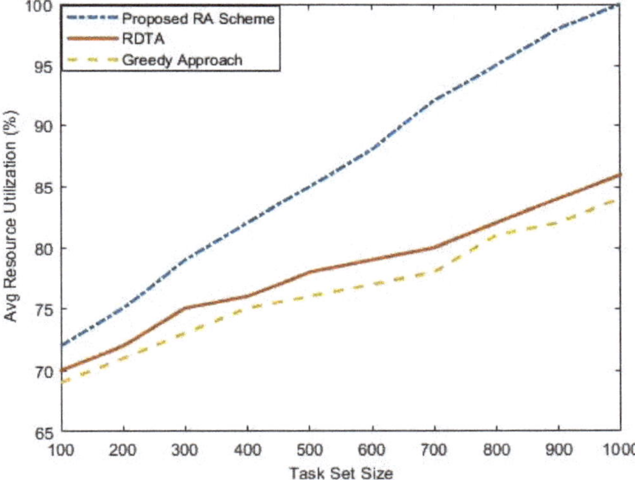

Figure 7. Effect on resource utilization.

The proposed algorithm is reactive in the context of deadlines missing by the tasks due to the efficient utilization of cloud resources, testing more and more *PNP* points, and task grouping on the basis of similar data files demands which skip the re-fetching of the same data files from the storage resources on the same computing resource again and again, which adds transfer time overhead to the makespan calculation. It is also obvious that the RM technique gives higher priorities to the tasks with shorter periods. In our case, since we assume that periods of the tasks are equal to the deadlines, so tasks with short deadlines have higher priorities.

The proposed RA methodology incorporates the scheduling points test as feasibility criteria for determining initial feasible resources, which is slower in performance as compared to the other scheduling techniques. This performance gap is due to testing task feasibility on all scheduling points for all tasks in a task set to offer more opportunity for finding feasible points. The other techniques, such as the iterative one, have the advantage of skipping large number of scheduling points in the feasibility analysis and concludes the task's feasibility very earlier. The proposed approach also does not consider the resources power consumption and the energy perspectives when operating with high power to execute more tasks within deadlines.

6. Conclusions and Future Work

This paper presented the problem of resource reservation for smart systems in the cloud computing environment. The real-time systems deadlines generated by smart devices were respected by modifying the task model by incorporating data constraints. A task grouping technique was introduced that reduces the priority levels and execution time. The scheduling and resource allocation decision is driven by the need to improve the traditional performance parameters, such as resource utilization, and decrease the total application execution time and cost. In a cloud computing environment, the providers are incentivized by profit motives; while consumers would have a limited budget and would, therefore, like to execute the application at resources that provide services within the budget. In such an environment, both provider and consumers aim to improve their utility. This research is user-centric, which selects computing and data storage resources in such way that the makespan and cost is minimized while keeping the deadlines of the real-time system intact. The results were obtained through mathematical formulations by modifying the original task model to incorporate the task's data files requirements. The proposed resource allocation scheme was compared with RDTA and Greedy approaches and celebrated results were achieved.

As a future work, it will be interesting to rank the cloud resources on the basis of different criteria such as resource computational powers, storage capacities, imbalance workloads, and cost and allocate them using machine learning techniques. It is also desirable to include energy-efficient perspectives in the proposed research when resources operate with high power to execute more tasks within deadlines.

Author Contributions: conceptualization, M.S.Q. and M.B.Q.; methodology, M.B.Q.; software, M.S.Q. and M.B.Q.; validation, M.S.Q.; formal analysis, M.F. and M.Z.; investigation, M.B.Q. and M.Z.; resources, M.S.Q.; data curation and writing–original draft preparation, M.S.Q. and M.B.Q.; writing–review and editing, M.Z. and S.A.; supervision and project administration, A.S.; funding acquisition, M.S.Q. and M.F. All authors have read and agreed to the published version of the manuscript.

Funding: The APC of this research was funded by University of Central Asia, Kyrgyzstan.

Conflicts of Interest: The authors declare no conflict of interest.

Nomenclature

Notation	Description
T	Task set
r_k	Release time of task k
e_k	Execution time of task k
d_k	Deadline of task k
CP_y	Computing power of resource y
PNP	Positive-negative points set
DF_k	Data files set required by task k
EET_{ky}	Execution time of task k on resource y
Υ_x	Task group x
GU_{Υ_x}	Group utilization of task group x
TU_i	Utilization of task i
CD	Compute and storage resource pair
CR	Computing resource set
t_P	Positive point
t_N	Negative point

TT	Total execution time
DR	Data storage resource set
R_w	Response time of the storage resource w
f_{kz}	File z needed by task k
$FT_{f_{kz}}$	Transfer time of the file $f_k z$
$card(T)$	Cardinality of task set T
dr_w	Data storage resource w

References

1. Venugopal, S.; Buyya, R. An SCP-based heuristic approach for scheduling distributed data-intensive applications on global grids. *J. Parallel Distrib. Comput.* **2008**, *68*, 471–487. [CrossRef]
2. Min-Allah, N.; Qureshi, M.B.; Alrashed, S.; Rana, O.F. Cost Efficient Resource Allocation for Real-Time Tasks in Embedded Systems. *Sustain. Cities Soc.* **2019**, *48*, 101523. [CrossRef]
3. Martel, S. How Cars Have Become Rolling Computers. Available online: https://www.theglobeandmail.com/globe-drive/how-cars-have-become-rollingcomputers/article29008154/ (accessed on 2 June 2020).
4. STATISTA. Number of Internet of Things (IoT) Connected Devices Worldwide in 2018, 2025 and 2030. Available online: https://www.statista.com/statistics/802690/worldwideconnected-devices-by-access-\technology/ (accessed on 25 May 2020).
5. Amazon Elastic Compute Cloud (Amazon EC2). Available online: http://aws.amazon.com/ec2/ (accessed on 5 June 2020).
6. Li, J.; Qiu, M.; Ming, Z.; Quan, G.; Qin, X.; Gue, Z. Online optimization for scheduling preemptable tasks on IaaS cloud systems. *J. Parallel Distrib. Comput.* **2012**, 72, 666677. [CrossRef]
7. Armbrust, M.; Fox, A.; Griffith, R.; Joseph, A.D.; Katz, R.H.; Konwinski, A.; Lee, G.; Patterson, D.A.; Rabkin, A.; Stoica, I.; et al. Above the clouds: A Berkeley View of Cloud Computing. Available online: http://www.eecs.berkeley.edu/Pubs/TechRpts/2009/EECS-2009-28.pdf (accessed on 12 June 2020).
8. Amadeo, M.; Molinaro, A.; Paratore, S.Y.; Altomare, A.; Giordano, A.; Mastroianni, C. A Cloud of Things framework for smart home services based on Information Centric Networking. In Proceedings of the IEEE 14th International Conference on Networking, Sensing and Control (ICNSC), Calabria, Italy, 16–18 May 2017.
9. Stergioua, C.; Psannis, K.E.; Kimb, B.G.; Gupta, B. Secure Integration of IoT and Cloud Computing. *Future Gen. Comput. Syst.* **2018**, *78*, 964–975. [CrossRef]
10. Chen, H.; Zhu, X.; Guo, H.; Zhu, J.; Qin, X.; Wu, J. Towards energy-efficient scheduling for real-time tasks under uncertain cloud computing environment. *J. Syst. Softw.* **2015**, *99*, 20–35. [CrossRef]
11. Lee, Y.S.; Son, Y. A study on the smart virtual machine for executing virtual machine codes on smart platforms. *Int. J. Smart Home* **2012**, *6*, 93–106.
12. Qureshi, M.B.; Alqahtani, M.A.; Allah, N.M. Grid Resource Allocation for Real-Time Data-Intensive Tasks. *IEEE Access* **2017**, *5*, 22724–22734. [CrossRef]
13. Zorkany, M.; Hussein, M.; Kader, N. Real Time Operating System for the Internet of Things, Vision, Architecture, and Research Directions. In Proceedings of the World Symposium on Computer Applications & Research (WSCAR), Cairo, Egypt, 12–14 March 2016. [CrossRef]
14. Dickerson, R.; Gorlin, E.; Stankovic, J, Empath: A continuous remote emotionalhealth monitoring system for depressive illness. In Proceedings of the Wireless Health, San Diego, CA, USA, 10–13 October 2011.
15. Hishama, A.A.B.; Ishaka, M.H.I.; Teika, C.K.; Mohameda, Z.; Idrisb, N.H. Bluetooth-based home automation system using an android phone. *Jurnal Teknologi (Sci. Eng.)* **2014**, *70*, 57–61.
16. Pavana, H.; Radhika, G.; Ramesan, R. PLC based monitoring and controlling systemusing WiFi device. *IOSR J. Electr. Commun. Eng.* **2014**, *9*, 29–34.
17. Khalid, A.; Aslam, S.; Aurangzeb, K.; Haider, S.I.; Ashraf, M.; Javaid, N. An efficient energy management approach using fog-as-a-service for sharing economy in a smart grid. *Energies* **2018**, *11*, 3500. [CrossRef]

18. Mckinsey, By 2025 Internet of Things Applications Could Have 11 Trillion Impact, 2015. Available online: http://www.mckinsey.com/mgi/overview/in-the-news/by-2025-internetof-thingsapplications-could-have\-11-trillion-impact/ (accessed on 28 May 2020).
19. Soliman, M.; Abiodun, T.; Hamouda, T.; Zhou, J.; Lung, C. Smart Home: Integrating Internet of Things with Web Services and Cloud Computing. In Proceedings of the 2013 IEEE 5th International Conference on Cloud Computing Technology and Science, Bristol, UK, 2–5 December 2013; pp. 317–320.
20. Son, J.Y.; Park, J.-h.; Moon, K.-d.; Lee, Y.-h. Resource-Aware smart home management system by constructing resource relation graph. *IEEE Trans. Consum. Electr.* **2011**, *57*, 1112–1119. [CrossRef]
21. Davidson, J. *Apple Homekit: The Beginner's Guide. Van Helostein*, 1st ed.; CreateSpace Independent Publishing Platform: Scotts Valley, CA, USA, 2017.
22. The Ambient. Available online: https://www.the-ambient.com/guides/samsung-smartthings-guidesmart-\home-163 (accessed on 20 May 2020).
23. Google Home Review (2018): The Smart Speaker that Answers almost Any Question, The Guardian. Available online: https://www.theguardian.com/technology/2017/may/10/google-homesmart-speaker-\review-voice-controll (accessed on 10 June 2020).
24. Liu, C.L.; Layland, J.W. Scheduling algorithms for multiprogramming in a hard real-time environment. *J. ACM* **1973**, *20*, 40–61. [CrossRef]
25. Qureshi, M.B.; Alrashed, S.; Allah, N.M.; Kolodziej, J.; Arabas, P. Maintaining the Feasibility of a Hard Real-time Systems with Reduced Number of Priority Levels. *Int. J. Appl. Math. Comput. Sci.* **2015**, *25*, 709–722. [CrossRef]
26. Min-Allah, N.; Khan, S.U. A hybrid test for faster feasibility analysis of periodic tasks. *Int. J. Innov. Comput. Inf. Control* **2011**, *7*, 1–10.
27. Sha, L.; Goodenough, J.B. *Real-Time Scheduling Theory and Ada, CMU/SEI-88-TR-33*; Software Engineering Institute, Carnegie-Mellon University: Pittsburgh, PA, USA, November 1988; p. 15213.
28. Suciu, G.; Vulpe, A.; Halunga, S.; Fratu, O.; Todoran, G.; Suciu, V. Smart Cities Built on Resilient Cloud Computing and Secure Internet of Things. In Proceedings of the 2013 19th International Conference on Control Systems and Computer Science, Bucharest, Romania, 29–31 May 2013; pp. 513–518.
29. Liu, J.W.S. *Real Time Systems*; Prentice Hall: Upper Saddle River, NJ, USA, 2000.
30. Ye, X.; Huang, J. A framework for Cloud-based Smart Home. In Proceedings of the 2011 International Conference on Computer Science and Network Technology, Harbin, China, 24–26 December 2011; pp. 894–897.

MDPI
St. Alban-Anlage 66
4052 Basel
Switzerland
Tel. +41 61 683 77 34
Fax +41 61 302 89 18
www.mdpi.com

Energies Editorial Office
E-mail: energies@mdpi.com
www.mdpi.com/journal/energies

www.ingramcontent.com/pod-product-compliance
Lightning Source LLC
LaVergne TN
LVHW070056170726
843515LV00007B/1547

* 9 7 8 3 0 3 6 5 1 6 2 7 1 *